Making It Clear

156 essays on business, management, leadership, human nature, projects, information technology, and life in general

By Harwell Thrasher

Published by Harwell Thrasher, Kissimmee, Florida.

The majority of the material in this book comes from previously published books and articles written by Harwell Thrasher. Articles were published on the web site www.makingITclear.com (no longer in existence), with copyright dates 2003-2024. Web site content is used with permission of MakingITclear, Inc.

ISBN: (paperback) 979-8-9861009-2-0

ISBN: (ebook) 979-8-9861009-3-7

Library of Congress Control Number: 2026920916

LCCN Harwell Thrasher: Kissimmee, Florida

Dedication

To my late wife, Sharon, who let me be alone when I needed to write, provided encouragement when I needed it most, gave me helpful feedback on every article even when it didn't interest her, and loved me despite my flaws. I miss you, darling.

Making It Clear

Introduction

In the ten years between 2003 and 2013, I tried to make sense out of the world by writing about situations and ideas in my business life and, in a few cases, in my personal life. I justified the writing by saying that it would help position me as a "trusted advisor," and therefore lead to consulting business. But honestly, it probably helped me even more than it helped my readers, because it enabled me to understand what I had done right in my various corporate jobs, and what I could have done better. I've always heard that the best way to understand something is to have to explain it to someone else, and that's what I did in my writing.

The writing started as an email newsletter, but it evolved into a web site called makingitclear.com. The "it" in "makingitclear" was intended as a play on words: I initially focused on the Information Technology world. But it seemed like many of my readers came from the general business world, and so the scope of my writing broadened over time. The readership peaked when I wrote about new managers, what characteristics they needed, and how they can be successful. That information is included here, along with everything else I wrote.

I honestly don't know whether to call these things "articles" or "essays." As a I understand it, articles convey information, while essays convey opinions and personal experiences. I called these "articles" at the time, and I continue to use that word in this book when I cross-reference one article from another one. But there is a lot of opinion and personal experience included too. So whatever you want to call them, here they are. I hope you find some value in what I've written.

Hashtags

The articles are included in the chronological order that I wrote them, but most people won't read them in that sequence. Feel free to skip around. To help you, I've added hashtags at the beginning of each article. The hashtags include:

#ACQUISITIONS is for articles concerning M&A (Mergers and Acquisitions). These articles cover some of the things I learned during my due diligence work. But they also apply to product selection decisions and to corporate partnerships.

#ALIGNMENT is for articles relating to attempts to align IT with the business, and vice versa. But in most cases, they are relevant for attempts to align any department or organization with the overall business.

#CAREERS is for articles relating to career planning, hiring, firing, and promotions. This includes articles on what it's like to be a manager, and how to become better at your job.

#IT is for articles that relate specifically to Information Technology organizations.

#LEADERSHIP is for articles that deal with organization direction-setting, corporate politics, and general navigation at higher management levels in an organization.

#MANAGEMENT deals with the fundamentals that managers need to know, and how to deal with various problems that arise.

#PERSONAL is for articles that relate to some personal transition I was going through at the time. They may be useful to you if you're going through any sort of life transition.

#PROJECTS is for articles that should be of particular value to project leaders, or to people trying to get projects approved.

#STRATEGY is for articles that talk about what strategy is, and how to create and follow a better strategy.

While hashtags are probably more value to someone reading this book on an electronic device—because you can search for the next hashtag—you may also find them valuable as you skip through the pages, as a way to focus your reading on certain articles that reflect your own current interests.

Why this information is timeless

Although these articles were written a number of years ago, I've found that while the technology might change, people remain pretty much the same. I've gone through the articles one by one and added some "2026 Commentary" to talk about their relevance in today's world. Most of the main points in the articles are as true today as they were when I wrote them, because they focus on people more than technology.

1. How to Get People to Change—the Human Side of IT Projects

Originally published on APRIL 8, 2003 in #IT, #LEADERSHIP, #PROJECTS, #STRATEGY

According to a recent study by the Standish Group, only 29% of Information Technology (IT) projects are completed on time and within budget. I believe that a large number of the remaining 71% of projects get in trouble because they don't adequately consider the human side of the projects. Good CIOs recognize that Information Technology is ultimately about people and change—software and hardware are just a means to an end. Implementing new systems requires people to change. The change, in fact, is often the most difficult part of a project. This article provides a perspective that may help you facilitate change, and in some cases may help you understand why intended change isn't happening.

There are three things required to get a person to change:

1. The person must be dissatisfied with the status quo.
2. There must be a clear, well understood vision of what life for the person will be like after the intended change, and that vision must be perceived by the person as positive.
3. The first steps for change implementation must be clear to the person.

As an analogy, think of a big boulder that has to be moved. You can apply pressure to one side of the boulder, and eventually, with a lot of effort, it will move. But think instead of a boulder on the side of a hill. The slope of the hill represents the perceived difference in appeal between the current situation (the uphill side) and the future vision (the downhill side). The boulder "wants" to roll downhill—gravity compels it. But inertia and friction cause the boulder to stay where it is until the first steps are taken to push the boulder in the right direction. Once the boulder starts to roll downhill, however, the momentum for change builds, and the boulder continues to move in the downhill direction as long as the slope (the quality of the vision, and its "goodness" relative to the status quo) remains steep enough. It takes significantly less effort to roll a boulder down a hill than it does to move the boulder on flat ground—you just have to start it rolling, and it continues on its own.

Similarly, people want to move toward a better life, but inertia and fear of change cause them to continue to do what they have always done. If you make the first steps clear and easy, however, then the people will begin to change. Their enthusiasm for the change will depend on the perceived benefits of the vision over their perceived status quo. Note the two uses of the word "perceived" in that last sentence. If you don't do a good job of explaining how good the future is, and if you don't convince them that they should be unhappy with the status quo, then the perceived difference won't be enough—no matter what the reality is.

There's a group effect too. Once a few boulders start to roll downhill, you can get a landslide effect that takes other boulders with it. The analogy is apt: once a few key individuals start to change, their enthusiasm promotes change among their peers.

All three requirements—dissatisfaction, vision, and clear first steps—have to be met. The absence of any one of them will prevent change from occurring. People who are satisfied with their current situation have no reason to change, and will be reluctant to try. People who don't understand the future vision have nothing to move toward, no matter how dissatisfied they may be (Unfortunately, perhaps a more common corporate problem is not the lack of clarity of a vision, but the confusion that results from too many conflicting visions.). And even if someone is dissatisfied and has a clear picture of what's better, nothing will happen until the first steps are made clear.

2026 Commentary:

I checked the Standish Group result every few years, and the numbers never got significantly better. In later years, the Standish Group added Quality and Scope to their measurements, so that "successful" projects had to meet goals of Quality, Cost, Schedule and Scope, and the success rate dropped even more. For more about Quality, Cost, Schedule, and Scope ("QCSS" for short), see my first book, *Boiling the IT Frog: How to Make Your Business Information Technology Wildly Successful Without Having to Learn Anything Technical.* Here's a quote from Chapter 9 of that book: "When you define the requirements for a project, you can't specify quality, cost, schedule and scope; only three of the four can be required, and the other variable is dependent on the process being used for the project."

Obviously, there are more issues than just human-related problems, including use of new, unproven technologies, and "estimates" that were driven top down by overbearing executives rather than bottom-up from the people who would actually have to do the work. But I still believe that people issues are at the heart of most unsuccessful projects: people who don't want to change, systems designed by outsiders who don't really understand the need, and failure to get the necessary support from the real users of the proposed systems.

2. Driving Information Technology—Is the CIO Just a Chauffeur?

Originally published on MAY 13, 2003 in #IT, #ALIGNMENT, #LEADERSHIP, #STRATEGY

In last month's newsletter, I said that Information Technology (IT) is about people and change, and that software and hardware are just a means to an end. However, some of you are operating under the impression that the IT organization *itself* is just a means to an end. To use an analogy, some of you believe that the IT organization is like a chauffeur, whose job it is to:

1. Drive the car to the destinations chosen by the passengers,
2. Maintain the car in good working order,
3. Keep up with changing routes, and occasionally suggest a different route to reach the destination faster.

If your IT organization has a chauffeur mentality, then its job is to:

1. Move the company's technology to a destination chosen by the company executives,
2. Maintain the technology infrastructure in good working order,
3. Keep up with changing technology, and occasionally suggest a different route to reach the destination faster.

You might ask, "Well, what's wrong with that?" Here's what's wrong: it's managing—not leading. If your IT organization has a chauffeur mentality, then:

1. The company is not able to get the maximum benefit from information technology. The chauffeur isn't involved in picking the destination. Neither is the IT organization included in the process of picking the best places to apply technology in the company,

even though the IT organization contains your technology experts. How is the CEO supposed to know the best use of technology for the company? They have other things on which to focus.

2. There is no coordination of technology decisions. The chauffeur isn't involved in the choice of transportation—whether to use a car, bus, train, or plane, or a combination of all four. Neither is the IT organization involved in the choice of technology. Maybe the sales organization picks a CRM package (which may, for example, require a Wintel platform), and the production organization picks an ERP system (which may run best on Unix), and the human resources organization picks time reporting software. Then the IT organization is faced with the job of integrating the results of these divergent decisions: both the data integration *and* the hardware/software integration. Were they the right decisions? Maybe or maybe not, depending on whether the integration trade-offs were adequately factored into the technology choice process. Wouldn't it have been better to make those trade-offs explicitly rather than by default?

3. The IT organization becomes a commodity, and is therefore easily subject to outsourcing. If you don't have need of a full-time chauffeur, it's easy to outsource to a limo service. There isn't much of a downside to the outsourcing decision, as long as the limo service has reliable people who can follow directions. Similarly, outsourcing a "chauffeur mentality" IT organization doesn't have much of a downside either. If IT just follows directions anyway, then why not outsource, either within the U.S., or off-shore? The risk is that when the company outsources IT, it gives up the ability to use IT strategically. Outsourcing IT may reduce cost, but it also greatly reduces the possibility of IT strategic benefit.

Is the CIO in your company a high-priced chauffeur? Then either replace the IT organization with the technology equivalent of a limo service, or get someone as CIO who can provide the technology advantage your company can use against its competition. If you're a CIO, then you're either strategic, or you're gone.

2026 Commentary:

This was written in the days when CEOs weren't very technical, and so they weren't particularly good at anticipating the possible impact of new technologies on the business. But even now, I think the number of CEOs who really anticipate technology impact can be counted in the minority. So the analogy still applies. It's common for the technology leader in a business to take orders from the CEO (or even someone lower in the organization), and have little or no input into the process of best applying technology innovation for the good of the business.

3. Three Technology Surprises in Acquisitions, and How to Avoid Them

Originally published on MARCH 10, 2003 in #ACQUISITIONS, #ALIGNMENT, #LEADERSHIP, #MANAGEMENT, #STRATEGY

In simple terms, due diligence is the elimination of future surprises. For an acquisition, due diligence is the research required to make sure that a year or so later, the buyer won't look at the acquisition and say "If I had only known then what I know now." There are lots of different aspects of due diligence, but they all come back to this concept of eliminating surprises. Financial due diligence verifies the financial records of the "target" company. Human Resource due diligence looks at employee contracts, agreements, benefits, lawsuits, and pending lawsuits. Information technology due diligence looks at the information technology used by the target company, including hardware, software, and the people who deal with it.

In each case, the objective of the due diligence is not to ensure that there is an absence of problems (although that would be desirable). Instead, the objective is to identify any problems that exist so that those problems can be factored into the acquisition and valuation decision. If the problems are severe enough, then the acquisition may be called off. In some cases the problems can be remedied before the acquisition takes place. In many cases the acquisition will proceed, possibly at a decreased offer price. However, in all cases, the buyer will be going into the acquisition with their eyes open, with a full understanding of the amount of work required to make the acquisition a success.

Three Possible Technology Surprises

There are three primary surprises associated with information technology as part of an acquisition, and technology due diligence focuses on preventing all three. The three surprises are:

1. The target company doesn't have clear title to the information technology being used.

2. The systems don't do what the buyer thinks they do, or the systems won't handle what the buyer wants to do with them in the future.

3. The amount of money and effort that may be required to deal with future systems needs, including integration of the target company's systems with the buyer's systems, is more than expected.

Let's look at each possible surprise in more detail.

1. The target company doesn't have clear title to the information technology being used.

This surprise can be prevented by classic due diligence – go through the paperwork provided by the target company to verify that the company has receipts and licenses that prove their ownership of all of the information technology they use. It sounds easy, but it usually isn't. Most companies will have receipts for their hardware, and so you can verify hardware purchases and lease terms. Software, however, is a different story. Every software vendor defines its software licensing differently, and the details of the license can change over time. To properly verify licensed software, you have to get an inventory of every software product on every computer, examine the license agreements and certificates provided by the target company, and then compare the inventory information, agreements and certificates to the latest license information from each vendor. Most companies are out of compliance with software licensing to some degree, and even those who are in compliance usually have difficulty in finding all of the documentation to prove it.

What about custom software? Or licensed software that has custom adaptations? Work could have been done by employees of the target company, or by independent contractors or consultants doing work for the target company, or a combination of the two. The key is to determine whether any of these people or companies who worked on the custom software retain any intellectual property rights (IPR) to the work they did. If the target company doesn't have signed documents saying that the IPR remains with the target company for all of the work done, then there's a potential problem. If clear title can't be obtained for the software, and the buyer wants to use that software after the acquisition, then there is a risk of future payments, a lawsuit, or even a requirement to discontinue use of that software.

2. *The systems don't do what the buyer thinks they do, or the systems won't handle what the buyer wants to do with them in the future.*

This surprise is difficult to avoid without getting due diligence help from someone experienced in technology due diligence. It is human nature to put things in the best light, and therefore any description of system capability presented by the target company will always be very optimistic. Most buyers want to hear good news, and so they interpret things positively as well. Unfortunately, most technology is described using strangely exotic "technology geek" language that only another technology geek would understand. The future surprises come mostly from the subtle details that a non-technology person wouldn't be likely to understand. Examples include an e-commerce web site that hasn't been implemented in a secure manner, or a system that hasn't been designed to scale up to handle higher transaction volumes.

If the target company has anything beyond basic accounting systems and off-the-shelf uncustomized software, then there is high risk of a future systems surprise. Custom systems usually evolve over time to meet the changing needs of a company, and so systems that were originally developed to meet one need are often inappropriately used to meet additional needs. The systems are like a rubber band – each additional requirement stretches the limits of the systems further, and eventually the new requirements will cause the rubber band to break, requiring major systems development work. Proper technology due diligence can assess the current systems condition of the target company, determining which systems are closest to the breaking point, and which systems are unlikely to be able to handle the new requirements imposed by the buyer.

3. The amount of effort and money that may be required to deal with future systems needs, including integration of the target company's systems with the buyer's systems, is more than expected.

Even if the systems used by the target company work perfectly, most buyers want to integrate some of their own systems with the systems of the target company. At a minimum the buyer will probably want to bring together some of the financial systems: general ledger, accounts payable, accounts receivable and payroll. Other integration is likely if there is any synergy expected between the two companies. For example, supply chain, customer service, order processing, ERP, and Human Resource systems are frequent candidates for some level of integration.

Strictly speaking, the resource required to integrate the systems of the target company and the buyer is outside the scope of the acquisition cost. Yet it is important for the buyer to understand the total cost that the business will incur as a result of the acquisition.

Therefore it is often useful to put together a preliminary estimate of systems integration cost, so that integration surprises can be avoided.

How Much Technology Due Diligence is "Enough"?

It is impossible to eliminate all risk in an acquisition, and there is no amount of due diligence that will eliminate all possible surprises. Each case is unique, but it's important to focus the limited resources available on the areas with the highest likelihood of risk and the highest potential cost.

The amount of technology due diligence required varies tremendously depending on the situation. A low amount of due diligence will be required for a target company that has very few automated systems, especially if the buyer intends to replace all of those systems with the systems of the buyer (e.g., moving the target company to the buyer's general ledger system). A high amount of due diligence effort will be required when buying a software company whose systems will be used for mission critical aspects of the buyer's business. Most due diligence efforts fall somewhere between these two extremes.

Recommendation

Evaluate the risk associated with each of the three surprise areas, and decide how much you need to invest in order to reduce the risk to an acceptable level. An investment in technology due diligence before you close the deal can greatly reduce buyer's remorse a year or so later, and it may enable you to negotiate a better price for your target company. Minimize the surprises!

Disclaimer

This article is for informational purposes only. It is important to consult with an attorney regarding specific situations your company encounters.

2026 Commentary:

The biggest changes in technology since 2003 are drastic increases in use of outsourcing, software-as-a-service, cloud technology, and Artificial Intelligence. These technology advances make due diligence even more difficult because they make it easy to hide software and service ownership and licensing issues behind Internet and intranet interfaces. The basic issues are the same, but it takes a lot more digging to exhaustively track down all of the software and service dependencies of the target company.

4. Seven Ways Information Technology Can Improve Your Business

Originally published on MARCH 10, 2003 in #IT, #ALIGNMENT, #LEADERSHIP, #MANAGEMENT, #PROJECTS, #STRATEGY

How can you focus your use of information technology for increased business benefit?

One of the things often missing from Information Technology (IT) strategic planning is a focus on doing the best thing for the business. Unfortunately, most IT strategic plans are more focused on the fine points than on the bigger picture – they're very detailed about systems and software development, but they frequently omit critical information about how all this IT expense is going to have a tangible benefit to the company. As a result, many businesses are not taking maximum advantage of their IT budgets because those budgets aren't focused on maximizing the IT investment for the business.

This article describes an approach that can be used to answer the question, "Where can Information Technology best contribute to the business?" By following this approach within your company, you can help focus your use of IT for increased business benefit.

Seven Areas of Improvement in the Business

There are seven areas in which a business can be improved by IT:

1. Reduced expense
2. Increased revenue
3. Improved margin
4. Improved quality

5. Improved customer acquisition
6. Improved customer satisfaction
7. Increased speed

Let's look at each area of improvement in more detail.

1. Reduced expense

This is the benefit of IT that first comes to mind for most people. The original IT projects were put in place to "automate" processes in order to reduce cost, often by replacing large numbers of people with a few computers. Reducing expense is still a valid way for an IT project to pay its way, but this is becoming one of the least strategic IT investments. Most companies have already automated the obvious areas for savings long ago.

Still, if your business is not competitive in your industry because of a higher than normal expense level, then this may be the most important way that IT can contribute to your business success.

Examples of IT projects which reduce expense include the classic automation projects for payroll, general ledger, accounts payable, and accounts receivable. IT projects also reduce expense by automating any process that requires heavy use of clerical people, or any process that uses highly repetitive well-defined procedures.

2. Increased revenue

For most businesses, increasing revenue offers higher payback than reducing expense. This is especially true in businesses which have relatively fixed costs, in which incremental revenue dollars flow directly to the bottom line. It's even more true when revenue can be added by increasing sales to existing customers (sometimes called "same store sales"), eliminating the necessity of additional customer acquisition costs.

Many e-commerce web sites in the dot com era were justified on the basis of increasing revenues, but most of them just provided an additional sales channel for existing customers, and actual revenues didn't increase. To truly increase revenues, IT systems have to attract new customers, increase sales volume from existing customers (possibly by making it easier to order), or increase cross-selling to existing customers, thereby increasing overall sales volume across multiple product lines.

In addition to e-commerce web sites, here are some other examples of systems that might increase revenues:

- Data Warehouse systems to help identify customers who are most likely to buy as a result of certain types of promotion, or who would be more likely to buy as a result of cross-selling efforts
- In service companies, software that provides a new service offering. For example, a bank might use software to provide a new credit card that counts purchase amounts toward a cash-back bonus or frequent flyer miles. A retailer might do the same sort of thing with a "frequent buyer card" supported by new information technology.

3. Improved margin

Improving margin isn't just a matter of improving both revenue and cost. To improve margin an IT project has to change the whole business model in order to change the incremental relationship between a dollar of revenue and the cost associated with that revenue dollar. In most cases such an IT project would have to be part of a business reengineering effort – something that is going to radically change the underlying business processes, and therefore the IT systems as well.

Where IT projects are able to play a significant role in a fundamental reengineering of the business model, there is usually significant payback to the business. Examples are web-based reservation systems for the hospitality industry, phone-based stock trading (using voice recognition technology), and web-based print ordering for digital cameras. Each of these examples has resulted in fundamental changes to the business model.

4. Improved quality

An IT project can improve the quality of your products or services if the project implements process improvements that would be too time-consuming or difficult to do without using computers. Many of those old "automation" IT projects provided as much or more benefit from improved quality as they did from reduced cost, because they reduced the amount of human error. Improved quality is listed here as a separate area of business improvement because it's frequently overlooked for IT projects. Yes, by improving quality, we can end up increasing revenue, improving customer retention, and even reducing cost, but quality improvement should be considered as a goal in itself.

Examples of quality improvement IT projects are:

- Sophisticated software to track spending patterns of credit card users, in order to identify likely fraudulent use of credit cards, and even to trigger a call from customer service to ask a customer whether the customer would like to have their credit limit raised
- Analysis of customer call logs to get an early warning on quality issues
- Use of modeling software to better understand and predict product failures, and to take action before products fail

5. Improved customer acquisition

Since more customers will usually lead to more revenue, it seems reasonable that IT projects which improve the customer acquisition process should provide benefit to the business. Many companies pursue benefits in this area, but get distracted by secondary goals like improving the productivity of the sales force. Remember that if a project in this area doesn't increase sales or add customers, then it isn't a success even if it makes things better for the salespeople.

Key to improving the customer acquisition process is getting an understanding of where the bottlenecks are in the current process, and then identifying IT projects that can help to eliminate or reduce those bottlenecks. Sales force automation is one of the project areas that can sometimes produce this kind of benefit, but also consider IT projects that reengineer the way that customers are acquired. Improvements to customer acquisition can improve the speed of customer acquisition (reduce the time to close sales), increase the number of customers that can be handled by a sales rep, or reduce the cost of acquiring a new customer.

6. Improved customer satisfaction

Improving customers satisfaction is one of those things that is obvious, and yet very difficult. If customers are more satisfied then your revenues will grow, either through repeat purchases or through word-of-mouth advertising. Costs are likely to be lower as well, since you won't have to spend as much to deal with customer complaints and returns.

Quality is one of the contributors to customer satisfaction, but there are IT projects which can improve customer satisfaction directly. The key phrase is "making it easy." How do we make it easy for the customer to deal with us, get questions answered, place orders, and generally interact with the business? Information Technology can help by providing web access for customers, and by making information available to customers by phone or by fax. IT can help customer service reps make it easy to deal with the customer by providing all of the information that a customer service rep could want, including a history of prior interaction between the customer and the company. Customers want to be treated as individuals, with the business understanding their likes and dislikes. And IT projects can provide the tools to make it possible to track customers as individuals, even when there are thousands or millions of them.

7. Increased speed

Everything is happening faster these days, and expectations have changed as a result. It wasn't so long ago that people expected letters to be answered in a week or so; now we expect emails to be answered within minutes. The saying "time is money" has never been more true. If you can speed up the sales process, the manufacturing process and the delivery process, then revenue will be recognized sooner. If you can improve the speed and accuracy of the ordering and fulfillment process, then you can reduce inventories and thereby reduce carrying costs.

IT projects that are associated with reengineering a business process can obviously increase the speed of that business process. But certain types of IT projects are focused purely on increasing speed. Consider, for example, the use of "integration broker" technology which makes it easier to exchange data among systems within your company, and between your company's systems and the systems in your partner's companies, including your suppliers and your customers. By making it easier to set up data connections, the technology makes it possible to set up new business relationships faster, and to more easily and quickly integrate your company's systems with the systems of other companies. Use of this technology is just one example of an IT project that can increase the speed of your business.

An Important Observation

Here's an interesting question about the seven areas of improvement to the business: "How many of these areas have easily measurable payback"?

Unfortunately, the answer is one – maybe two (#1 reduced expense and maybe #2 increased revenue). This is the underlying issue behind one of the biggest problems in trying to get the IT organization to make the maximum possible contribution to the business. Most companies still think of IT as a way to automate things. Automation projects can be cost-justified quite easily because they save money. Sometimes other IT projects can be justified because someone is willing to stand up and commit to increased revenue as a result of the project. But the other five areas of improvement are more indirect in their effect on the business, even though in the long run those other five areas may be more important. And because their benefits are more indirect, projects in the last five areas don't tend to get proposed, and don't get approved when they do get proposed.

Recommendation

To derive maximum benefit from your IT budget, you should:

- Look at each of the seven areas in turn, and think about what benefit there would be to your company (even if you can't easily measure it) if you could make improvements in that area.

- Have brainstorming sessions to separately focus on each of the seven areas, proposing possible IT projects that would make improvements in that area.

Depending on the relationship between IT and the business, you may want to conduct both separate (business and IT separately) and joint (business and IT together) sessions. And don't get hung up over whether a project is in area "a" or area "b"—just list it wherever it comes up, even if it comes up in multiple areas.

- Within each of the seven areas, prioritize the IT project lists that emerge based on your best understanding of business benefit and "what's important for the company", and regardless of whether or not you can quantify the benefit.
- Merge the lists for the seven areas, prioritizing the resulting combined list, again basing the priority on "what's important for the company", regardless of whether or not you can quantify the benefit.
- Compare this list to your current IT project list, and try to understand the difference. If a highly ranked item on the new list isn't on your current IT project list, then why? If a highly ranked item on the current IT project list is lower ranked on the new list, then why?
- Adjust your IT project list and your IT strategy to focus on the real needs of the business, instead of just the projects that are easiest to justify.

Conclusion

I once heard a story about a woman who left her office late one night and encountered a man in the parking lot. The man was on his knees under a street lamp, fumbling around like he was looking for something. The woman walked up to him and asked him what he was looking for, and he replied, "I dropped my car keys." The woman bent down and helped him look for a few minutes, then thought to ask, "Where did you lose them?" The man replied, "Over there in the corner." The woman, somewhat taken aback, asked him, "If you lost your keys over there in the corner, why are you looking for them here?" The man answered, "Because the light is better."

A lot of companies justify IT projects based on easy-to-measure savings associated with cost reduction. But does that mean that those are the best projects to do? Or are we doing them just because "the light is better" and they're easier to justify?

Go through a full evaluation of all possible IT projects that could help the business. And then put your money on the projects that will offer the most benefit to the business, regardless of how easy it is to measure the return.

2026 Commentary:

I think all of this is still true, although there are some obvious technology changes since 2003. The Internet is now the primary customer interface for most customer service, either through a browser or by using a smartphone app. And "integration broker" technology still exists, but I don't think it has turned out to be the big thing it looked like in 2003.

5. IT Doesn't Matter?

Originally published on JUNE 10, 2003 in #IT, #LEADERSHIP, #STRATEGY

There is an article in the May, 2003 issue of *Harvard Business Review* with the disturbing title "IT Doesn't Matter." Yet despite its inflammatory title, the article is highly supportive of today's information technology, comparing it to the likes of electricity, transportation and telecommunication in its importance to business, and declaring it "the backbone of commerce." The essence of the article is that IT has reached a point where, in the author's opinion, all companies are so equally matched in their IT capabilities that no one company is able to derive any strategic advantage from IT for very long. Any leader into new territory is quickly matched by its competitors within a matter of months. Furthermore, the leader is negatively rewarded for its original thinking by having to pay the price for innovation, while its competitors enjoy the lower price associated with duplicating the leader's idea.

According to the author, Nicholas G. Carr, there are thus no significant new opportunities for strategic advantage in IT, and therefore "IT doesn't matter" as a strategic business weapon. Instead of using information technology for strategic initiatives, Carr recommends that companies focus on cleaning up their act in IT to provide for higher stability and efficiency within their own IT operations, concentrating especially on the areas of IT security and the prevention of system outages.

While there are aspects of the author's argument with which I agree, I can't help but think of IBM's Tom Watson and his famous observation in 1943 that "I think there is a world market for maybe five computers," or DEC's Ken Olson and his famous comment from 1977, "There is no reason anyone would want a computer in their home". These men were proved short-sighted by subsequent events, and so could Mr. Carr.

Yet what concerns me more is the article's implicit reinforcement of a theme that is becoming all too common among big business in today's economy. In spite of his blind spot to personal computers, Ken Olson was well-respected in the technology industry during the glory days of DEC, and he was known within DEC for his strategic advice to "Do the right thing" no matter what the situation. I'm afraid big business has now rearranged those words to say "Just *avoid* doing the *wrong* thing."

Where is that spirit of innovation that has propelled our technological development into the twenty-first century? Have CIOs become a group of caretakers for the technical infrastructure? Some have, I'm sure. Yet I firmly believe that there are still those of us who want to use technology in new strategic ways to radically change the way that business works, and to fundamentally improve the perception of business by its customers, thus increasing revenue and profit. To us, and to the companies to which we dedicate our efforts, IT truly *does* matter. The technology itself may not be strategic, but the way in which IT is *used* is very strategic.

Where do *you* stand on this issue? Sure, you're trying to make your infrastructure so reliable that it becomes virtually invisible. But are you doing anything more? Does IT matter in *your* company?

Notes:

1. The statements from Watson and Olson seem ridiculous today, but their words were heavily influenced by the thinking of their peers, and their thinking was constrained by self-imposed unconscious limitations on their strategic vision. They were proved wrong by subsequent events, but only because other men and women changed the rules of the game. I believe that continually changing the rules of the game is the most important aspect of

technological progress, and one of the key ways in which IT can be strategic. Use IT to change the rules!

2. Notice that I said "use technology in new strategic ways" instead of "use new technology." I think there is more opportunity for profitable innovation in new uses of existing technology than there is in the typically unprofitable use of new "bleeding edge" technology.

2026 Commentary:

Although the Carr article attracted a lot of attention at the time, I doubt anyone remembers it now. New technologies go through cycles: First, they get a lot of attention, but the early adopters get burned due to failures and high costs. Eventually the technology is perfected enough where more mainstream companies can adopt the new technology for good purposes and with little fanfare. “Cloud” technologies went through that whole cycle, and are now taken for granted. Artificial Intelligence is still in its fad stage, but will eventually be taken for granted as well.

6. What does Folding a Suit have to do with IT?

Originally published on JULY 8, 2003 in #IT, #ALIGNMENT, #LEADERSHIP, #MANAGEMENT, #PROJECTS, #STRATEGY

Like many of you, I travel a lot. Some of the travel requires me to wear business suits, and I've had to learn how to pack a suit coat so that it's wearable when I unpack it. Years ago I learned the secret, but it recently occurred to me that the secret of packing a suit coat can be applied to Information Technology as well.

The secret of packing a suit coat is not to avoid wrinkles, as you might think. Instead, the secret is to make sure the wrinkles end up in places where wrinkles won't be noticed.

Why does this relate to IT? Because it's a lesson in perfectionism and prioritization. We all want our systems to be perfect, but we know that we don't have the time or the budget to make that possible. So the secret of having our systems appear to be more successful is to make sure that the system limitations end up in places where the limitations won't be noticed.

Sounds obvious, doesn't it? Then why don't we do it? Think about the typical everyday user of one of your systems. For online systems (web or not), what screens does that person spend most of their day using? For batch systems, what reports does the person use the most? Have you focused the majority of your systems improvement effort on making things better for these most common uses of your systems? On improving their usability? On improving their system performance and reliability? That's where the productivity payback is, because that's where most of the business labor goes: into the systems that are used the most.

But a common trap in IT is to get distracted by the so-called "corner cases," the very uncommon situations where due to a certain combination of variables, weird buggy things happen in software. We tend to spend an inordinate amount of IT dollars on fixing these problems, "removing the wrinkles," when we might be better off in some cases just moving the wrinkles to where they won't be noticed.

Got a low-importance system that requires periodic down-time for maintenance? Don't waste money on replacing the system with something that doesn't require down-time—just schedule the down-time for a low-use time of the week, and let the users know. They won't care if they don't use the system much.

Got a software system that won't properly handle a few extremely infrequent but bizarre combinations of variables? If it's not a critical situation, then put in a workaround (e.g., suspend the transaction in question for manual intervention) instead of wasting resources on something that only happens once in a blue moon. Sure it's a "wrinkle," but who will notice!

The type of person who is attracted to the IT profession tends to be a perfectionist. That can be a plus sometimes, but it means that we always want to deal with 100% of a problem. Unfortunately, the world doesn't need a 100% solution to a few problems as much as it needs 80% solutions to more problems.

When there's not enough time or money to solve 100% of a problem, it's usually better to focus on the 80% of the problem that matters the most to the most people. You can't avoid wrinkles, but you can make sure they end up where they won't be noticed.

And, oh by the way, if you haven't learned the secret of folding a suit coat yet, turn the page to see my instructions.

2026 Commentary:

This is still true, whether we're talking about apps or browser-based systems. And it ought to be true of AI systems as well. You don't want an AI system that pretends to know the answers to everything, even though the underlying data isn't there to support it. Have your AI system configured to own up to situations where there's not enough data to draw a conclusion.

7. The Secret of Folding a Suit Coat

Originally published on JULY 8, 2003 in #PERSONAL

As I said in the previous post, the secret of folding a suit coat is to make sure the wrinkles end up in places where wrinkles won't be noticed. On this page I'll show you how to fold a suit coat using this secret approach.

First, let me say that I'm going to describe this from a male point of view because I've never had the experience of folding a woman's suit for packing. I know that a woman's suit might include a skirt instead of pants, but otherwise I'm pretty sure that this secret applies to women's suits as well.

Everybody starts pretty much the same way: by folding their pants over the hanger, like this. (Women, I guess you do pants or a skirt here.)

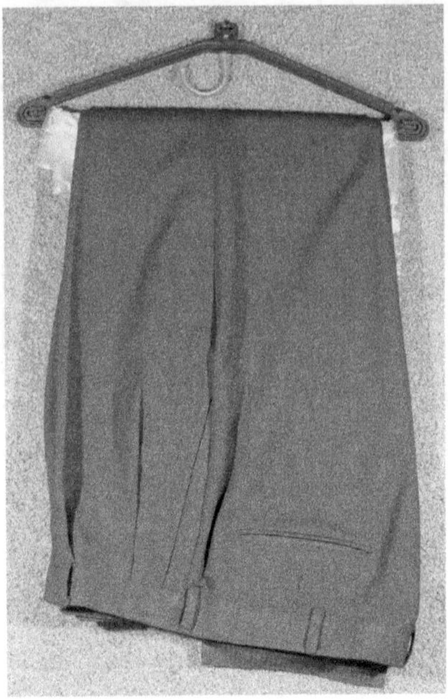

Next you put the coat on the hanger over the pants. But most people who haven't learned the secret do it like this.

To see this a little better, let me move the sleeves out of the way.

This is where most people go wrong. They put the coat on the hanger the way they wear it, with the coat buttons lined up with the button holes. They think that this approach will minimize wrinkles, but it usually just puts extra wrinkles along the sides of the coat, and it makes it fairly likely that you'll get wrinkles on the front of the coat as well.

Here's how to fold the coat instead. Fold the front of the coat all the way over, with one side of the coat (on the left side of this picture) snug up against the pants.

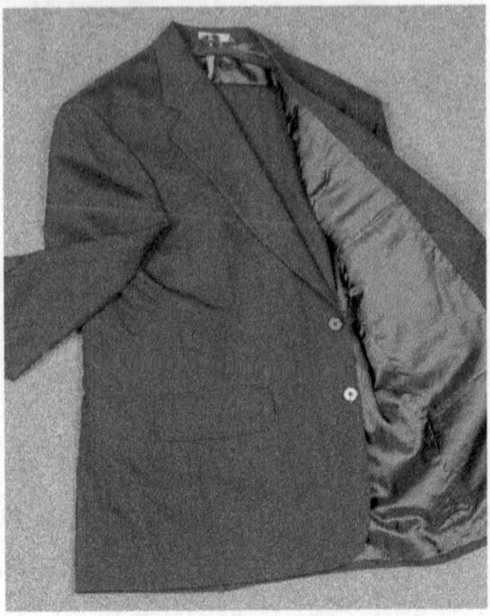

Now fold the other side of the coat across, snug up against the pants on the other side (on the right side of this picture).

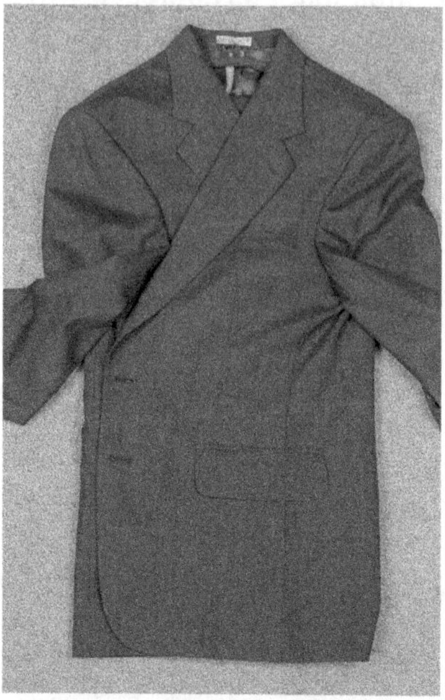

Now fold one of the sleeves down across the front of the coat and straighten out the sleeve. There will be wrinkles under the armpit, but people won't see them. You may also get a few wrinkles at the shoulder, but they'll look natural. Do the same thing with the other sleeve.

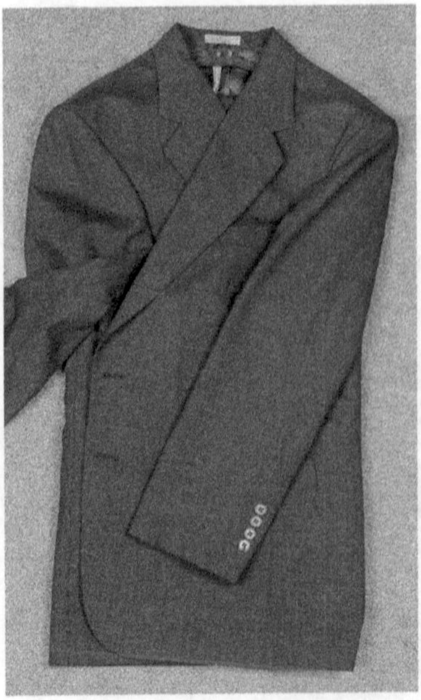

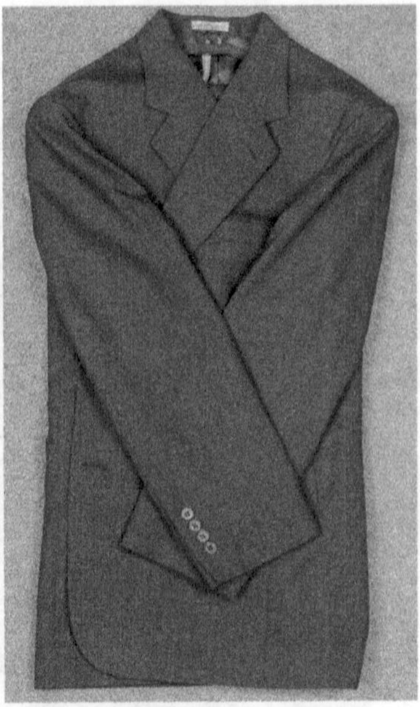

I like to put a plastic dry cleaning bag over the coat and pants at this point before I pack them, but that's optional. I'll leave it off in the next few steps so you can see what I'm doing.

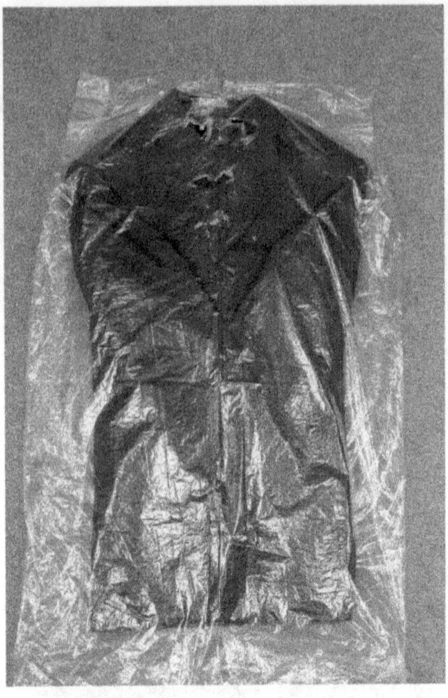

Put the coat and pants into the suitcase so that the coat will fold backwards instead of forwards. If you get wrinkles they'll be on the back where they're less likely to be noticed.

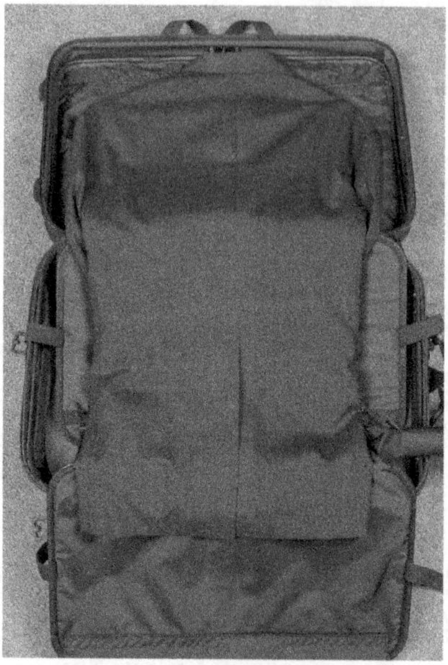

Fasten the coat in (I'm using a Rollaboard Suiter suitcase here. Rollaboard is a registered trademark of TravelPro USA.)

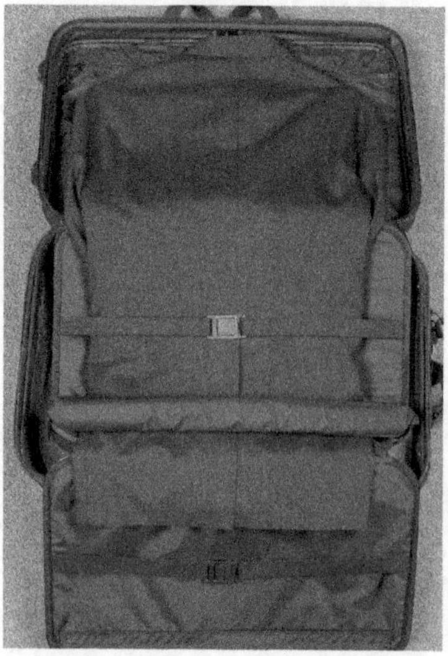

Fold it up.

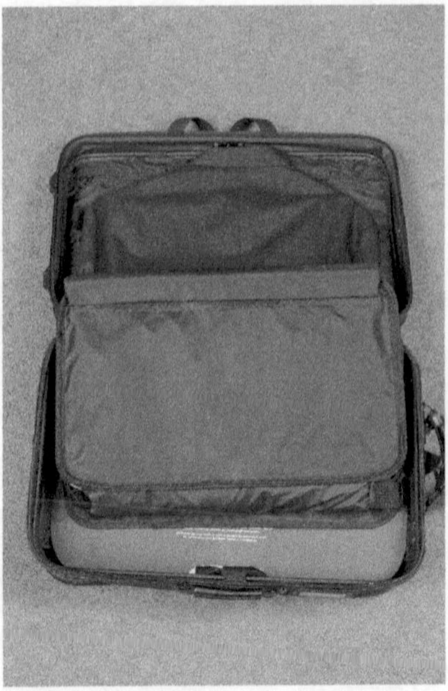

Fasten the straps to secure the folded coat.

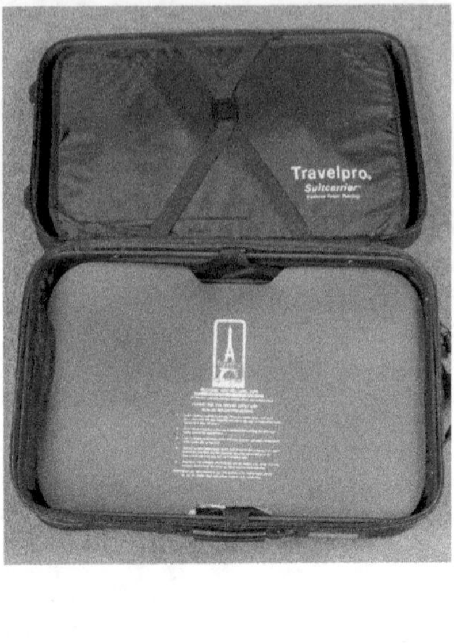

2026 Commentary:

This is probably less relevant now that people are dressing more casually, but the technique still works.

8. Web Services aren't the Answer...but use them anyway

Originally published on AUGUST 12, 2003 in #IT, #STRATEGY

Web services aren't the answer, but you should use them anyway. I'll tell you why.

Let me start with a quick definition of web services. When you use web services, you allow software applications to communicate with each other using three technologies: HTTP, XML, and SOAP. HTTP is the primary protocol used for data transmission over the Internet. Using HTTP means that applications can communicate with each other if they're in different parts of a company, in different companies, and even in different countries. HTTP is allowed through most firewalls because it's the same protocol used by web browsers. Web services use HTTP because it's the universal common denominator for communicating across the Internet.

XML is a data format that includes the definition of the data along with the data itself. Sending data in XML is like sending a book with a complete dictionary of the language as well a classroom guide on how to read it. Using XML for data transmission means that the sending and receiving applications don't have to agree in advance on how to store the data.

SOAP (Simple Object Access Protocol) is a method for packaging the XML data into a request that a receiving application can understand, and to which a receiving application can respond. It's like a work order that anyone can interpret: it says "I want you to do this, and when you're done, here's what I want the result to look like."

Combine HTTP, XML and SOAP into "web services" and you have a way for any two applications to communicate with very little prior work. The approach even works between applications which use entirely different architectures, like Java-based systems and Microsoft .NET-based systems. That's why it has been hailed in the technology press in terms that make it sound like the Holy Grail of IT.

While web services are undoubtedly a good idea, they are not nearly the be all and end all that the press has described them to be. As a child, did you ever have one of those sets of boxes where you open a box to find another box, and then open the second box to find another, etc., etc., until you finally end up with a very small interior box? All Web Services do is open up another box, but the next box down in the layers of boxes is much more challenging.

Web services make it easier for two applications to communicate with one another. But opening up most applications to outside access makes it painfully obvious just how confusing and proprietary those applications are. Most application-specific data requires an understanding of the application software in order to interpret it. In spite of industry adoption of relational databases as a standard, many databases are far from normalized. And what's worse, the accuracy and validity of data in most databases is extremely uneven: some data is highly accurate, but other data is hardly verified at all, much less kept up to date.

Using web services to arbitrarily connect two applications is a little like buying a car from an ad in the newspaper without seeing the car in advance and without having anyone test the car. You might get something worthwhile, but it's hard to know.

After you open up the web services box, the next "box within a box" is the need for the standardization of data to make it more understandable and more reliable. Maybe we should even "grade" our data elements so that someone reading the data can know what data they can trust and what data to ignore. Until we solve this data problem, web services aren't the answer; they just make it easier to communicate with another application that isn't very trustworthy.

Should you use web services? They definitely make inter-application communication easier, particularly across diverse architectural platforms. But they're also valuable because they call attention to the data validity problems we all face. So go ahead and use web services. After all, you never would have found the second box if you hadn't opened up the first one.

2026 Commentary:

I never hear anything these days about web services and SOAP, probably because other buzz words have taken their place. But the underlying issue with the trustworthiness of data is still very, very real. And this applies as much to AI as it does to web services. A sophisticated interface based on untrustworthy data still gives a bad result, even though it may look pretty.

9. Get Off the Train, and Join the Fleet

Originally published on SEPTEMBER 9, 2003 in #CAREERS, #MANAGEMENT, #PROJECTS

I remember the first time I was in a management role, more than 25 years ago at Digital Equipment Corporation (DEC). I was a bit anxious because I had been put in charge of people who had been my peers. Then I surprised my new direct reports by asking them questions no one in management had ever asked them before.

I met with each of them one-on-one and asked them, "What do you want to do? Where do you want to go with your life and your career?" Sometimes these questions are asked as part of a job interview, where we ask the questions to assess the ambition level of the job applicant. But here I was asking the questions for an entirely different reason.

I've always had the belief that people and organizations have a natural direction, with different directions for each individual person and organization. Most companies subscribe to the "get on board the train" approach, in which the company sets a direction, defines the jobs that it needs (the boxcars in the train), and then finds the people to fill the jobs. This approach simplifies administration, particularly in large companies, but it doesn't take maximum advantage of the capabilities of the employees. It also creates some strange models for individual behavior. Workers can be motivated to some degree by the "vision" of the company's direction (the train's destination). But when that vision is too abstract, unfocused or distant, the employees are more likely to be motivated by more mundane activities like jockeying for a better boxcar on the train. As a result, much of the employee energy is wasted on things which don't contribute to the company's direction or to achieving the personal life goals of the employees.

The most important omission in the "get on board the train" approach is the lack of attention paid to individual employees' natural direction. But consider an alternate approach, which I'll call "join the fleet." Imagine a company as a fleet of ships that sail the oceans. Then imagine that the company wants to carry a large amount of cargo from port A to port B. The cargo quantity is too large to be carried by a single ship, so the cargo is put into containers that can be carried by multiple ships.

As with the train approach, the company has a destination. But our container carriers aren't as rigidly constrained as boxcars. Smaller, faster ships can get their cargo to the destination sooner. Sailboats can tack back and forth to take advantage of variable wind direction. Ships can even make intermediate stops on their way to the destination, taking on supplementary cargo to optimize their own performance. Some ships may reach the destination and go back for more cargo. Others will make one trip and move on to other things.

Applying the analogy to employees, the "fleet of ships" approach takes advantage of the inherent differences among employees. Each employee is given high-level tasks ("cargo") that are consistent with the employee's long-term direction. Employees carry out the tasks in a way which is optimized for them as individuals, working at different speeds, and simultaneously maximizing their own performance (and their own lives) as well as their contribution to the company (the "fleet").

So what happened twenty-something years ago when I asked people about their own personal goals? One employee told me he wanted to work toward going home to his native Greece. We contacted the DEC office in Greece to find out what skills they would require, and arranged his work assignments to help him develop those skills. Another person said he was very interested in gaining technical proficiency, and that operating systems had always fascinated him. So I moved him toward increasingly more technical tasks. A third person said she wanted to travel more. We worked out her responsibilities to do more support of DEC's remote manufacturing plants, and she got to travel as part of that support role.

I'm not saying that everything always falls naturally into place when you try to align company goals with employee personal goals. It often takes ingenuity and redefinition of jobs. But when it works, it's fantastic: the employees are enthusiastic, and they're doubly motivated because they know that their goals and the company's goals are aligned. Productivity skyrockets because high goal alignment means that employees can focus on the tasks at hand without being distracted by their "other life."

So get off the train and join the fleet! You'll be amazed how much everyone can benefit.

2026 Commentary:

I look back at this approach now and wonder whether I'm the only one who ever tried to work this way. I never hear of anyone else trying this, even though it worked extremely well for me.

I think that job definition goes through cycles. Before machines and computers, people were the only resource. Jobs were often incredibly repetitive and boring, and there wasn't a lot of flexibility in the way that jobs could be done (see a related article on bureaucracy, "23. Julia Roberts, Training Wheels, and Bureaucracy.") During the time referenced in the article, there was a lot more flexibility in the way software was written, and so it was possible for people to take individual approaches to the work. Now it seems that AI tools are removing the flexibility, and in many ways we're going back to jobs becoming repetitive and boring. That's too bad.

So I'm not sure that this approach would work in today's environment. But I think it would be nice if it does.

10. 18 CIO Secrets for Information Technology (IT) Effectiveness

Originally published on OCTOBER 6, 2003 in in #IT, #ALIGNMENT, #LEADERSHIP, #MANAGEMENT, #PROJECTS, #STRATEGY

Do you want ...

- A more satisfied boss?
- Less stress?
- More successful projects?
- Infrastructure savings?
- Enhanced leadership?
- Low risk solutions?

Here are some proven approaches for reaching these goals.

A more satisfied boss

Communicate more clearly. Use business terms when talking to business people, and avoid technical jargon. Talk about business benefits instead of system features and capabilities. Have key presentations reviewed by an outsider to ensure clear and effective communication.

Focus on what's important to the business. IT is a means to an end rather than an end in itself. What are the critical issues in your business? How can IT contribute to the resolution of those issues? Be specific.

Provide more value from IT. Value can be improvement in revenue, expense, or profit, or it can be a time-based improvement. For example, because of IT, perhaps your company can react faster to changing conditions, or it can process transactions faster. Can you summarize the value in dollars that you have provided to your company in the last year?

Less stress

Get a sounding board—someone you trust with preliminary ideas. Your best choice is someone outside your company who can provide a more impartial and objective view of what's happening.

Have resources to help you when you need help. Think of your IT organization as more than just the people reporting to you. Build up an "extended family" of advisors, supporters, and available supplemental workers. This lets you deal with the usual resource limitations in a better way.

Have someone who can guide you in the areas in which you're not as strong. Know your strengths, and spend most of your time in those areas. Recognize your weaknesses, and have people available to help you in your weak areas. You won't become great by wasting your critical time on areas in which you're merely mediocre; focus on leveraging your strengths.

Use outside arbitration when there is a technology dispute within your organization that you don't want to arbitrate yourself, when impartial judgment is important, or when you don't have the required expertise. Also use an outsider when the disputants have to work together after the decision; it gives them someone to blame.

More successful projects

Make sure you're doing the right projects. It's not enough to just do projects well, you have to be sure that you're focusing on the most important projects for your business. The right project done adequately is better than the wrong project done well.

Identify why your previous successful projects worked so well. Is your success repeatable? Find out what went right in successful projects, and take steps to ensure that it happens again. Hint: it's not just the people – processes are usually more important.

Determine why your previous unsuccessful projects didn't work so well. Forget blame – the important thing is to understand the process problems that contributed to undesirable results. You'll find it easier to get to the truth if you use an indirect approach: ask "What could be done differently next time to make the project more successful?" That elicits the answers you want without embarrassing people.

Shut down less beneficial projects. In some cases the best thing you can do for a project is to shut it down. Think of it as thinning out the new growth to make room for an overall healthier crop.

Infrastructure savings

Develop a set of standards, and then follow those standards. An inordinate amount of infrastructure cost is caused by supporting too many platforms. Implementing standards can reduce your platform diversity and save you money.

Identify specific ideas for infrastructure savings, and then implement the ideas. It sounds easy, but it's seldom done. Most infrastructure focus is on increasing uptime instead of decreasing cost. Ironically, lower cost can lead to higher up-time.

Enhanced leadership

Have a vision of the future that generates employee enthusiasm. Leaders are trusted individuals who have a vision that people can understand and support. You're probably trusted, but do you have a clear vision? Have you communicated it within your company? Why not?

Have a clear direction and strategy. This is the next level below the vision. It translates the vision into reality by showing specifically how you're going to achieve the vision. A strategy is actionable.

Have an understandable set of principles to serve as the foundation for decision-making. Principles are the primary way to successfully delegate decision-making. Together with vision, direction and strategy, they define your success.

Low risk solutions

Go for base hits instead of home runs. Sure those big projects are impressive, but they're also incredibly risky. It's far safer to break up big projects into a series of smaller projects, each of which offers some tangible business benefit. This not only reduces project risk, it also increases customer satisfaction, helps build trust, and is more adaptable to changing requirements.

Never do a project that will take longer than half the average tenure of a company executive. To be safe, it's better to deliver something to the business (something significant that the business can use) at least once every six months – three months is better. If you're using new unproven technology, then do a low-cost proof of concept first. And only work with vendors who guarantee their work.

2026 Commentary:

Even today, I think all of this CIO advice is good. But the advice is difficult to follow if you're thinking short term, or if you're following orders instead of actually implementing a strategy. See the earlier article, "2. Driving Information Technology—Is the CIO Just a Chauffeur?"

11. 3 Keys to Service Success

Originally published on OCTOBER 14, 2003 in #MANAGEMENT

What do capability, motivation, and expectations have in common? All three are essential for a successful service organization, whether that organization is in Information Technology or in any other field.

Capability

Let's use a help desk as an example. For such a service organization, capability includes the basic skills necessary to be able to answer questions and help customers. Some of these capabilities are technical, and employees have to constantly update their skills through training and self-study. Other capabilities are more intangible but just as important: things like "people skills" and the ability to deal with an irate customer.

For other types of service organizations, the capabilities can vary. But in most cases, the capabilities still focus on two major areas: knowledge of the "what" in order to answer questions, solve problems, and offer advice; and knowledge of the "how" in order to convey the relevant information, problem-solving and advice in a way that is appreciated by the customer.

Motivation

The second key to service success is motivation. People who have the capabilities but not the motivation will ultimately hurt your organization. Even with the right people skills, unmotivated employees will still reveal their lack of motivation in other ways. Maybe the service rep won't go the extra mile to please the customer or to truly solve a problem instead of a symptom. Perhaps the unmotivated manager will tolerate inadequate work performance. Sooner or later that lack of motivation will become known to the customer, and the customer will switch to a different, more highly motivated service provider. Customers have human needs; we all prefer to work with someone who cares about us and wants to take care of our needs.

Expectations

So far we have capability and motivation. And interestingly enough, many service organizations can do just fine on just these two keys. If an organization has enough resource to exceed customer expectations, then those expectations are never an issue. But when cut-backs take place in a service organization, and the organization is forced to try to provide the same service with fewer and fewer people, the third key becomes more and more important.

There are multiple dimensions of the word "expectations." First, there's the need for a service organization to understand the expectations that already exist within their customer base. This usually requires some sort of measurement: maybe a survey, or possibly something less formal, like interviews with selected customers.

Second, there's the need to "set expectations for the customer." That's a common phrase, but it's not so easily achieved. Setting expectations can be explicit, with a published list of what you'll do and how long it will take, or it can be implicit in situations where you consistently deliver the same result in the same way, and so the customer knows from experience what to expect. The implicit approach works well with simple, short-duration transactions, like a telephone directory assistance service in which expectations are set by how long it takes to answer the phone, and then how quickly a number can be provided. For more complicated, longer-duration services, a more explicit approach is usually preferred.

However, "setting expectations" is only successful when the expectations you set come close to meeting the expectations of the customers. For example, if you want to provide four-week turn-around on fixing a user's PC, then you'll find yourself in direct conflict with customers who think (quite rightly) that they can't do their jobs without a working computer. Thus you'll have to look for other alternatives, like providing a loaner PC, or swapping out the user's PC for another PC immediately.

If you're providing a service internally within a company, and you're working on a budget that doesn't allow you to meet reasonable customer expectations, then you're looking at guaranteed failure. The customers can't go elsewhere to meet their needs, and you don't have enough resource to keep your customers satisfied. There's only one way out of this dilemma: work with the customers to financially justify an investment to increase the service levels. If your internal customers are unproductive as a result of your not having the resources to meet their needs, then that lack of productivity has a cost to the business, and that cost can usually be translated into a financial justification. If you can't make this approach work, then I suggest you start looking for another job in another company, because you're not going to survive long where you are. It's not fair for you to be asked to meet reasonable customer expectations with unreasonably low levels of resource, and you're certainly not going to enjoy your work.

Capability, motivation, and expectations are the three keys to success in a service organization. Ignore them at your own peril!

2026 Commentary:

My telephone directory assistance example is pretty dated (does that service still exist?), but otherwise the content still rings true. Some technology advances like chatbots and web-based FAQs complicate the current situation, but from the customer perspective, things are the same. The customer has expectations, and the customer will measure you based on those expectations, whether the service is delivered by chatbot, AI, or in person. Capability and motivation are still key issues for customer-facing jobs. And indirectly, capability and motivation are key for people who design, implement and manage automated support systems.

12. On Time at the Wrong Restaurant

Originally published on NOVEMBER 11, 2003 in #CAREERS, #IT, #LEADERSHIP, #MANAGEMENT, #STRATEGY

A friend of mine struggled with bad weather and worse traffic to make his way across town, arriving just in time for a scheduled lunch meeting. Unfortunately, he had misunderstood his calendar, and he was at the wrong restaurant.

When he told me about it the next day, it struck me how his predicament is an appropriate metaphor for the way many of us do Information Technology. We work extremely hard, we struggle against all sorts of obstacles from the technology itself and from the people in our company, and yet, when it's all done, we find that we've reached the wrong destination. We find that we misunderstood the business requirements, or maybe we pursued the solution to a symptom instead of digging deeper and finding the root problem.

I've seen this happen hundreds of times: work done on optimizing a system that was soon thrown out and replaced; effort expended on an application that the customers didn't want; time wasted on improving the human interface of screens that are only viewed once in a while by system administrators. We get so focused on problem-solving and roadblock removal that we don't stick our heads up once in a while and see if we're still headed in the right direction. Forget the forest and trees analogy; we're below the tree level down in the dirt.

Why does this happen? Sometimes it's because we become focused on "resource utilization," that magic phrase which means that it's a crime to actually have an employee idle for a day (See "126. A Manager's Biggest Burden, and 5 Ways to Deal with It"). We find things for the people to do, even if the things are not really all that important, and before you know it, a minor task has escalated into something major.

But more often it's because we do what we're told, even if we don't understand why, and even if we push back and try to put things into a broader context. Because in most cases the day-to-day success of a CIO isn't measured by what they accomplish; it's measured by how what is accomplished is perceived by their superior. Hype, spin, and positioning can cover up a multitude of sins.

My job is to help companies think strategically about IT. But the two biggest issues I find are that either (a) the business isn't thinking strategically about itself, or (b) that no one connects the IT strategy to the business strategy (See "106. 8 Reasons Businesses Can't Connect Business Strategy and IT Strategy"). When either of these things happens, the result is that IT direction setting becomes a process of prioritizing random requests rather than positioning those requests in the context of a broader strategy.

Without the strategy, it's as if we turned our restaurant choice over to bunch of six-year-old kids. They don't think about what's the best nutrition for their bodies—they just want to go to the place that has the best kids' meal toys. Similarly in the business world the people making random requests for IT help are often going after "fast food projects": projects which provide a quick sugar rush in terms of visible short-term results (notice I used the word "results"—not "benefits"), but which probably don't contribute much to the long-term needs of the business.

IT Strategy is a whole lot more than prioritizing projects. It's knowing where the business wants to go, what's holding it back, and what can be done to remove the limitations that are preventing the business from becoming bigger, better and more profitable. It's like nutrition in the sense that you have to understand the whole body need, and then put together meals or snacks (smaller projects) which make the body stronger. And yes, you have to do some "glamour projects" once in a while to placate your boss and make your IT group look good, just like you probably ought to take your six-year-old to McDonald's every now and then. But "glamour" is not a word that I've ever heard associated with being in the IT business.

Mostly, IT is "three steps forward, one step back." It's trying to keep focused on what's important even when you're being bombarded by short-term requests and broken technology. It's trying very hard to nourish your company, with many of the exact same rewards that come from trying to nourish your family: "Oh ma, do we have to eat meatloaf again?!?"

And if we're lucky and diligent, we'll do what our company needs to have done, and we'll end up on time at the *right* restaurant. Because you know what happens if you show up on time at the *wrong* restaurant? You have a nice lunch alone.

2026 Commentary

Nothing has changed since I wrote this, which isn't surprising, since human nature hasn't changed. And while the article is written from an IT perspective, it actually applies to the way companies work in all departments. And it applies to governments as well.

13. IT is a lot like Gift Giving

Originally published on DECEMBER 9, 2003 in #IT, #ALIGNMENT, #LEADERSHIP, #PROJECTS, #STRATEGY

It's that time of the year when many of us are thinking about finding appropriate holiday gifts for loved ones. The other day it struck me that there are similarities between the gift giving process and the profession of Information Technology (IT). Here are my thoughts about the philosophy of gift-giving as it relates to IT:

1. Gifts are all about the recipient.

The first rule of finding a gift is to find something that the gift recipient will appreciate. Good gifts may be something that the recipient needs, but the best gifts are the things that thrill: gifts that cause exclamations of pleasure and make the face of your loved one light up with joy.

Those thrilling gifts are tough to figure out for most people. You have to really understand the person who will receive the gift: know what they enjoy, know their secret desires, understand their fantasies. This level of understanding requires conversations about their philosophy of life, their hopes and dreams, and how they see their life changing over the upcoming years. It's not enough to know their present; you have to get a feel for their past and future as well.

There is a message here for IT organizations. In my experience, most IT failures are due to miscommunication. IT strategy gets out of sync with business strategy. Projects fail due to misunderstandings about requirements. The IT customer asks for one thing, but really means another. Somehow the words get turned around, maybe by the speaker, maybe by the listener, and most likely by a combination of the two. It's a lot like Aunt Sally saying she wants a record for Christmas when she really means a CD. Her terminology may be out-of-date, but her audio equipment isn't. You would know that if you knew more about Aunt Sally.

And that's the point, isn't it? You have to know your customers well in order to communicate effectively. Know your business to focus your IT strategy. Know your project beneficiaries to be able to fulfill their needs. It's all about the recipient.

2. Gifts have to be viewed as valuable.

Gifts don't have to be expensive, but they have to be viewed as valuable to the recipient. Value can mean that the gift serves a purpose; some people really do use salad spinners. Or value can mean that the gift will be cherished: a great baby picture or a finger painting by your four-year old.

The key here is that value is defined by the person receiving the gift—not the person giving the gift. An ideal gift is one which costs the giver very little but provides enormous value to the recipient. Sound familiar? We're talking about cost/benefit here.

Yes, just as IT projects are invariably held up against the yardstick of cost/benefit or return on investment (ROI), the same thing applies to holiday gifts. And just as with holiday gifts, the ideal IT project costs very little but provides enormous benefit to your business.

3. Gifts need to be timely.

Gifts of baby clothing carry the most extreme penalty for being late; the clothes just won't fit a fast-growing infant. But no one likes it when they receive a birthday gift a month after their birthday, or a holiday gift after the holiday is over. Technically the gift is still the same; if it's valuable when it's on time then in most cases it will be valuable when it's late. But emotionally it's not the same.

It's that way with IT projects too. When the project first starts out, the projected completion date is usually pretty arbitrary. You pick a date that sounds reasonable, and then you try your best to keep things on schedule. But the longer the project goes on, the more "cast in concrete" the completion date becomes. It's repeated so many times that people begin to believe in it. The expectation for the delivered project is like kids waiting for Santa Claus; they know it's going to be great when the presents arrive, and they count down the days as the date draws closer and closer.

How would your kids feel if Santa postponed his visit for a month or two? That's how IT customers feel when you miss a date for a project deliverable. Was the date arbitrary? Yes, but arbitrary dates don't stay arbitrary—they become part of customer expectations.

A final observation

In thinking about the comparison of gift giving and IT, it occurs to me that a lot of people spend more time picking out the perfect gifts for their relatives than they do picking out the perfect IT project for their business. It's like taking your kid to a toy store. Kids don't examine their budget, prioritize their needs, and pick the toys that will provide the most return for their investment. No, they just dive in and start begging for the first toy they find that appeals to them.

How many of you have technology strategies that were developed the same way?

2026 Commentary

It's odd how the reference to records and CDs has reversed now, with many audiophiles using vinyl records instead of CDs. And of course with music subscription services, the whole concept of buying a record or a CD is no longer applicable for many people. Nevertheless, I think the meaning still comes across. And I think the idea of thinking of IT projects like Christmas gifts is still appropriate.

14. When to Outsource and When to Offshore

Originally published on JANUARY 13, 2004 in #MANAGEMENT, #PROJECTS, #STRATEGY

The subject of outsourcing is in the news a lot lately, particularly when the outsourcing is done to a location outside your country. Based on my experience, a company should consider outsourcing when one of the following criteria is met:

1. The vendor can do the job better than your company, at a reasonable cost.
2. The vendor can do the job just as well as your company, but the vendor is less expensive.
3. The vendor can't do the job as well as your company, but the vendor is less expensive, and the vendor does a "good enough" job.
4. The vendor may not do the job as well as your company, but the need for the job is uneven, cyclical, or unpredictable. As a result, it would be easier and less expensive to have a vendor do the job than to maintain a workforce in your business to deal with the uneven demand.

When shouldn't you outsource?

Here are some outsourcing disadvantages:

1. You increase risk because you create a dependency on a single point of failure over which you have very little control. Risk is even higher if the vendor is dealing with your customers (like outsourced customer support), or if the vendor is handling an

aspect of your business that differentiates your company from its competition.

2. The process becomes more rigid, and is more difficult to change. An outsourcing business can only be profitable if the vendor has set up extremely well-defined processes. You may have to conform to the vendor's processes, follow the vendor's rules, and give up the flexibility that you had with your own resource.

3. You have to invest in a one-time expense to negotiate the outsourcing contract, work out all of the processes that you will use in relating to the vendor, determine the details of how the vendor will be measured and compensated, and handle the transition to the vendor's services. In a large outsourcing deal, this is a huge undertaking.

Most outsourcing starts out as a financial decision, but it's important that you go beyond finance to evaluate risk and return. Outsourcing your cleaning service may have very low risk, but outsourcing the development of a key company product is a whole different story. It's important to put together a risk mitigation plan as part of the early investigation into outsourcing, including the answers to questions like:

1. What are the risks? Specifically what could go wrong?
2. What are the consequences if each of those things happens?
3. What could you do to minimize the probability of those things happening?
4. What would you do if they happen anyway?
5. If things don't work out with this vendor, what options do you have? How difficult would it be to move the job to another vendor? To bring the job back in-house?

If the risks are too high, or you're not prepared to handle the consequences of things going wrong, then you shouldn't be looking at outsourcing.

Why Offshore?

I'll define offshoring as outsourcing work to a vendor outside your own country. I can think of only three reasons why you might want to outsource your work to an "offshore" vendor:

1. It's less expensive. This is the primary reason for offshoring in the IT world. Outsourcing to a country with a low cost of living or with a favorable exchange rate can be considerably less expensive than outsourcing to someone local.
2. The vendor may offer better quality. I don't hear this reason so much any more, although it was common in the electronics industry at one time. Some offshore software vendors claim to have better quality, but I haven't seen any statistics about actual work done by these vendors.
3. You can take advantage of time zone differences. For example, to provide customer support in off hours, or to have programmers implement your designs during the night to be ready the next morning.

Note that these three reasons don't replace the initial four outsourcing criteria—they just add another dimension to the decision. First you should look at whether you want to outsource, then you should look at whether you want to outsource to an offshore vendor. If it's too risky to outsource in the first place, then an offshore vendor won't change the situation. If an outside vendor can't do a job as well as your own people, then you either say "no" to outsourcing or you figure out whether the vendor can do a "good enough" job. What's "good enough"? That depends on the situation. Dell recently changed one of its offshore vendors because customers complained about service from offshore workers. In Dell's case, it wasn't good enough.

Additional Disadvantages of Offshoring

All of the disadvantages of outsourcing apply to offshoring, but offshoring introduces a few more things to watch out for:

1. Language barriers can cripple an offshoring relationship. If you thought it was tough for a business user to work out the details of a process with an IT person, wait until they try to do it with an IT person who speaks English as a second language. Be sure that any offshoring agreement makes it clear who will be talking to whom, and that language proficiency requirements are spelled out.

2. Currency fluctuation can change your costs unless you make sure your pricing is defined in your own country's currency. And even if your price is protected, that just shifts the currency fluctuation risk to the vendor, who has to absorb the cost. If your vendor becomes unprofitable, then the vendor may try to change the contract, or even go out of business.

3. Your agreement is at the mercy of international politics. Things we take for granted, like safe locations for work, are not as guaranteed in certain countries. Trade embargoes and wars between nations could impact your offshore vendor, or even cut off your access to the vendor. One offshore vendor I worked with had a "disaster recovery plan" that called for all of the workers to get in their cars and drive to a neighboring country when their government collapsed. Problems like this make power outages appear trivial in comparison.

4. Legal agreements may be difficult to enforce. You can go after the local representative, but they probably don't keep very much capital in your country. So be very careful to avoid pay-

ing in advance for work unless money is held in a trustworthy in-country escrow account.

Conclusion

Think of offshoring as an extreme form of outsourcing. All of the same outsourcing rules apply, and there are a lot more risks. But the much lower cost of offshoring has fundamentally altered the mathematics of the outsourcing decision, tipping the balance of the equation away from keeping work in-house. That's why offshoring is in the news: it's very attractive to consider cutting some of your costs by 50-90%, even if it means eliminating local jobs.

Note: This article was incorporated into my book, *Boiling the IT Frog,* as Chapter 15.

2026 Commentary

All of this is still true. And an additional disadvantage of offshoring has emerged recently: the possibility that tariffs might raise the cost of the imported goods or services.

15. The Opposite of "Project Risk" isn't "No Risk"

Originally published on FEBRUARY 10, 2004 in #MANAGEMENT, #PROJECTS

I've got a speaking engagement in a month or so where I'm going to talk about "How to Reduce Risk in IT Projects" (for full text of the presentation, see article #50.). In thinking about what I want to say in that presentation, it occurred to me that "risk" is an interesting word. We define the word as the uncertainty of bad things happening, and one of our goals in project management is to minimize that uncertainty and prevent the bad things. But what about the uncertainty of *good* things happening? What do we call that? Serendipity? Good luck? Good fortune? Happenstance? And why is it that we focus on preventing the bad things but we don't try to make good things more certain?

Look at it another way. We usually consider a project to be successful if it achieves the planned result with the expected quality and if it finishes on time and within budget. Realistically, this success is pretty rare; according to the Standish Group, only 34% of U.S. IT projects in 2002 were completed on time and within budget, and $55 billion was wasted due to canceled projects or cost overruns, out of $255 billion spent on U.S. IT projects. Yet it seems to me that we ought to be striving for more: projects which *exceed* expectations, provide *superior* quality, finish *early*, and come in *under* budget.

Do I expect this to happen very often? No, but I think we're making a mistake in not taking the right steps toward this extraordinary result. I believe that many of our projects are late because once we create a project plan, we never try to do better than the plan. We try to bring the project in on time, but *never early*. If a task is completed early, then we figure we have some breathing room and we don't start the next task right away. In many cases, we *can't* start the next task right away because no one assumed that we would be able to finish the first task early, and so things aren't ready for us to start the next task. If we adopt this attitude for all tasks which finish early, then the best possible outcome is a project that's on time, and the project will be late if the least little thing goes wrong.

Similarly, our projects run over budget because we try to bring them in *on* budget but not *below* budget. If we're under budget in one part of the project, then we tend to spend the money on other lower priority items that probably aren't required. Again, the best we can hope for in any budget category is to be *on* budget, and therefore most projects run *over* budget.

Swing through the Ball

In case you're getting the wrong impression, I'm not advocating "sandbagging" your project estimates, ensuring a favorable outcome by deliberately adding more time or money to your project plan than you think you'll need. But if your people have the attitude that they can take as much time to finish the project as was planned, then there is no incentive to beat the time. It's kind of like the idea in golf (as well as baseball and tennis) of "swinging through the ball." You don't just hit the ball and then stop your swing on impact. Instead, you keep swinging even after you've made contact with the ball, and this gives you more power to make the ball go farther and more control to make it go where you want it to go. In projects, you shouldn't just create a project plan and consider it the best that you can do. You should treat the project plan as a reference point and then try to do better than the plan. "Swing through" the project plan by trying to beat it.

How to Make This Work

Let's take a look at the way we should schedule our tasks for a project. For each task, most people estimate the most likely effort and duration, and then they provide some reserve time or "slack time" after each task to allow for possible uncertainty. Suppose instead we estimate the lowest *reasonable* effort and duration[1], and we define the project tasks accordingly. Then we should provide a "slack *task*"[2] after each major task to pad it out to the most likely effort and duration, and we should remove these slack tasks from the schedule as their corresponding tasks are completed. Are we kidding ourselves by doing this? No, we're just giving a message to our employees that they should try to achieve the lowest reasonable effort and duration for each task, even though the overall plan is based on the most likely effort and duration. Mediocrity shouldn't be a goal; it's what you get when you don't achieve your goals.

Ok, how about the slack time that we typically add to a task above and beyond the most likely effort and duration? Take it out of the plan, and instead of scheduling the slack time at the task level, you should include an aggregated slack time at the end of the overall project. If you include slack time at the task level, then

---

1 Notice that this is the lowest *reasonable* effort and duration, and not the lowest *possible* effort and duration. The estimate should be challenging but not so challenging that employees give up when trying to achieve the estimate.

2 A "slack task" is a dummy task that acts as a placeholder in a project schedule. It's like slack time, except slack time has a duration but no effort, while a slack task has both duration *and* effort. The duration in this case should be the difference between the lowest reasonable duration and the most likely duration for the task corresponding to the slack task. Similarly, the effort should be the difference between the lowest reasonable effort and the most likely effort for the task corresponding to the slack task. The effort should be assigned to the same resource(s) used for the task corresponding to the slack task.

people will assume it's for their use as breathing room and they'll take advantage of it, often to the detriment of the project schedule. If you include it at the end of the overall project, then it can be used as a buffer for task overruns on the project's critical path where it's needed the most.

Let's look at another aspect of good project management, the risk mitigation plan. A good risk mitigation plan has a list of risks, a contingency plan for each risk, a well-defined trigger event that will cause the contingency plan to be invoked, a "risk owner" who is responsible for tracking the trigger event, and a list of steps we're going to take to make each risk less likely. But again all of the risks are negative events, so the contingency plans are designed to get the project back on schedule—not to bring the project in early or under budget. Have you ever seen a project plan that includes a "serendipity encouragement plan" or a "good luck plan"? The terms seem absurd because the idea is so uncommon that we don't have a good name for such a plan, but the idea is to list the positive events that might happen and what we're going to do in the project to make them more likely. For example, user training might go faster if the system is easier to use, so list the things we will do to ensure that it's easier to use. Or costs might be lower if we can reuse some existing servers, so list the things we'll do to make that reuse more likely.

Conclusion

We have naively assumed that the opposite of "risk" is "no risk," but that's not the case. The opposite of "risk" (negative uncertainty) is "good luck" (positive uncertainty), and we ought to be taking steps to make good luck more likely. As Seneca said, "Luck is what happens when preparation meets opportunity." If you anticipate possible areas where "good luck" things can happen, prepare for them, make them more likely, and create your opportunities, then your projects will always be much better than just mediocre.

2026 Commentary:

I haven't kept up with project management literature, so I don't know how typical this approach has become. If it hasn't become popular, I'm guessing it would be due to political pressures. Including anything called "slack" in a project plan tends to attract attention from people who don't understand the problem. The natural inclination from people who don't manage projects is that slack shouldn't exist. Of course they're wrong.

Maybe using this approach requires keeping "two sets of books": one with slack managed the way I describe, and another set of plans that hide the slack.

16. Experience, Knowledge, Wisdom, and Better

Originally published on MARCH 9, 2004 in #LEADERSHIP, #CAREERS

I'm writing this on my birthday. Birthdays are a pretty arbitrary measure of aging—the earth has made some number of revolutions around the sun since I was born. Time passes, and our experience grows. But it's not our experience that counts in life as much as what we do with it. Do we learn from it? Do we apply it to help others? Does our experience give us knowledge and wisdom?

Experience and knowledge by themselves are not enough. Knowledge is just information that's made meaningful. You sometimes hear the phrase "full of useless knowledge" describing people who know a lot of meaningful information about topics that are regarded as irrelevant by most of us. But what makes a topic relevant? Something is relevant if it can be put to use, if it can be applied to a situation to make things better.

That word "better" seems to be the key. We want to make the world a *better* place to live. We want to make things *better* for our children and grandchildren. The key to life is progress toward better things. Sometimes progress is slow, requiring our patience and dedication; and sometimes progress is fast, requiring our control and focus.

Things don't always get better; sometimes they get worse in spite of our efforts. But then we try to turn things around, making them better again.

I mentioned wisdom in the opening paragraph, but what is wisdom? I think it's knowing what "better" is. It's knowing that sometimes things that seem better in the short-run aren't really better at all, and that we need the perspective to strive instead for what's better in the long-run. Wisdom is being able to differentiate between subtle shades of "better" to pick the best goal and the best path to achieve the goal.

Wisdom is taking all the knowledge and experience we've gained over the years and putting it together into an understanding of what's right and wrong, important and unimportant, better and worse. But it goes two steps further:

1. Wisdom applies that knowledge and experience to help others choose personal and business goals that make things "better" for them.
2. Wisdom helps them achieve those goals by enabling small course corrections on the route toward goal achievement, when changing conditions cause us to alter the meaning of "better."

Most of our life is spent achieving goals, but we aren't making progress if things aren't becoming better. Be sure your goals are the right ones. Spend a little time on your birthday, or any other day, thinking about whether you're headed in the right direction to make things better. If not, then apply some of your wisdom (and the wisdom of others) to pick different goals or different ways to achieve them.

2026 Commentary

Right now our country is going through a period when the definition of “better” is widely contested. I hope I live long enough to see people gain the wisdom to resolve those differences.

17. The world is non-linear—use it to your advantage

Originally published on APRIL 13, 2004 in #MANAGEMENT, #PROJECTS

We live in a non-linear world but it's funny how we keep forgetting that fact. Eating twice as much for dinner won't make up for missing lunch. A runner who can do a 4 minute mile can't do 10 miles in 40 minutes. If it takes two painters one day to paint a room, then sixteen painters can't do the job in an hour. And no matter how much they might try, nine women can't have a baby in a month.

So why is it that everyone assumes that Information Technology is full of linear relationships? I've seen companies try to cut project time in half by doubling the project staff, only to find that the project takes longer than the original estimate because of all the extra training, coordination and rework. I've seen companies cut their support budget by 10% and double their customer complaints. And it's not unusual to pay for a computer that's twice as fast as your current one only to get a mere 20% performance improvement.

I think our love affair with linearity goes back to our early schooling, when we were given simple math problems like "If comic books are ten cents [this is *early* schooling] and John has five dollars, how many comic books can John buy?" Or the old classic "If a train leaves the station at 6:00 PM going 30 MPH, …."

But where were the footnotes on those problems? Footnotes that should have said "Warning: The real world is much more complicated. Quantity discounts apply. Some items are out of stock. You have to pay sales tax. Trains have to slow at various points on the journey …." The list goes on and on.

I remember being told in college that every curve looks linear if you look at a small enough segment of the curve. And that's where we usually make our mistakes: we truly believe there is a linear relationship between two variables, like labor and time, or money and time. Then we extrapolate based on that linear relationship beyond all reasonable expectation. But more labor requires more people interaction and time for communication, and so exponential relationships between labor and time are more common. And more money leads to more justification, more meetings, more red tape, and therefore less effective use of the money.

The Up-Side

Is there an up-side to non-linearity? Can we actually use non-linearity to our advantage? I think a lot of the recent software development methodologies are taking advantage of the up-side. Techniques like "extreme programming" and "agile programming" focus on small teams doing rapid development efforts aimed at quick deliverables. Instead of a long drawn-out project, large deliverables are chunked down to a series of small projects, each with its own focused team. In some cases the user function is delivered in phases, starting with a bare minimum, and then with each phase adding a little more function and detail.

Performance on these small team projects is stellar. Deliverables are crisp, and with short project durations there is little scope creep from start to finish. Morale is high—everyone knows what they're doing, and they focus on the result. The projects move too fast for team members to get bored. Red tape is at a minimum, and because of the small team, communication is easy, and rework is minimized.

Even if requirements are somehow misunderstand at the start of the small project, it isn't as much of an issue as it would be with a large project, since you haven't gone far enough in the wrong direction to make it a big issue to change your way. There is less wasted effort in a series of small projects. Compare a large project deliverable to a long sea voyage across an open ocean toward a far away destination. With a large project you only "shoot the stars" to determine your position (sanity check your deliverables) a few times during the voyage, so the currents (the gap between what the business needs and what you think you're doing) could take you very far off course. With smaller projects you "shoot the stars" with every small incremental deliverable, and so course corrections can be made much more frequently. As a result you'll waste far less effort by breaking up that long voyage into a series of smaller trips.

Conclusion

The world isn't linear, but the trick is to make non-linearity work *for* you instead of *against* you. The larger the project, the more non-linearity works against you. But use one of the new techniques, break up your large projects into smaller projects, and the non-linearity can be used for your benefit and for the benefit of your company.

2026 Commentary

When I wrote this in 2004, this was not common practice in software development the way it is now. But the smaller-project approach is equally useful for projects outside of IT, and its use for those non-IT projects isn't as widespread as it should be.

18. IT is Moving toward Property Management

Originally published on MAY 11, 2004 in #CAREERS, #IT, #ALIGNMENT, #STRATEGY

Last month I had the unique opportunity to help a large university plan its future curricula for its undergraduate and graduate degrees in computer information systems. The university recognizes that Information Technology is changing, and wants to make sure that its students are being prepared for the real world. To this end the university has asked various people to help them see things from an outsider's point of view.

The exercise made me think about my own schooling, and how things have changed since that time. When I first entered MIT as an undergraduate over thirty-five years ago, the only computer-related degree was in the electrical engineering department, and that degree focused on designing and building computers—not using them. I started in that program and rapidly became disillusioned; I then moved to the management department because it seemed to be more focused on the business use of computers. I had to know at a high level how computers work, but I certainly didn't need to know how to design computers in order to use them.

Now it seems we have moved another level away from the computer itself. Off-the-shelf software packages from SAP, Oracle, PeopleSoft, and other similar companies have drastically reduced the need to write software from scratch. And in the few cases where proprietary advantage might be gained by having custom-written software, outsourcing and offshoring have become the dominant methods for creating that custom software.

The net effect is to make the knowledge of how to design software almost as irrelevant to many future IT college students as knowing how to design computers was to me thirty-five years ago.

Real world IT people are moving away from writing software. Instead, they spend their time buying off-the-shelf software packages and the hardware to run them; they customize and adapt the software to their business needs; they interconnect and interface the purchased software and hardware; and then they keep their infrastructure running smoothly and efficiently.

This new definition of the IT role in business reminds me very much of another business role: real estate property management. This is the organization that runs your buildings, keeps your offices at the right temperature, repairs leaky plumbing, keeps the grounds attractive, controls the security to the building, and generally gives you a nice, clean, comfortable and safe place to work. Think about the property management role and its similarity to IT. The property management organization manages infrastructure. They have to buy, maintain, and replace large expensive complicated assets like elevator systems and HVAC (heating, ventilation and air conditioning). They maintain and upgrade a network of pipes and ducts throughout the building. They handle changes in requirements caused by your evolving business.

Furthermore, as building technology has advanced, the property management role has become more and more computerized. It's not unusual for an extreme temperature registered by one of the building sensors to trigger an alert to someone's pager in the middle of the night. The remote person can then connect to the building HVAC system over the Internet, see a graphical display of the building temperatures and HVAC settings, and make changes to various settings to re-balance the system. This is exactly the same kind of approach that IT network or server engineers use now to deal with computer problems that occur at odd hours.

Any new technology goes through phases. In the "frontier" phase, there's a lawless environment where it's "every man for himself" because there are few support systems to help, just like the Wild West in the 1800s. Then as the use of the technology grows and becomes standardized, and as support systems become common, things become much more practical and ordinary.

Computer information systems went through the frontier phase up until around the 1990s, but gradually during the 1980s and 1990s the systems became more and more standardized, and off-the-shelf software began to proliferate. Sure, there's still room for new frontiers, especially in the communications and medical computer fields. But for most systems in most businesses, we're out of the frontier and into the practical and ordinary. Life in IT in most companies is about improving business processes by applying technology—not about developing exotic state-of-the-art software.

So where does that leave students in future computer information systems programs? I think it provides a choice of directions. For some students who are more business focused and who are interested in making things incrementally better, the focus should be on many of the same types of skills that are taught to real estate property managers: the tools of the trade, how to work well with customers, how to select and deal with vendors, how to measure and achieve success, and how to manage money and large projects. For other students who want to work on the next software frontier, the attraction of designing and developing software will be strong, and they should be given the opportunity to learn more, just as our electrical engineers continue to learn new and better ways to design computers.

Eventually, I think the degree titles will change. Computer information systems will break apart into two degrees: one in "software engineering" and one in "management of information systems" (or maybe "information systems leadership"). Check back with me in 10 years and we'll see if I'm right.

2026 Commentary

Well it's 22 years later. I think I was generally correct about the way that education would shift, but it seems to have broken into a lot more categories than I anticipated. What I called "software engineering" now covers security and data protection, artificial intelligence in all of its forms, and various hardware/software combination devices for robotics, geographic positioning and tracking systems, and intelligent appliances. What I called "management of information systems" now covers what I talked about, plus data management and analysis, as well as niche IT uses like investment analysis and arbitrage.

19. Hidden Consultants within your Organization

Originally published on JUNE 8, 2004 in #MANAGEMENT

You've all heard the old joke about a consultant being someone who uses your watch to tell you the time, and then steals your watch. There's some truth to the story: consultant recommendations are often the same things that your employees or customers have been telling you all along. But while you will listen to a consultant, you don't listen to your employees and customers. Why is that? Why do companies pay more attention to consultants than they do to employees or customers? And what should you do about it? But let's start with an even more important question: *why should* you listen to employees and customers?

Why listen?

The best reason for listening to employees and customers is that they have a detailed understanding of your company's problems. They're close to day-to-day operations so they see what's happening and what's wrong. When you implement their ideas they're committed to success because of their personal involvement. The result is a fast implementation of change, with a high probability of success.

So why don't companies listen to employees and customers? Based on my own experience, here are the primary reasons why companies don't take advantage of their hidden consultants:

No Clear Summarization

First, we don't hear recommendations from employees and customers in a clear summarized way. We interact so much with these people that the recommendations aren't separated out from everyday comments. A profound statement is dropped casually into a conversation on another subject, and we miss the importance of the statement.

When I'm working as a consultant doing interviews with employees and customers, I often hear significant statements, but I notice them because I'm listening for them. My brain is focused on gathering important information, and so I'm able to separate out the irrelevant stuff from the important things. Most people don't listen that way on a day-to-day basis.

Then, after I've heard an important statement and verified it with others, I'll figure out the best way to convey the statement to my client. Sometimes the issue with accepting a recommendation isn't so much the recommendation itself; it's how the recommendation is presented. Important truths have to be presented in a way that makes the client see the light without taking offense. Employees and customers don't often use appropriate summarization and presentation techniques, and so we reject their recommendations.

Bias

Second, we think employee and customer opinions are biased and therefore unreliable. We think they are trying to advance their own personal agendas. For example, an employee just wants to do that project because it would make their job bigger. Or a customer wants us to improve our service because their won't take responsibility for problems in their own company. Sometimes personal bias will color an opinion, but we don't take the time to sort out the motivations and get to the truth of the matter. Instead, we just tune out the employee and customer comments, throwing out valid suggestions because we think the source is biased.

Reluctance

Third, we have made it pretty clear to employees and customers that we don't want their advice, and as a result, they are reluctant to offer it. They see us criticize ideas and shoot them down, and they see us label idea creators as rabble-rousers and troublemakers. In such an unsupportive environment, they have determined that it's best to keep their heads down and their ideas to themselves.

Other reasons

Here are some other reasons I've run across:

- Some managers don't want to acknowledge that their own employees can be more knowledgeable about a subject than the managers are. The managers forget that the employees (a) are usually closer to everyday problems, and (b) have had a life before working for this manager, and so they have other experience to bring to the table.
- Managers sometimes feel that giving an employee a strong say in an issue will be viewed as "giving up control." We forget that we aren't in control anyway. At best we're leading and steering, and certainly we're accountable, but the employees who do the work actually have control over the process—not the managers.
- There's a feeling of "you get what you pay for," so we feel that a low-paid employee can't provide as good an opinion as a high-paid consultant. This is a narrow viewpoint, but it feeds the families of many consultants.

How to use your hidden consultants

So what can you do to take advantage of these hidden consulting talents? Here are some suggestions:

1. Help your hidden consultants learn how to focus. Provide training for your employees and customers in techniques that help them find the root cause of a problem, determine possible solutions, and put together a plan to solve the problem.
2. Provide a way to get feedback from employees without you being biased by the source of the feedback. Create a method for employees to submit suggestions and ideas anonymously, but with a way to subsequently identify the suggester if you want to provide a reward.
3. Identify someone (internally or externally) who is good at summarizing and presenting. Have that person summarize employee and customer feedback and present it in the way that an outside consultant would.
4. Have a program in which selected employees can be "consultants for a week." Having these employees think like consultants takes them outside the day-to-day process, if only temporarily, and gives the employees the opportunity to identify issues and recommend solutions. Sometimes this approach is even more effective if the employees act as consultants for different departments than their own.
5. Help your employees to learn how to differentiate between a "reason" (why you are a certain way) and an "excuse" (why you stay that way). And make sure that you understand the difference yourself.

Conclusion

I make my living as a consultant, so I obviously don't want you to stop using my services. But ultimately my goal is to help companies be more profitable and become better places to work, and most companies are missing a huge opportunity for self-improvement. By taking advantage of the ideas generated by your own employees and customers, and by focusing those ideas on providing benefit to your business, you can reserve the use of outside consultants for the things we're best at:

- Providing skills and expertise that don't exist within your organization, and
- Helping your organization develop better processes for optimizing your own skills and expertise.

You have a huge pool of hidden consulting talent within your organization. You just have to focus it and use it.

2026 Commentary

Artificial Intelligence has added another tool for analyzing business problems and providing solutions, but don't be deceived into thinking that AI is more trustworthy than employees and customers. AI is basing its "expertise" on whatever database it's been trained on. It might come up with useful suggestions, but you can get generalized useful suggestions from reading any business book (like this one).

I view AI as just another tool in the toolbox. Employees and customers are still better informed about current processes and systems, and where they work and don't work. So use AI and business books to come up with ideas if you like, but sanity check those ideas by discussing them with employees and customers. It's the employees and customers who have to live with the result.

20. Why ROI Isn't Working

Originally published on JULY 13, 2004 in #ALIGNMENT, #MANAGEMENT, #PROJECTS, #STRATEGY

ROI (Return on Investment) is the most common and popular method for project ranking, both in IT and elsewhere. But ROI isn't working in most companies, and as a result, businesses are making bad project decisions. In this article I'll explain why ROI isn't working. Then in the next article I'll tell you how you can improve your use of ROI, and I'll talk about an important additional project selection criterion that should be used in conjunction with ROI.

Here are some of the reasons ROI isn't working:

Biased proposals

In most companies, project proposals are put together by the people who want the projects to be approved. It might be a marketing or operations department who benefits from a project, or it might be the IT group itself. In any case, proposals are going to be deliberately biased to make the projects look good because the people writing the proposals know that they have to compete against other project proposals which are similarly biased toward their own projects. Making projects look good isn't a problem, but when these same people do the ROI analysis for their projects, the results are far from fair and objective. ROI exaggeration becomes a game, and under the rules of this game, if a benefit *can* be claimed, then it *will* be claimed, whether or not there is any probability of it actually occurring, and whether or not it will have real financial benefit to the firm. For example, if a new system has the potential to save an hour per week for 200 secretaries, then you can be sure that the average hourly wage of a secretary (with benefits) will be multiplied by 200 and then by 52 weeks to chalk up some annual financial benefit. Never mind that the hour will probably be absorbed by other tasks, and that no financial savings will actually accrue to the company.

On the other hand, if the secretaries have to be trained on how to use the new system, and if there's a learning curve involved, then you can be assured that this training time will be minimized in the estimate (maybe an hour or less per secretary). In addition, the labor costs associated with people being trained will likely be ignored, and the learning curve will be neglected altogether. The general rule for preparing the ROI in a proposal is to maximize possible benefits and to minimize possible expenses, thus inflating the ROI beyond all reality.

Omission of Secondary Costs

But it gets worse. In their eagerness to get their project approved, the proposal writers will conveniently forget that aspects of their project will increase costs in other areas. This is especially likely if a project resource is shared with other projects or other departments. Things like increased network traffic or server usage will probably be ignored. And even more likely to be forgotten are the tertiary costs such as increased support dollars for those secondary higher resource demands.

Neglect of Transition Issues

Of course, no project will get its full payback on the first day that a new system is installed. Not only will there be a transition period while users learn a new system, but in most cases there will be a short-term decline in productivity and quality while people make mistakes and then figure out how to deal with a new environment. Yet most ROI proposals show only two worlds: the one before the project is implemented, and the vision of the perfect new world after the project is completed. The costs of the transition period are usually ignored.

Minimization of Risk

Risk is usually forgotten in ROI projections. This is particularly ironic since risk is such a large part of project management once a project gets approved. Nevertheless, it is rare that risk is considered in ROI since an ROI projection boils down to one number, not a series of numbers. Have you ever seen an ROI calculation that says "there is a 50% probability of a $100 million savings, a 40% probability of no savings, and a 10% probability of a $50 million loss"? I doubt it, since ROI proposals usually show only one outcome: the successful one. Any problems with risk are considered "implementation issues" and won't be addressed until the project is started.

Also conveniently omitted from many project proposals is the increased risk to the business of becoming dependent on a new and less reliable process. This is the risk of the new system itself, which is over and above the risk of whether or not the project can be successfully completed.

Missing Prerequisites

The ultimate bit of gamesmanship for a true ROI game player is to carefully break up a project into pieces, with a separate ROI for each piece. Keep most of the costs in the earlier project pieces, or if you're a grand master, then bury some of the costs in an infrastructure project that also includes infrastructure improvements for other projects outside your department. All you really care about is the final project piece, which now has minimal cost and incredibly high benefit because all of the major costs have been buried in other prerequisite projects. Through careful project segmentation and manipulation, you can arrive at a great-looking project with a massive ROI that is guaranteed to be approved. Of course, actually doing the project is going to require prerequisite aspects of other unapproved projects, but you can blame that problem on IT or some other convenient service department, so you don't need to mention it in the project proposal. After all, shouldn't it be up to the other departments like IT to get their own projects approved? That's not your problem as a proposal writer, is it?

No Follow-up Measurement

Ok, so the project gets approved. How many companies go back after the project is complete and actually measure the costs and benefits to calculate the real ROI? I've never seen anyone do it (If you have, I'd like to hear about it). And even if the costs and benefits are measured, there's a high likelihood that the numbers will be distorted and therefore unusable. Labor savings are minimized by other process changes that weren't anticipated when the project was proposed. Costs are shared with other projects and so cost accounting is difficult. And underlying volume assumptions have changed—up or down—and this affects the measurement. It's very difficult to say whether the project ROI was actually achieved, and so the proposal writers who inflated the estimates get away with their crime, and they will go on to use similar tactics in additional projects.

Conclusion

ROI isn't working because the highest ROI doesn't go to the best project—it goes to the most creative exaggeration of ROI by a proposal writer. But in spite of its deficiencies, I'm not saying that ROI ought to be discarded as a tool. In the next article, I'll give you some recommendations for improving the accuracy of your ROI calculations, and for selecting the projects which can provide the most benefit for your company.

Note: This post has been combined with another post on ROI from August, 2004, "How to Improve ROI and your Project Selection Process," and incorporated into Chapter 5 of my book, *Boiling the IT Frog.*

2026 Commentary

Human nature hasn't changed, so this is all still true.

21. How to Improve ROI and your Project Selection Process

Originally published on AUGUST 10, 2004 in #ALIGNMENT, #LEADERSHIP, #MANAGEMENT, #PROJECTS, #STRATEGY

In my previous article I explained why ROI isn't working in most businesses. Based on comments I received from readers, I want to quickly point out that I don't believe that the "game players" I mentioned are bad people, in spite of my use of the word "crime" in the article. I believe that everyone falls into that game playing mentality when confronted with a process that requires it. It's the process that's at fault—not the people. The people just learn that in order to get a project approved, they have to exaggerate the ROI, and this provides the basis for the problem.

So how do we fix the process? The core of the problem is an inherent bias in figuring ROI, so let's try to eliminate the bias. Here are some techniques you can use:

1. Have a single group of people who determine the projected ROI for all projects, or who at least review the ROI calculations for projects to make them consistent. Eliminate the advantage that imaginative exaggerators have in doing project proposals.

2. Ensure that ROI proposals include *all* costs, including all prerequisite projects, infrastructure requirements, and all secondary and tertiary costs. If a project is to be implemented in phases, then include the ROI for each phase as well as the cumulative ROI for all prior phases, and make sure that it's clear to everyone that a later phase cannot be accomplished without an earlier phase, and therefore cannot be approved without the earlier phases.

3. If a project depends on a prerequisite that is also a prerequisite for other projects, then either (a) break out the costs of the prerequisite that are a specific requirement for this project, and include them in this project's ROI calculation, or (b) go ahead and approve the prerequisite and take its costs out of the ROI calculations for all of the projects on which it depends.
4. Differentiate in proposals between real dollar savings which affect the bottom line and theoretical savings (e.g., an hour saved per week from each of 200 people) which may be absorbed by the business.
5. Specifically include a section in each proposal on the transition from old system to new: how the transition will occur, how long it will take, what will happen, and how productivity and quality may be impacted during the transition. Factor those productivity and quality costs into the ROI.
6. Include consideration of risk by doing scenario planning in the proposals that require it. Identify the risks in a project, including probabilities of various outcomes. Explicitly identify potential issues, and discuss the steps that will be taken to minimize or eliminate them. Develop multiple ROI calculations based on various likely scenarios, and then build a single composite ROI number based on a probabilistic weighting of the various ROI numbers. For example, if there is a 50% probability of a $100 million savings, a 40% probability of no savings, and a 10% probability of a $50 million loss, then the composite ROI number is $100 x .5 + $0 x .4 - $50 x .1 = $45 million. This may still be a project worth approving, especially if clear steps are taken to minimize the 10% probability of a loss.
7. To the extent possible, go back after each project to determine the actual ROI that resulted from the project. This should be done as part of a generalized project close-out process, which

should also discuss things that should be done differently in the next project.

8. Even better, get user commitment for savings *before* the project starts. For example, if the project is going to eliminate the need for one clerk in each office, then go ahead and reduce the office budgets in advance to reflect the savings. If the project succeeds, then the savings have already been locked in. And now every office has an additional incentive to make the project successful, because they're already committed to the savings.

I've used this last approach successfully in two different companies. In both cases, advance commitment was made to headcount reductions in anticipation of savings from a new system. Because the commitment was made far in advance, the companies were able to meet the headcount reduction goals through normal attrition without any layoffs. And the system ROI was easily achieved because the savings were built into the budget.

The Missing Project Selection Criterion

By using the techniques listed above, you can reduce the bias in ROI estimates. But consistent ROI estimates aren't enough; an underlying problem with using ROI for project selection is that the project with the best payback isn't always the best thing for your business. Your business has a strategy for advancing in the marketplace, and the best projects are the ones that help implement that strategy—not necessarily the ones with the highest ROI. Choosing projects based on ROI may contribute to the bottom line in the short-term, but this approach will not position your company for the future.

To improve the project selection process, add another criterion to project selection: strategic alignment. There are a lot of ways to do this; here are a few:

- Use an ROI multiplier. For example, if the project is in strong alignment with the strategic direction, multiply the ROI by 1.5 to get an "adjusted ROI"; if the project is in moderate alignment, multiply ROI by 1; if the project goes against strategic direction, then multiply ROI by 1/2 or 1/4. Then rank projects on the resulting adjusted ROI.
- Use strategic alignment as the primary criterion. Only consider projects that are in strong or moderate alignment with strategic direction, regardless of their project ROI.
- Use strategic alignment as a project budget allocation tool. Set aside a certain amount of the project budget (say, 75%) for projects that are aligned with strategic direction. Then spend up to 25% of your project budget for unaligned projects *if* their ROI is higher than any aligned project.

You can work out the specifics of *how* to use strategic alignment in your project selection process; the important thing is that it is taken into consideration in some way in your process.

Conclusion

ROI is probably the best known project ranking tool, but it's seldom used consistently and correctly. We're kidding ourselves when we say that we use ROI to rank projects on financial returns. The reality is that calculating project ROI is a game in most companies, and the best game players get their projects approved.

However, you can change the rules of the game to make ROI a better reflection of reality. And you can ensure that your projects are strategic by measuring their alignment with strategic direction. The result will be a better choice of projects for your business, and a higher percentage of successful projects overall.

Note: This post has been combined with another post on ROI from July, 2004, "Why ROI isn't Working," and incorporated into Chapter 5 of my book, *Boiling the IT Frog*.

2026 Commentary

There are some good ideas here, but there are political issues in implementing them. Everyone in a company has their pet projects, and they get annoyed if the process interferes with their projects getting approved. Applying these rules to *most* of the projects will be more likely to gain approval within your company.

22. Acquisitions 101: Why Companies Acquire Other Companies

Originally published on SEPTEMBER 14, 2004 in #ACQUISITIONS, #ALIGNMENT, #LEADERSHIP, #MANAGEMENT, #STRATEGY

If you haven't been involved in an acquisition, then wait a few months; it eventually happens to all of us. In preparation for that event, it's useful to get an understanding of why companies do acquisitions, so that you can anticipate the effect that an acquisition will have on you. I've been lucky enough to represent the acquiring company in more than twenty acquisitions, and so the information in this article is based on real experience. Yet I don't claim to know all industries or all cultures, and so my understanding is obviously limited to what I've observed.

3 Reasons for an Acquisition

In my experience, a company acquires another company for one or more of three reasons:

1. Vertical integration
   To take control of a key supplier, product or service on which the company depends; or to gain the profits that are otherwise going to these suppliers, products or services
2. Horizontal integration
   To fill in "holes" in the product line, to add complementary products, or to add customers in new markets
3. "Synergy"
   That magical situation in which the combined whole is greater than the sum of the two companies separately

Underlying all of these reasons is the assumption that the company wants to grow. The quest for growth is (unfortunately, in my opinion) common to all public companies and most private ones. Investors expect returns from the companies in which they invest, but an increase in stock price only comes from increased revenue per share or, to a lesser extent, increased profitability per share.

Now, looking at the three reasons, we see that the first two types of acquisitions can be done while keeping the two companies at arm's length—by continuing to run the acquired company as a totally separate business unit. When acquisitions are done that way, the impact on an information technology (IT) organization is pretty much the same as the impact of having to deal with typical partnerships or third-party relationships: mostly interfaces and data feeds.

But I've never seen an acquisition where synergy didn't come up. Attempts to achieve synergy can cause major disruption for IT and almost every other organization in both companies. Since synergy is one of the reasons for almost every acquisition, and because it's the primary reason in a good number of acquisitions, let's take a closer look at what synergy is all about. Here are three common synergy scenarios:

Synergy Scenario 1: More revenue, less expense

In this scenario, it is expected that together the two companies will have greater revenues, but that expenses will be lower. Revenues are expected to be higher due to cross-selling between the two customer bases; combining the sales forces; combining marketing and advertising; and due to an enhanced brand. Expenses are expected to be lower due to economies of scale; combining operations; consolidating systems and computers; eliminating duplicate functions and organizations; and in some cases by "centralizing."

Synergy Scenario 2: Target business is losing money

This is a variation on the first scenario. The "target" company (the one being acquired) has higher expenses than its revenues. But by taking advantage of the techniques listed in scenario 1, the combination of the two companies is expected to be not only profitable, but *more* profitable than the parent company doing the acquiring.

Synergy Scenario 3: Take the customers

Under this scenario, the parent company doesn't really want any of the operations of the target company; it just wants their customers. This type of acquisition is common when the parent company is in a business that already provides all of the products or services that the target company's customers use, and the acquisition is done to add revenue to the parent without going through the cost of acquiring the customers individually.

In a typical acquisition of this type, say in the banking industry, some of the target company's customers jump ship and refuse to become customers of the parent (the expected percentage of lost customers is factored into the acquisition price). Customer conversion is the key factor here: if customers can be converted to use the parent's products and services without any disruption or effort on the customer's part, then most customers will be expected to make the switch.

How does synergy affect IT?

There are a number of questions that an IT organization has to answer in order to deal with an acquisition. First there are the due diligence questions that should be answered before the acquisition takes place. What synergy scenario applies to this acquisition? If it's scenario 1 or 2, then answer the following questions:

1. Does the target company really have the technology they say they have? Do they own the rights to it (intellectual property rights, current third-party licenses)?

2. Can the combined company benefit from technology economies of scale? If so, where and how?

3. How easily can data and systems in the two companies be integrated or interfaced to meet company requirements?

If synergy scenario 3 applies (take the customers), then answer these questions:

1. How difficult will it be to replace customer-facing products, systems and services for target customers with the products, systems and services of the parent?

2. What migration path should be used for target company systems and data? How long will the process take (long migrations guarantee more lost customers)?

3. Will the existing parent systems support the increased volume? If not, then what will have to be changed?

Then after the acquisition takes place, there are more questions that have to be answered for all scenarios:

1. How will roles and responsibilities change? You'll have to get a feel for the new people with whom you'll deal: What are they like? Can you trust them?

2. What technology economies of scale will you actually implement? What systems will be combined? How will data centers and servers be affected? How can this be done to maximize payback while minimizing disruption to the business?

3. How will data move among the various systems in the two businesses? What integration or interface tools will be used to make the data flow more efficient?

And finally, there's the question that you ought to be asking yourself whenever you try to take advantage of economies of scale: What's the likelihood that the acquisition will fail and will have to be spun back off? If we combine systems and organizations, what problems will this create if the business has to be split up again later? Is it worth it?

Conclusion

Acquisitions are a fact of life in business today. Most acquisitions are done in the expectation of some sort of synergy, and the acquisitions that fail usually fail because that synergy wasn't achieved. With the proper due diligence before an acquisition, you can set more reasonable expectations. And with the proper planning, you can achieve the synergy that your company expects and desires. Acquisitions are like anything else: they're great if done well and awful if done badly.

2026 Commentary

I mention that "an increase in stock price only comes from increased revenue per share or, to a lesser extent, increased profitability per share." There is actually another way to increase stock price that wasn't being done as much when I wrote the article: the company can buy back its own stock. However, companies that buy back their own stock are typically doing so as a tax strategy: it gives the stockholders a tax-deferred capital gain instead of a taxable dividend. So this doesn't really affect the message in the article.

While the article is written from the IT viewpoint, the same logic can be applied to any other department affected by an acquisition. Are there savings that can be obtained through combining the relevant departments of the parent and target companies? How easily can that combination be done? And how easily can that combination be undone if the acquisition fails?

Also, the article is written about acquisitions, but the concepts apply equally to mergers, with one additional issue: Who is in charge of each company area that gets merged? This introduces political issues that complicate the process.

23. Julia Roberts, Training Wheels, and Bureaucracy

Originally published on OCTOBER 12, 2004 in #CAREERS, #ALIGNMENT, #LEADERSHIP, #MANAGEMENT

What do the three things in the title of this article have in common?

Julia Roberts

In the movie *Mona Lisa Smile*, there's a scene in which a Wellesley instructor played by Julia Roberts shows the class a painting, and asks them whether or not it's any good. Up until this point all of the art they've seen in the class has been well-known, and so the group is confused. They frantically search through their textbooks, looking for some mention of this painting. The Julia Roberts character is making a point: the appreciation of something as good or bad is inside of you as an individual, and cannot be obtained by simply following the "rules" for good art. We need to be more than just people who follow rules; we need to develop a sense of individuality and self.

Training Wheels

When a young child first rides a bicycle, the bike is often equipped with training wheels to make learning easier. The theory is that the child will get a feel for the basic mechanics of pedaling and steering before having to cope with the additional chore of balancing the bike. Of course the bicycle can be used as transportation while it has training wheels: the child can pedal from one place to another. But the full capabilities of the bicycle cannot be exploited while using training wheels—you certainly never see anyone in a bicycle race using them.

Bureaucracy

When the word "bureaucracy" was first invented, it was supposed to be a good thing. Corporations were trying to cope with growth in situations where it was no longer possible to do enough hands-on management to ensure that the company implemented processes consistently across the enterprise. Bureaucracy was a standard set of processes and procedures which created that consistency even when implemented with different people, in different locations, and in different cultures. But the strength of bureaucracy—consistency—is also its biggest weakness; a bureaucratic environment can't cope with changing conditions or with situations for which the rules weren't designed.

There's a rule-of-thumb in software design that the difficulty and cost of fixing a problem goes up by a factor of ten at each level in the software creation process. Thus a problem caught in the design stage is ten times easier to fix than the same problem caught in programming, and a hundred times easier to fix than the problem caught in system test. Release the software with the problem, and the cost is now a thousand times higher than fixing the problem at the design stage.

Apply this rule-of-thumb to process implementation, and you get something like this: A process standardized as a philosophy is ten times more flexible than a process standardized as guidelines, a hundred times more flexible than a process standardized as procedures, and a thousand times more flexible than a process standardized as hard and fast rules.

Bureaucracy is hard and fast rules. See the problem?

What We Should Do Instead

We would do far better to equip our employees with a good understanding of our philosophy and strategy of business, and the fundamental skills to make their own decisions on the detailed day-to-day implementation of the principles behind these philosophies and strategies, than to construct a set of rules that "must be followed, no matter what." Businesses can't win the competitive race using training wheels, nor can employees expect to know what's good or bad based on the rules in a book.

You can't be best by being bureaucratic, because you're throwing away the capabilities and initiative of your employees by artificially limiting them. But if you use a principle-based approach instead, you can provide broad direction to your employees while still focusing their efforts in the direction you want to go. As a result, the employees are happier, more productive, and more capable of reacting to changing business conditions. And your organization and business will be more successful and better able to satisfy its customers.

There's still another reason why you ought to avoid bureaucracy. When faced with a conflict between bureaucratic rules and an ethical choice, a person often follows the rules even if it results in unethical behavior, because of our ingrained sense of "doing what we're told." Avoiding bureaucracy can help our businesses be more ethical. Make sure your employees do what's right—not what the rules say.

2026 Commentary

I still agree with what I wrote, but it occurs to me that having a philosophy-driven business doesn't necessarily mean that it's an ethical business. Some companies (and governments) have what I would consider to be unethical philosophies. However, even in those situations, it's better to be philosophy-driven than bureaucratic. At least everyone will know where you stand, and they can choose to participate or not.

24. Use Their Terminology—Not Yours

Originally published on NOVEMBER 9, 2004 in #CAREERS, #IT, #ALIGNMENT, #LEADERSHIP, #MANAGEMENT, #PROJECTS, #STRATEGY

A few months ago I was a speaker in front of a group of CIOs, discussing some of the issues facing IT organizations. One of the CIOs asked me what he could do to better communicate his problems to his business users, who seemed to have trouble understanding the difficulties associated with making changes to software. I suggested that he rephrase the problems in the users' terms instead of using the traditional IT terminology. This is fairly common advice, but I'm always amazed at how few people actually use this approach. As IT people we keep falling back on the language we use with our peers, instead of trying to understand the business user's point of view, and explaining things in user terms.

In the last couple of months, I've seen two interesting examples of explaining IT in user terms. At Southern Company, CIO Becky Blalock had a breakthrough in communication when she explained IT in terms that are more understandable at a power generation company. She told her business users that just as Southern Company uses coal to generate power and then distribute the power to its customers, the Southern Company IT organization uses data to generate information and then distribute it to the business users. The business users got the point, and they began to understand some of the similarities between power distribution and the distribution of information.

At The Home Depot, Chief IT Architect Barbara Sanders uses a similar approach to define the stages that a new technology goes through before it's implemented in a live production environment. Her architecture organization issues "building permits" to pilot projects for new technologies, and then delivers a "certificate of occupancy" when the pilot is proven and the technology is ready for general release. Everyone at Home Depot understands exactly what that means, and there are far fewer questions about why a new technology isn't yet in use.

I've used similar approaches myself. At an architectural firm that specialized in designing college campuses and the buildings that go with them, I explained a plan for a future intranet by showing a campus metaphor for the intranet. Users enter the intranet through the "quad" in the middle of the campus, and then enter various "buildings" (subsystems), always coming back to the quad to navigate from building to building. I identified virtual classrooms for training, a "history department" that keeps records of the buildings the firm has designed in the past, a virtual library of white papers and presentations, an engineering library of details for Computer Aided Design (CAD), and even a transit station for "public transportation" to take the intranet user to selected areas of the Internet. By translating difficult technical concepts like home pages, databases, and gateways into terms with which the architects were familiar, I was able to help the firm understand the benefits of an intranet, as well as some of the issues they would encounter during construction.

For a lower level of explanation, I've often tried to relate IT issues to things that are familiar to the average person. For example, I've frequently used a plumbing analogy to help explain why a system change is so difficult. I describe a hypothetical house-building project, and ask why it might be a big problem if the homeowner changes the location of a bathroom after the wallboard is up. The answer is obvious to most people, yet the concept of a difficult system maintenance change (like expanding a purchase order number) is in no way intuitive to these same individuals. Using the plumbing analogy makes the issues clear.

Similarly, most people seem to be able to relate to an example of trying to add four additional floors to a house that has a foundation designed for a single floor; they know that such a change will necessitate some major foundation work. But it's usually extremely difficult to directly explain why, for example, a system designed in Microsoft Access for a single user can't be easily expanded to be available for hundreds of users on an intranet.

For additional perspective, think of the problems that doctors have in telling their patients about complex diseases and their treatment. The best doctors are able to get their message across using examples from cars, plumbing, electrical circuits, construction, and other generally understood processes. Other doctors just throw medical jargon at you and expect you to figure it out. Which kind of doctor inspires more trust? Which kind of doctor would you rather have do your surgery? Amazingly, we base our opinion of a surgeon's competence not on their surgical abilities, but on the ability to communicate with the patient.

IT is the same kind of thing. Since business users can't judge our technical ability, they base their judgments of us on our ability to communicate clearly. Communication is an important part of every project, but it's especially important when your business users don't understand what their IT organizations do for them. The next time you have to explain something, try using an analogy that your users can better understand. You'll find that you communicate better, and you'll get a lot more support.

2026 Commentary

Although a few of the examples are a bit dated, I think the message still comes across.

25. Don't Get Stuck in a Learning Stage

Originally published on DECEMBER 14, 2004 in #CAREERS

I've heard that we go through a progression of stages as we learn a new topic. Before starting to learn a topic, we're so oblivious to the subject matter that we're not even aware of what it's about. We're in Stage 1: *we don't know what we don't know*. Gradually, we become aware of our own limitations in knowledge, and we go to Stage 2: *we know what we don't know*. Then, as our knowledge increases, we enter Stage 3: *we know what we know.* At this point we have the knowledge, but it's new enough to us that we consciously have to think about using it. Finally, as we integrate the knowledge into our day-to-day lives, we go to Stage 4: *we don't know what we know.* The knowledge becomes part of us, and it's buried so deep inside us that we don't even have to think about it to use it.

The progression sounds reasonable—it rings true with my own experience. But I'm convinced that some people get stuck in some of these stages.

Stuck in Stage 1: we don't know what we don't know

There are those who are stuck in Stage 1. They are fundamentally ignorant about a topic, yet they think they know everything about it, and can best be described as "often wrong but seldom in doubt." People like this who aren't aware of their own limitations are dangerous. Working with them is like walking on a frozen lake in the spring—we never know where the ice will be so thin that we fall through into the frigid water below.

If you have to work with Stage 1 people, try to get a second opinion for everything they recommend in the topic area. If *you* are the kind of person who gets stuck in Stage 1, then start figuring out what your strengths and weaknesses are, and learn to tell the difference. Ask yourself, "What can I *learn* that will help me *do* what I need to do?"

Stuck in Stage 2: we know what we don't know

Then there are the people who get stuck in Stage 2. They get overwhelmed by their lack of knowledge, just as someone might be overwhelmed by trying to count the grains of sand on a beach. They lose perspective, and they forget their strengths by overshadowing them with all of their perceived weaknesses. Some of the Stage 2 people become knowledge junkies, always trying to read more, learn more, get a better understanding.

Not long ago, a university professor told me about one of his students who had graduated and gone on to a professional career. The former student kept calling the professor to get a good reference book or article for this problem or that. The professor was lamenting to me that he couldn't make the former student understand that it's time to just *do* something, and to quit trying to do so much research.

The student was stuck in Stage 2, frightened by all of the things in real life that aren't taught in college. What the student was forgetting is that no matter how good we are or how prepared we are, we're going to make mistakes. If you're stuck in Stage 2, then you need to leverage your strengths and move on. Use what you know, learn about some of the things you need to learn, but don't get overwhelmed by all of the things you don't know.

Stuck in Stage 3: we know what we know

Being stuck in Stage 3 is a bit different. It's usually the result of a lack of self-confidence. Most people who learn a new skill will eventually get comfortable with it. If you don't get comfortable, then either you don't really have the skill yet (keep trying), or the difference between your skill level and the skill levels of others is so great that you're overwhelmed.

It helps to remember that the others who are more skilled have a lot more experience than you do. Don't expect miracles overnight —just keep trying to get better and better every day. Ask yourself, "What can I *do* that will help me use what I've *learned*?"

Stuck in Stage 4: we don't know what we know

What does it mean to be stuck in Stage 4? After all, Stage 4 is the point where we've integrated the knowledge into our life, so isn't that the end stage?

Unfortunately, no it isn't. Human beings are on a never-ending journey to constantly increase their knowledge and skill throughout their lives. After Stage 4 comes Stage 1 again, just at a different level of proficiency or for a different subject. Being stuck in Stage 4 is like being stuck in Stage 1: we've progressed to a certain level, and we've stopped learning. It may not seem quite as bad as being stuck in Stage 1, but the result is the same: no forward progress.

Conclusion

Learning comes in stages that help us learn, integrate the knowledge, use the knowledge, and then learn some more. If we focus on the learning without deriving any benefit from the learning, then we're not progressing as individuals. If we focus on the doing without continuing to learn, then our progress stops as well. Improvement in our lives depends on both learning *and* doing in the proper mix.

Don't get stuck in a learning stage: keep making things better for yourself, for your family, and for your business.

2026 Commentary

A couple of thoughts here. First, we're in a world that's full of misinformation and disinformation. Sometimes that makes it difficult to tell whether or not we're competent at a given subject. These days you can find a YouTube video on just about any subject, yet those videos may be wrong about the subject you need, or the video may give such an oversimplified view of the subject that you won't appreciate the subtleties. Other Internet sites, podcasts, TV channels and influencer opinions may give you equally bad information. It's important to seek out truth from sources that deserve your trust.

Second, Artificial Intelligence sources and chatbots not only contribute to possible misinformation, they also can falsely verify information you may provide. You have to remember that most AI sources these days get their information from a collection of Internet sources that may themselves be unreliable. Don't believe an AI source because you are led to believe that "computers are always right." The old computer principle of "garbage in, garbage out" still applies. AI sources may sound correct, but that doesn't mean their information is valid.

26. Why Middle Managers are Important

Originally published on JANUARY 11, 2005 in #MANAGEMENT, #CAREERS

Middle managers don't get much respect. All of the glory goes to the CEOs and senior executives, who in turn focus their own occasional reward programs on the "worker bees." Middle managers play a vital role in most organizations, but it's a shame that many middle managers don't understand their role, or see their jobs only as stepping stones to higher management.

In my experience, there are four vital aspects of the middle manager job:

1. Focus

- Concentrate effort where it's needed
- Focus the work of the organization on what's relevant and productive
- Motivate employees to do this focused work
- Don't put effort where it's not needed
- Prevent work on things which aren't relevant or aren't productive

2. Align

- Coordinate efforts of the organization with peer groups or with other organizations

3. Support

- Support the employees of the organization by eliminating roadblocks and barriers to success

- Get the tools and resources that the employees need
- Provide a good working environment
- Resolve any personal disagreements among the employees in order to maintain a positive working experience for everyone
- Represent the employees upward in the organization, so that outstanding employees can be recognized for their talents and potential
- Help a few employees to realize that they are better off working elsewhere (Yes, I consider it support when you don't try to fit a square peg into a round hole, but instead help find a square hole elsewhere.)

4. Build

- Build the organization and its capabilities by hiring as needed and by training the employees who are there
- Improve the individuals within the organization by increasing their talents and capabilities
- Advance the careers of the individuals who work for you

Where do middle managers fail?

This all sounds obvious, right? So where do middle managers go wrong?

I see most failing middle managers ignoring the first aspect: focus. They react to things that are thrown at them instead of keeping the overall perspective that is necessary to truly prioritize the work that needs to be done. There's a big difference between important and urgent, but many middle managers see them as synonyms. As a result, they work in continual crisis mode, moving from one crisis to the next without ever thinking about the purpose of their organization and how it is supposed to contribute to the company's bottom line.

The second most common failure I see is in the area of support. The managers somehow get the impression that the employees are supposed to support the manager instead of the other way around. Yes, I admit that there are reciprocal aspects of support: employees need to support their managers in certain ways. And yes, some employee positions, like administrative assistants, are truly there to support a manager. But think about it this way: is the middle manager actually contributing any value to the company's bottom line except through the employees? Isn't the middle manager just there because there are too many employees to have them all report directly to the next manager level up? "Span of control" is the term we always hear, but the real reason for having middle managers isn't "span of control"—it's "span of focus and support." If a manager has too many direct reports then the manager can't do a good enough job of focusing their efforts and supporting their needs. That's why middle management jobs exist.

The middle manager needs to view him/herself like a magnifying glass using the sun's rays to start a fire. The sun rays are always there, but the magnifying glass focuses the power of the rays to create combustion. Similarly, the employees might be there, and they might have skills and capabilities, but the middle manager focuses those skills and capabilities to achieve specific results that are beneficial to the company. Can the magnifying glass start a fire without the sun's rays? No—try it some time on a cloudy day. Similarly, middle managers are nothing without their employees.

What's "Middle"?

I've talked so far about middle managers, but when you think about it, these four aspects of a middle manager job apply equally well to senior executives and even CEOs. The big difference is the addition of a fifth aspect of the executive job: leadership and direction-setting. The senior executive doesn't just focus work—they decide what objective the work will accomplish. But in all other respects, the job of the senior executive is identical to that of the middle manager. Focus is at a high level, but for that reason it's all the more important. Alignment is also at the company level, maybe aligning a business with other companies in partnerships and joint marketing efforts. And support and building are absolutely identical to the same roles performed by middle managers. The game is the same, but the stakes are higher.

Conclusion

Middle managers should think less about themselves and how they're going to advance within the company, and more about their employees and how they can focus and support their efforts. Ironically, by making their employees more productive, the middle managers actually do a better job of advancing their own careers.

2026 Commentary

I still agree with this. And there are a couple of other articles on “span of control” that appear later in the book: “70. The Right Span of Control Isn't a Number,” and “72. How Does Outsourcing Affect Span of Control?”

27. Logic isn't always the Logical Choice

Originally published on FEBRUARY 8, 2005 in #CAREERS, #LEADERSHIP, #MANAGEMENT, #STRATEGY

When we come into this world as babies, we believe that the earth revolves around us, and from the way that most parents treat their newborns, I guess that's true to some extent. As we grow out of babyhood, we gradually become aware of other people, and our concept of existence evolves into a view that the world is made up of people, and we're people just like everyone else. Unfortunately, that's incorrect. Sure, we're all people, but it's not true that everyone is the same.

Now it's obvious that we don't all look alike, and we don't have the same interests and talents. But somehow we retain one last misconception about people that goes deeper than looks and talents: we believe that we can persuade others in the same way that we like to be persuaded.

A long time ago, at the very start of my career, I was fortunate enough to be able to take a course called "Positive Power and Influence," and I have derived tremendous value from that course ever since that time. The idea of the course was that there are four basic ways that we can influence others, but that we rely primarily on just a few of those styles—typically those influence styles that we find appealing in persuading ourselves, or those styles which were used on us by our parents when we were children. The course helped me explore all of the styles, and learn how to use the ones with which I was less familiar.

The four influence styles taught in the course were:

1. Assertive persuasion—presenting the facts and a logical argument

2. Common vision—painting a vision for a better future, and inspiring others to pursue that vision
3. Participation and trust—relating to others on a personal level, and having them follow your direction because they value and trust you
4. Rewards and punishments—the carrot and stick approach

Because I had an IT background, of course my primary style was assertive persuasion—the use of logic (if you're in IT, then you probably rely on this style as well). And since at that time I hadn't quite grasped the concept that others might like to be persuaded differently, I used that style in almost all of my communication: upwards to higher management, sideways to my peers, and downward to my subordinates. Sometimes it worked, but often it didn't, and I was left wondering how I could be more effective in presenting my logical argument. When I took the course it opened my eyes to the fact that my influence failures weren't all related to logic failures; most of them were due to a poor choice of influence style for the particular situation.

The concepts I've just talked about may sound familiar to some of you who have taken courses on how to alter your sales approach to accommodate different types of buyers. It's a similar idea: some people want to buy logically, based on a comparison of features, while other people want to trust their salespeople and believe that they are selling the best product for the customer's needs. And still other people just want to have the salespeople treat the customer like royalty, regardless of the benefits of the product. There is no one "right" sales style to use in every situation—just different styles that should be used depending on the situation and the particular customer.

I've heard this described another way as a variation on the Golden Rule. The Golden Rule, of course, is to "do unto others as you would have them do unto you," and it's good at pointing out that you shouldn't be self-centered and treat people as inferior. But when you think about it, it doesn't go far enough. It still makes the babyhood assumption that we're all the same. What it ought to say is to "do unto others as *they* would have you do unto *them*." In other words, treat them the way *they* want to be treated—not the way *you* would want to be treated." That's really the underlying idea behind the different styles in the "Positive Power and Influence" course: you have to find out how others want to be persuaded, and then adjust your style to persuade them.

Whether you're leading them to a bold new future, or just trying to get them to follow some new process, the issues are the same. In order to get someone to do something that is foreign to them, or that is perceived as undesirable, you have to use an appropriate persuasion technique. In some cases, you will have to use multiple persuasion techniques to accommodate each of the personal styles in your organization.

For an IT person who deals with very logical and consistent computers, this may be an uncomfortable concept. But remember: Technology is all about people, processes and change. The hardware and software are just a means to an end. If you can't persuade others, then you can't be successful in IT, or for that matter, in any other profession.

2026 Commentary

It's easy to forget this idea of influencing others using the style that they want. I keep falling back on my own preference, and I have to frequently remind myself of the other influence style possibilities. But this whole idea is so important!

28. If You're Stuck, Get a Jiggler

Originally published on MARCH 8, 2005 in #ALIGNMENT, #LEADERSHIP, #MANAGEMENT, #STRATEGY

In Stephen Covey's latest book, *The Eighth Habit: from Effectiveness to Greatness,* he includes the following Q&A:

Q: "In your experience, what is the best question to ask people when you hire them?"

A: "In my experience, the best question is to say 'Starting with your earliest memory, what did you really like doing and did well?' Then push that through grade school, junior high, high school, university and work assignments afterward, until you start to see a pattern of where people's real talents and strengths are—where their real voice is."

Before you read any further, think for a few minutes about how *you* would answer such an interview question.

I've thought about how *I* would answer the question, and I've found my own pattern. All of my favorite memories are related to learning something new, teaching other people how to use it, and leading a team to put it into action. One of my friends from long ago used a World War II analogy, saying that I'm a "take the beach" person as opposed to a "hold the beach" person. I think he was right; I can certainly look at my work experience and find a large number of positions that I started from scratch.

We made a similar distinction between types of people when I worked at Digital Equipment Corporation (DEC). It was the 70s, the big growth period for minicomputers, and DEC was adding manufacturing plants at a rapid rate to accommodate all of the orders for computers and peripherals. DEC found that the person they send to start up a new plant is not the right person to leave in charge of the plant after the startup period. The startup person has to be good at improvising, creating something out of nothing, and getting everything to the point where it works. But it requires a different kind of person to take something that works and tweak it to the point where something good becomes better. If you leave the startup person in charge, then things will never settle down: someone who is comfortable with big changes will continue to make big changes, or will get bored when the big changes aren't needed.

Much has been written in the management self-help press about the differences between incremental change and radical change. The phrase "outside the box" was invented to describe the type of thinking that's necessary to create radical change—to see fundamental flaws in an approach and to (forgive the overused expression) "change the paradigm" to make things significantly better. But the trouble with a lot of the management press is that they don't recognize the value of both sets of skills: not everyone should be an outside-the-box thinker; you need inside-the-box thinkers to manage the day-to-day operation and to "keep the ship afloat."

Even Michael Hammer in his reengineering books recognizes that companies go through alternating periods of incremental change and radical change. No one can tolerate constant radical change, and in big companies it can shake the company apart. You need a quieter period of incremental change after each radical change, so that the organization can absorb the big change and gradually bring its lower level processes into equilibrium. Thus there's a need for two types of managers: one type who manages outside the box, and another type who manages inside the box.

But how about the companies where everything is inside the box? Where all change is incremental change? Is that good enough? In today's world, I would say "no." A hundred years ago, this type of company behavior was the role model for industry. Companies who could institutionalize their processes and guarantee consistency over a number of years were hallmarks of success. But in today's global economy with competitors coming at you from all directions, there aren't very many businesses (I can't think of *any*) where absolute consistency is a virtue. Change is an inevitable part of keeping your company alive.

So if you've got an inside-the-box company, with inside-the-box people, what do you do? There's where a "jiggler" comes in. I first heard this term from Gerald Weinberg, one of my favorite authors and an inspiration for process consultants everywhere. The term comes from the way that people try to fix a toilet that's stuck and won't stop running: they jiggle the handle. The idea is that a jiggler goes into an organization and looks at it with a fresh set of eyes. The jiggler looks for processes that are stuck and need major change, or for new viewpoints that need to be considered. The jiggler provides the outside-the-box perspective to a company that may not have the right people to provide that perspective within their own organization.

By definition, the jiggler is one of the "take the beach" people, the type of person who would start up a manufacturing plant instead of just making it a little better. You don't necessarily want to have a jiggler full-time within your company (unless you've got a whole lot of jiggling to do). But it's the perfect role for a consultant: someone brought in for a specific purpose and then sent away until needed again down the road.

And there's another reason why an outside consultant is better as a jiggler. People inside a company—even the outside-the-box thinkers—are usually inhibited in their comments by their desire to have long-term personal success within the company. They're afraid that if they go too far in their suggestions, then their co-workers and managers will laugh at them, and their credibility will be ruined. So while internal jigglers are better than no jigglers at all, they will be self-limiting in the strength and depth of their ideas.

Ok, I admit it: I'm a jiggler. But this isn't so much a sales pitch as it is a description of a role that you ought to consider. No matter where your jiggler comes from, the important thing is to *use* one from time to time. You need someone to periodically introduce fresh ideas, provide a different perspective, point out things you need to do better, and even to point out things that you do well that you shouldn't be doing at all. Every so often you just need to jiggle your organization.

And, oh by the way, if you haven't made a major process improvement in the past year, then you're stuck—whether you think so or not—because I can almost guarantee that your competitors have made such a change. And if you're stuck, get a jiggler.

2026 Commentary

Are there other ways to "jiggle" a company or its business processes? I'm sure there are. One idea is to read a lot of different information on different businesses and industries than your own. Notice things that other companies have done that suggest the possibility of similar or parallel things that could be done to your own processes.

It's also possible that Artificial Intelligence sources could take a look at your processes and make suggestions. But bear in mind that any suggestions are based on the information that has been fed to the AI as part of its training. So make sure you independently verify any suggestions before trying them.

Another thought: Try new ideas on a small scale before implementing them on a wider scale. There may be advantages to some of the new approaches, but they may also bring undesired side effects that you don't anticipate. It's best to work out the problems in small scale implementation before you cause bigger problems that could hurt your company.

29. User Training is Like a Joke

Originally published on APRIL 12, 2005 in #ALIGNMENT, #MANAGEMENT

Shifts in frame of reference are the root of most humor. We all tell jokes, but we seldom recognize that most jokes are funny because they lead our thinking in one direction and then abruptly cause us to shift our thinking to a different direction. Jokes essentially accomplish mental sleight-of-hand by using the most basic of a magician's tricks: misdirection. Here's an example:

A man and a dog go into a bar. The man says to the bartender, "This here is a talking dog. If I can show you that he talks, will you give me a free drink?"

The bartender says, "Sure. But this had better be good."

The man says, "Watch this," and then asks the dog, , "What do you call the top of a house?" The dog barks an answer, "Roof!"

The man asks the dog, "How does sandpaper feel?" The dog barks out, "Rough!"

The man asks the dog, "Who was the greatest baseball player of all time?" and the dog answers, "Ruth!"

The bartender throws them out of the bar.

The man and dog walk down the street and ponder the exchange with the bartender. Then the dog turns to the man and asks, "DiMaggio?"

This joke raises an initial expectation that the dog will talk, then convinces us that it's all a hoax. The punch line shifts our thinking abruptly, making us realize that it was just bad questions that caused the problem—not the absence of canine talking ability.

Many jokes within the IT community celebrate system users' lack of understanding of fundamentals. We talk about the PC user who wants to replace a broken cup-holder (CD ROM drawer) on their computer, or a customer service rep who asks a PC user how many windows they have open and gets the answer, "The windows in my office can't be opened." I even saw an absurd example myself back in the days of 5 1/4″ diskettes, when we asked a customer to send us a copy of the data on a diskette and the customer mailed us a paper Xerox copy of the front, back, and *insides* of the diskette (the customer cut open the diskette to allow it to be copied).

We laugh about these situations, but the training issues are real. System users who are accustomed to working in a certain way will have difficulty when we change the fundamental principles underlying their work. It's like a joke: first you lead their thinking in one direction (the assumptions of the old system), and then you abruptly alter their thinking by changing the assumptions for the new system without telling them. But with a joke the punch line is funny—our system training just causes heartache for our users.

Last year my family bought a computer for my 84-year-old mother. It seemed to be a logical replacement for her worn-out typewriter, but although my mother has an in-depth understanding of how to use a typewriter (to use the phrase from one of my previous articles (#25), "She doesn't know what she knows."), the shift to using word processing software on a computer has been very difficult for her. Here are some of the assumption changes that she has confronted so far:

- Margins don't have to be set before you begin typing.
- You don't have to press the *Return* key at the end of each line.
- You don't have to pay attention to whether the page is full (Soon after she started using the computer, she called me to say that she'd filled her first page; how does she go to the second page?).

- Since you can edit text after you type it, you don't have to write a first draft by hand.
- It's not necessary to understand every aspect of every setting in the word processing software like you do for a typewriter.

All of these things are obvious to those of us who have used a computer for a while, but these are major hurdles for someone just starting out.

Our Training Approach is Backwards

Think about how most of our training programs will teach someone like my mother to use word processing software. Most of the training will focus on feature/function. The trainers will explain the menus, they will explain the various tool bar options, they will talk about how to perform certain functions like indenting a paragraph. But they will rarely take into account the user's current frame of reference and specifically list the changes in that frame of reference that will be required of the user in order to use the new system.

Instead, it will be up to the user to read between the lines and figure out the changes in thinking that have occurred between the old system and the new system. Most users won't do this, and so they will have to resort to rote memorization of a list of arbitrary feature/function descriptions. This is a much more difficult way to learn, and the users won't be as productive as they could be.

The Way Training Ought to be Done

When you're training people to use a new system, you should get an understanding of the users' current frame of reference *before* the training. Then you can start the training class by explaining to the system users:

1. The fundamental assumptions that have been made in the use of the old system that will change in the new system,
2. The specific user limitations of the old system that have been removed in the new system, and
3. The major fundamental assumptions in the old system that the users can still count on.

Once the expected changes in thinking are out on the table, then you can go on to explain feature/function, tying the feature/function changes back to the list of fundamental assumption changes that you provided at the beginning of the training session (i.e., "we used to do it this way; with our new assumption we'll do it this other way.").

The assumption changes are the most important part of the training; they're the new principles that have to be incorporated into the user's day-to-day processes. The features and functions are just the specifics on how those principles will be used. Specifics are important, but they're best presented in the context of how things have changed.

And that reminds me of another joke:

A famous lumberjack in the days of the double-headed axe was asked to be the guinea pig for a revolutionary new technology: the chainsaw. He took the new tool into the woods and tried it out, but initial results were disappointing. On the first day his lumber output dropped in half, and his supervisor attributed that drop to the learning curve with the new device. Gradually the lumberjack improved, but after a few weeks he was still unable to beat his output from his axe-wielding days.

Hearing about the numbers, and seeing a major marketing problem, the chainsaw manufacturer sent their product developer out to visit the lumberjack in the woods.

The developer introduced himself to the lumberjack, and asked to see the chainsaw. The developer looked the chainsaw over, found no obvious defects, and then pulled the cord to start the engine. The lumberjack looked startled and exclaimed, "What's that sound!"

If only the training for the lumberjack had started with an explanation of the changes in fundamental assumptions....

2026 Commentary

The references to CD ROM drives and diskettes are undoubtedly dated, but the concepts in the article still apply. I still see this feature/function emphasis in technology training, without any reference to paradigm shifts.

30. Fixing Broken Windows (not the Microsoft kind)

Originally published on MAY 10, 2005 in #ALIGNMENT, #LEADERSHIP, #MANAGEMENT, #PROJECTS, #STRATEGY

A few weeks ago there was an article in an Atlanta newspaper about George Kelling, the author of the book, *Fixing Broken Windows: Restoring Order and Reducing Crime in Our Communities*. Dr. Kelling has been called "the man who helped Rudy Giuliani turn around New York City" by refocusing some of the city's priorities. As I was reading the article about Kelling, it struck me that a lot of the same principles apply to an IT organization.

Kelling's studies have shown that crime goes down in areas of the city where there is a police foot patrol. That seems like a no-brainer, but what's not so obvious is that the decrease in crime comes from two things:

1. A better relationship between the police force and the community, and
2. Increased enforcement of laws for minor crimes like peddling, begging, graffiti, vagrancy, etc.

Better Relationship

The better relationship is probably the more important of the two factors. Putting police officers on the street personalizes the officers to the members of the community, and makes them approachable. When people interact regularly with another group on a person-to-person basis, they are more likely to understand each other's point of view. When they recognize common goals, they're more likely to take joint responsibility for achieving those goals, and they're more likely to work together to achieve them.

Where foot patrols are used, the members of the community begin to take ownership of the crime problems, and they don't just expect the police to deal with the crime by themselves. The community and the police force work together to make the city a better place.

Increased Enforcement of Laws Against Minor Crimes

The second cause of the drop in crime is increased enforcement of laws against minor crimes. What is surprising is that the enforcement of laws against *minor* crimes leads to a reduction in the number of *major* crimes. It turns out that there is a domino effect here. For example, if the sidewalk in front of a store becomes an annoyance to customers due to panhandling, graffiti or vagrancy, then the customers will go elsewhere to shop, and the store will eventually close and be replaced with a less legitimate business. Minor crime areas tend to evolve into areas for drug dealing, and the crime in that vicinity escalates. Stopping the minor crime changes a downward spiral of decay into an upward spiral of improvement. The business becomes more attractive to customers because they feel safer, and the area around the store benefits from the increased customer traffic to the business. A small safe area evolves into a larger safe area, and pretty soon even a place like Times Square can be safe enough to become an attraction for out-of-town visitors.

Kelling talks about broken windows in a building. If broken windows go unrepaired, then it's a signal that the building owner doesn't care. One broken window leads to more broken windows, and eventually the building becomes just another piece of urban decay. On the other hand, a small investment in fixing broken windows makes the spiral go the other direction. If you show that you care about something, then other people will care too.

Applying this to Information Technology

Most IT organizations don't start out as big organizations. In the beginning, some small business or department gets to the point where it needs information technology, and it hires IT people to help the business. Think of these people as the initial "foot patrol" because they usually have a good understanding of how the business works, and they work hand-in-hand with the business people who hired them.

But over time, two things happen:

1. As the IT need grows, the administration and management of IT becomes a bigger and bigger chore. Someone has to coordinate all of these individual IT foot patrol people, and that new IT bureaucracy widens the gap between IT and its user community.

2. As systems go into production, and more of the IT resources are shifted toward the care and feeding of the existing systems, the IT people move into a reactionary mode, responding to formal requests for resource instead of sensing the user needs as part of IT's day-to-day contact with the users. Today's all too common budget cuts and staff reductions make this reactionary mode even more likely.

Most police forces around the country (and probably around the world) are in this reactionary mode. You can call the police when you have a crisis (in the U.S., dial 9-1-1), but you don't have any day-to-day contact with police officers, and there's a tendency for the community to depersonalize the police department and blame it for crime.

Is it any surprise that the same thing has happened to many IT organizations? If you're part of the end-user community, then you probably don't talk to IT people on a day-to-day basis—you're just in contact when you have a complaint or when you're trying to deal with the bureaucracy of defining system requirements.

The Solution for IT

So how do we learn from what Kelling has done in New York City? What can we do in IT that is analogous to Kelling's "fixing broken windows" approach? In my experience, there are three areas for change.

First, have your IT people spend more unstructured time with their end-users, getting to understand them and their problems. This seems obvious, and yet very few IT organizations do it, with most IT people coming in contact with end-users only as part of a specific project. When I managed a systems group that supported manufacturing, I made sure that all of my programmers and analysts regularly spent time in manufacturing plants. When I supported a group that did tax filing, I made sure that all of my people spent some time working side-by-side with clerks filling out tax returns.

Where possible, have the IT people actually *do* some of the user work to get a sense of the user issues. Where it's not possible, have them buddy-up with a user and spend a day just getting a better understanding of what the users do and how it could be made better through the application of new processes and technology. You'll be amazed at the payback you'll get from this IT immersion in the user world.

Second, there is a way to directly apply the "fixing broken windows" method to IT. In every systems group I've ever managed, I've created a "Quick Response" (QR) team that specifically focuses on dealing with the minor system annoyances that will never make it through a formal project review process. We put size limits on the QR projects (e.g., nothing that takes more than a person-day), and the QR team has a limited budget, but within those limitations we let the users manage the Quick Response priorities.

The result is better camaraderie between IT and the users, and the users get a sense that things are better in the short-term. This makes an enormous difference in user morale and in overall perception of the IT organization. And it has an unusual side-effect in some cases: with a few minor tweaks, the existing systems are now more comfortable to the users, and so the users are less likely to demand unnecessary extreme solutions. Just as reducing minor crime can lead to a reduction in major crime, eliminating minor system annoyances can lead to a reduction in the need for major projects.

Third, have your managers do more walking around in the user parts of the building, in effect forming a foot patrol. When I was in school at MIT, my advisor, Professor Thomas Allen, told me about a project he had done to improve communication between rivaling departments. He had the business lock its bathrooms and give bathroom keys to the employees. But as a twist, he made sure that the employees of these two rivaling departments got keys to bathrooms that were located in their rival department's area—not in their own area. This simple action, which forced employees in the two departments to interact with each other multiple times a day, led to a breakdown of the communication barrier between the departments. And while I'm not recommending locking your bathroom doors, this example does make a point: just having periodic informal face-to-face contact between IT people and their customers will lead to better communication and better community spirit.

Conclusion

In these days of stretch goals, constant business change and downsizing, it's easy for IT to become a reactive organization. But look at what happens in communities when the police get overloaded and become reactive: an increase in crime and urban decay. You can fight lack of community in your business the same way that it was fought in New York City. Think about "fixing broken windows" and increasing foot patrols. It worked for New York City, and it will work for you.

2026 Commentary

Time has not been good to Rudy Giuliani, but the article refers to his tenure as the mayor of New York City. There has also been some criticism of this "broken windows" theory, but please don't apply any negative press for that theory to the conclusion I derive for IT. I know from experience that this approach works for IT, and I highly recommend it. And if you're not in IT, but you're in some sort of support role, then think about whether it might work for you as well.

31. The Information Technology Merry-Go-Round

Originally published on JUNE 14, 2005 in #CAREERS, #IT, #ALIGNMENT, #LEADERSHIP, #MANAGEMENT, #STRATEGY

The world is full of cycles. There are stock market highs and lows, periods of good weather and bad weather, even apparent cycles of good luck and bad luck. Many of the people-related cycles are caused by basic human nature. The bad part of a cycle starts when we hear bad things about a particular topic, and we talk about them. The talk turns into concern, and we begin to pay more attention to the problem. The more attention we pay to the problem, the more likely it is that other people will get concerned as well. And the more we're concerned, the worse things seem. The bad cycle hits its peak of "badness" when someone hears something good, and people begin to say more good things than bad. It doesn't matter whether the original bad thing was true or not, or whether the new good thing is true or not—what matters is how people talk.

The good part of a cycle works the same way, starting with good comments that escalate due to increased conversation. The good cycle ends when we begin to get skeptical or nervous: "Will the good times last?" or "How good can things get before they get worse again?". Then people begin to say bad things, and we go back to the bad part of the cycle again.

We hear about certain kinds of cycles every day. "The Dow was up." "We're having a warming trend." "The economy is down." But while we commonly apply this thinking to certain areas of our life, it's less common to think this way when we consider business processes that have longer cycles and aren't typically considered repeatable.

Centralization / Decentralization

Look at the cycle of centralization and decentralization. You may be surprised to hear the word "cycle" used for this subject, but it's a cycle nevertheless. Companies who centralize any part of their operations (like IT) will eventually begin to find fault with the centralization decision. User organizations throughout the business will start complaining that the centralized IT organization doesn't meet their needs. All of the benefits of centralization (chiefly related to economies of scale) will be forgotten as negative talk escalates. Unless the negative comments are suppressed by executive management (rare), the talk will grow into an uproar, and eventually the centralized organization will be decentralized, with or without the cooperation of the current manager of the centralized organization.

Peace will prevail for a while, and talk will die down. But then, gradually, the negative aspects of a decentralized organization will be noticed, and eventually those negative aspects will be the main topic of conversation. You can guess what will happen next—you've probably seen it if you've been in business very long. The advantages of the decentralized organization (chiefly related to better responsiveness to customer needs) will be forgotten, and a push will be made to cut costs and centralize. This cycle will repeat endlessly in a centralization/decentralization merry-go-round.

Outsourcing / Insourcing

Of course, centralization isn't the only long-term cycle that affects IT. Another related cycle is outsourcing/insourcing, and its popular variation offshoring/onshoring. Just like centralization, outsourcing is primarily driven by cost savings, but there is an inherent loss of flexibility and responsiveness to user needs (see my previous article, "14. When to Outsource and When to Offshore").

Therefore it fits the pattern that causes a merry-go-round: two conflicting goals that are very difficult to achieve simultaneously.

Unless an outsourcing firm demonstrates extraordinary responsiveness, or unless the process being outsourced is mostly invisible to employees and customers (e.g., payroll), then it's only a matter of time before business change, human nature, and escalating complaints force the cycle back the other way, and the outsourced business process is brought back inside the company. Then eventually cost savings will become more important again, the cycle will repeat, and the business process will be outsourced again.

This same argument applies to outsourced business processes that are sent offshore. But with offshoring there is one key difference: the current cost differential between insourced and offshore processes will require a very high level of user complaints before action will be taken. I doubt that this cost differential will be maintained in the long run, but we may not see the cost differential change significantly in *our* life-times. Thus an offshored process is not likely to be brought back in-house any time soon.

The "Jump at the Top" Career Strategy

Some people have decided to pursue a career strategy tied to one of the cycles I've described. So, for example, they might specialize in centralization of IT. They'll get hired into a company that wants to centralize IT (usually to replace an IT executive who's dragging their feet), and they'll stay with the company through the glory days of centralization until they begin to hear the first noise about decentralization. Then they'll quickly find a position in another company that's about to start its own centralization.

Should *you* follow this "jump at the top" career strategy? Maybe, if you feel strongly about a particular part of the cycle.

Should *you* hire someone who is following this career strategy? Absolutely not! Although the person may be very good at their specialty (centralizing, decentralizing, or whatever), the person lacks the experience in doing the specialty in a manner that allows it to survive for a long time in a company. It's easy to centralize IT —it's much harder to do it in a way that provides lasting benefit to the business with minimal backlash. You have to get the experience of managing through the aftereffects of a change to truly be able to do it right.

Recommendations

Here's my advice on what to do about business cycles:

1. Be aware of them. Know that there isn't a right way and a wrong way to do most things, but that the current *perceived* right or wrong way is a function of where your business is in the applicable cycle.
2. Take advantage of cycles when you can. For example, if current sentiment is moving toward outsourcing, then figure out the things that you think could truly benefit from outsourcing, and push them to be outsourced.
3. Don't fight current sentiment in a cycle. It's like being stuck in the middle of a raging river and trying to swim directly upstream—you'll always end up drowning. When stuck in a river, the best approach is to swim at right angles to the direction of the river, and to take advantage of the current to push you to shore. Similarly, in business the best approach is to take advantage of current sentiment to push for the things you feel are important.
4. Look for compromises which don't weaken your position. If you're centralized, then consider decentralizing certain processes that are more visible and which don't derive a great deal of benefit from economies of scale. If you're decentralized, then consider centralizing key processes like standards or high-volume purchasing. The strongest dam doesn't try to stop all of

the water—it lets some of the water through to benefit those downstream.

Conclusion

The world is full of cycles. It's frustrating to see business people lobby against something that you've put in place, but it's important to recognize that it's just human nature to find fault with things and to want something better. Sometimes it leads us in circles, but we can hope that each trip around the merry-go-round will be a little better than the last one. In the mean time, just enjoy the ride.

Note: if you're interested in a history of weird cycles caused by people's overreaction, you'll enjoy the book *Extraordinary Popular Delusions & the Madness of Crowds* by Charles Mackay.

2026 Commentary

I mention that the cost savings from offshoring tend to keep offshored processes there. But now we're seeing Artificial Intelligence processes such as intelligent chatbots provide a potential customer service experience that is even less expensive than offshoring. This AI cost savings may cause us to bring back offshored processes and let the merry-go-round continue.

For a related article, see "67. Questioning IT Realignment—Should IT be Centralized?"

32. IT Lessons from a Waitress

Originally published on JULY 12, 2005 in #IT, #ALIGNMENT, #LEADERSHIP, #MANAGEMENT

I went out to dinner last night to a place I've gone hundreds of times, and I ordered a salad that I've ordered many times before. The salad wasn't as good as it's been in the past: the lettuce was old, and the dressing was watery. When the waitress asked her usual question, "How is everything?," I politely told her the truth. Her reaction is one that I've seen many times in interactions between service people and customers. She responded as if I had said something entirely inappropriate, and she acted like she wanted to get away from me as fast as possible.

Of course, that's not the way it's supposed to be. The waitress is supposed to focus on the customer, listen carefully, make apologies, discuss what should be done about the problem, and perhaps offer alternatives. The waitress should bring the incident to the attention of her manager and the appropriate chef. Then the waitress is supposed to remember the incident for at least the duration of the meal, make an apology again at the end of the meal, and perhaps (depending on the incident and the restaurant) offer some form of compensation at the end: maybe a free dessert or a reduction in the bill.

"How is everything?"

So far you probably agree with me. But let's shift the scene to the interaction between an IT person and their customer. The first difference you'll see is that the IT person rarely asks, "How is everything?" Most IT people are so strapped for time that they can't seem to accomplish the tasks they've been given. They rarely take the time to converse with customers, and they seldom ask a customer how they feel about a system unless the conversation is an explicit part of a systems analysis task.

So if an IT person hears a complaint, it's probably because the customer feels strongly enough about the problem to make a special effort to raise an issue. That's bad for three reasons:

1. By the time the customer makes the complaint, there is a large amount of frustration and anger. Interaction with IT is strained, and the customer wants an immediate solution to a problem that probably doesn't have a quick solution.

2. Many problems and complaints will remain unvoiced. Some customers will just silently tolerate poor systems, poor processes and poor service. They'll never complain to IT directly, but they'll build up a lot of resentment toward the IT organization, and they'll complain to their co-workers and friends. To them, IT will become the enemy, and you'll never know it until it's too late.

3. Resentment builds in the IT organization as well. The IT people hear only negative comments, so they form a very one-sided view of the customer. Through IT's eyes, a customer becomes someone who complains.

Listening and Apologizing

Let's move on to the next step in the waitress example. There are two things that should go on in a problem discussion between a waitress and a customer: listening and apologizing. Through listening, the waitress should get an understanding of what the problem is, and how the problem impacts the customer. And in apologizing, the waitress should convey two messages: (a) that's not how things should be, and (b) steps will be taken to make things better.

Looking at a typical interaction between IT and a customer, we see that listening usually takes place but apologizing seldom does. And the listening, particularly in help desks, often takes the form of just listening long enough to categorize the problem into something that's been seen before. Once the categorization has been done (whether right or wrong), then listening stops.

Why don't we get apologies in IT? Perhaps it's because we don't understand what an apology is. An apology is not an admission of guilt, nor is it an acceptance of personal liability. It is instead a way of connecting to the customer on a human level, letting the customer know that you can relate to the pain they're experiencing, and that you truly want to make the pain go away.

Problem Resolution

Once the problem is understood and apologies have been made, we move on to the next step: problem resolution. Sometimes the resolution of a problem is an explanation instead of a solution. We don't see this much in restaurants, but we see it a lot in IT help desks. Someone calls in with a problem (e.g., "All the text in my document is double-spaced."), and we tell the customer how to use the system differently to achieve the desired result (e.g., "Use the 'Format Paragraph' menu option to change the line spacing."). In IT organizations, we tend to view this type of call as an annoyance, but there is a real underlying problem (e.g., "the system is too complicated, and system help isn't sufficient") that could be solved if we devoted the time and energy to solve it. That's why call tracking is so important for IT help desks: to determine the most common calls so that we can focus our efforts on solving the most common problems.

Ok, but what about problems that can't be resolved with an explanation? In a restaurant or a small company, the processes are simple, and it's easy to notify the right people about the problem. Whether they solve the problem or not depends on their own attitude, dedication, and resource availability. The key here is to avoid blaming people, and to instead look for a way to change processes to make things better.

Follow-up

Our last step in the restaurant example is to follow-up with the customers. The follow-up has two goals:

1. Convince the customers that you care about them, that you care about solving their problems, and that you'll take steps toward appropriate solutions.
2. Make the customers feel better about the restaurant so that they will return.

How often do you follow-up on a help desk call to make sure that the problem is resolved? How often do you follow-up on a customer complaint to see if the customer is happy? In IT organizations we tend to neglect the follow-up because of resource constraints and because, frankly, the customer has no place else to go. But in neglecting this step, we're doing a huge disservice to ourselves and our customers. Providing follow-up makes the customer feel better; it makes the IT person feel better (especially if the customer is pleased); and it gives us an opportunity to identify situations where an inappropriate solution has been applied, so that we can apply a better solution.

Recommendations

Here's what we can learn from the waitress example:

1. Plan for informal interaction between IT people and their customers. Give the IT people time to ask "How is everything?" and to really listen to the answer. If you do this well, then small problems can be resolved before they develop into big problems. You'll occasionally hear praise, which is a huge morale booster. And most important, you'll develop a better relationship between your IT organization and their non-IT customers.

2. Teach your people to *really* listen—not just categorize complaints. Categorization can still occur where it's necessary, but it should happen *after* the listening.
3. Teach your people how to apologize, remembering the two messages in the apology: (a) that's not how things should be, and (b) steps will be taken to make things better. Make sure that everyone understands that you can apologize for something without accepting personal blame for it, and without blaming anyone else.
4. In resolving problems, look for process issues instead of blame. Even if it's someone's "fault," the person made the mistake because of the process that was taught or because of the reward structure that provided the motivation.
5. Don't ignore problems which can be resolved through explanation. If enough people need an explanation, then there's an underlying problem that ought to be corrected.
6. Have a specific process that ensures follow-up on all problems (or a random subset if resource constraints prevent full follow-up). If a problem isn't resolved, then restart the problem-solving process.

Conclusion

I started by discussing a common situation in a restaurant which was handled badly, and I drew parallels to the relationship between an IT organization and its customers. I can make allowances for an occasional waitress lapse in a lower-priced restaurant, but in a higher-priced restaurant I wouldn't tolerate this sort of behavior. Don't our IT customers deserve to be treated at least as well as customers in a higher-priced restaurant?

2026 Commentary

I used the restaurant analogy as an anchor for discussing a general customer service issue I see a lot, both in IT and in other service organizations. When the article first came out, I had some criticism about the restaurant example. Some people felt that restaurant service people are not paid enough to deal with situations like this. Others cited examples where serving staff did bad things to the food in retaliation for the criticism.

My advice would be to pay less attention to the restaurant example, and more attention to the advice on how to relate better to your customers.

For a related article about complaints, see "115. How to Deal with Complaints—7 Steps."

33. Hiring like a Jigsaw Puzzle

Originally published on AUGUST 9, 2005 in #CAREERS, #MANAGEMENT, #STRATEGY

There are two basic approaches to hiring:

- Hiring like a jigsaw puzzle, and
- Hiring like an assembly line.

Hiring like an assembly line is more common. When you work on an assembly line you have very little variation in what you do. One particular task might be to put a bolt into a hole and then tighten the bolt. The holes are a standard size and you know the depth of the hole, so it's obvious which bolt to use. You just select a bolt that's the right size, and you slide it in.

Hiring like an assembly line works the same way. You have standard job definitions—maybe a few different kinds, but it's a limited number. Your task in hiring is to find a person who fits a particular job definition, hire them into the company (put the bolt into the hole), and do so in a way that ensures they stay in the position for a few years (tighten the bolt).

The nice thing about hiring this way is that it's simple. But what we tend to forget about assembly lines is that a lot of thought and planning goes into the design of a product to optimize it for production on an assembly line. Manufacturing process engineers spend their lives making things go smoothly, balancing the labor up and down the line, and defining each task in an optimum way. In contrast, we typically spend very little effort setting up job definitions. Most companies just borrow heavily from the definitions used in other companies.

Once the new workers are installed in their well-defined positions, you learn more about the people you hired. Not surprisingly, the people who share the same type of position are fundamentally different. If you're lucky, their individual strengths and weaknesses will balance each other. But it's more common when you hire using the assembly line method to have a bunch of people who share many of the same weaknesses and strengths. You'll have holes in your organization in the weak areas, and some of the overlapping strengths will be unnecessary.

Hiring like a Jigsaw Puzzle

Let's go back to that other method of hiring. When you solve a jigsaw puzzle, you usually start with some distinctive object that appears in the puzzle, and you put together that part of the puzzle first as a core from which to build the rest. You might also find the puzzle pieces with straight sides and assemble the edges of the puzzle. Then the search begins to expand the core of the puzzle. You look for individual pieces that have distinctive markings or colors, or you look for a specific shape. Gradually you assemble the puzzle, as each piece you fit into the core creates a different requirement for the next piece.

Now consider putting together an organization this same way. You start with a few core people; maybe they're the founders of the organization you're trying to staff, or maybe they're really good people who've been assigned to be the foundation for the new organization. You might also identify some "edge" people to start with—customers and supporting departments who have to interact with the people in your organization.

Based on what you start with in the core and on the edges, you figure out the next piece of the puzzle that you'll need. Just like with a jigsaw puzzle, you can build on the core in multiple directions at the same time, looking for multiple people who are needed to augment the skills of the core team in very specific ways. You can use a job description to help you, but it's just a starting point, and each description will have to be tailored to suit the individual requirements for this specific piece of the organizational puzzle.

Once you find a person who fits the requirements for an individual puzzle piece, you bring the person into your organization, and the strengths and weaknesses of this new person help determine the next piece of the puzzle. For example, maybe you are looking for a few project leaders with specific experience in your industry. You find a fantastic project leader who has extraordinary project experience, but only minimal industry experience. You need a second project leader anyway, so now you're going to look for someone who has more industry experience but who doesn't necessarily need the same depth of project leadership.

Conclusion

Extraordinary people are seldom hired using the assembly line method because they don't meet all of the detailed hiring criteria and so they get weeded out. These people have tremendous strengths, but most people who are very strong in one area have corresponding weaknesses in other areas, and these weaknesses eliminate them from consideration. The assembly line hiring method gives us more consistency in our people, but it also tends to give us people who are average in all respects but extraordinary in none.

The objective of the jigsaw puzzle approach to hiring is to build an organization of extraordinary individuals who can work together to exceed organizational goals. By tailoring our hiring to specific organizational needs and by taking into account the current people in the organization, we stand a better chance of hiring people who are outstanding in some important aspect of their skill set and experience. Hiring extraordinary people is the basis for an extraordinary organization.

2026 Commentary

This idea of hiring using the assembly line method has gotten worse as companies have started using Artificial Intelligence processes to filter resumes and applications. The result is to weed out superior people and to focus hiring on people who are skilled at fooling your system into thinking they're a good fit. I recently heard a story about a CEO who was concerned about his company's hiring practices and anonymously submitted a resume based on his own background. Sure enough, it was rejected without even an interview. As I recall, he asked for an overhaul of their hiring process as a result.

34. Preparing for your own Hurricane Katrina

Originally published on SEPTEMBER 13, 2005 in #LEADERSHIP, #MANAGEMENT, #STRATEGY

Disaster struck the southern United States last month as Hurricane Katrina did major damage to New Orleans and southern parts of Louisiana, Mississippi and Alabama. We don't yet understand the full impact of the storm in terms of lives lost, families disrupted, and the impact on the American and global economies. But we know that a key part of our responsibility as IT executives is to anticipate disastrous events like Katrina and be ready for them. Here are some of the things I've observed about the Katrina experience that are applicable to the IT arena, especially in the areas of business continuity planning and disaster recovery:

No one wants to follow the mediation plan if it's an inconvenience, but everyone chastises you afterwards for not pushing harder.

If Hurricane Katrina had swerved at the last minute and missed New Orleans, then I can guarantee that the press would be having a field day telling everyone how stupid it was to evacuate so many people. This is one of those "damned if you do and damned if you don't" situations that make it so hard to be in a position of responsibility. No matter how well you do, it isn't good enough in the eyes of some people. And if you're perceived as over-cautious in a situation where nothing happens, then the criticism will be just as fierce.

Pre-disaster exercises don't help if you don't apply what you've learned.

FEMA (the U.S. Federal Emergency Management Agency) conducted a week-long exercise in 2004 to help Louisiana emergency officials plan for the possibility of a hurricane very much like Katrina. But some of the processes used in the exercise were ignored when Katrina hit, including a process for the large-scale evacuation of people who don't have their own transportation.

Your contingency plans need their own contingency plans.

Part of the New Orleans contingency plan was to use the Superdome to shelter people who didn't have anywhere else to go. But the Superdome had to be evacuated when toilets backed up, the air conditioning broke down, and high winds ripped a hole in the roof.

No matter how much you plan, you still have to improvise when the disaster strikes.

There is no amount of planning that will anticipate every possible outcome, and there comes a point where additional planning makes no sense. You have to be prepared for surprises, and make sure that you have the right people in leadership positions to make the on-the-spot decisions that are required.

Insurance policies don't begin to make up for the loss of business and goodwill, and obviously don't make up for the loss of life.

Don't let an insurance company be your disaster plan. Think of an insurance policy as a safety net if everything else in your plan fails.

Contingency plans need to have a defined and published trigger event, and the contingency plans need to be executed when the trigger event occurs.

I believe that more lives would have been saved if each area of the coast had an evacuation plan with a timetable. For example, "If a category x hurricane is headed for this area, then y hours before its scheduled arrival, everyone must be evacuated except designated critical personnel. Here is how that will happen …." Without a trigger event, everyone holds out a little longer before acting, pushing beyond reasonable limits. This happened on a large scale with Katrina, as both federal and state agencies delayed before taking any action.

Any disaster has secondary and tertiary consequences that are difficult to anticipate.

Katrina caused localized gasoline shortages throughout the Southeast United States as panicked car owners rushed to fill their tanks. It's still not clear how badly the storm will hurt the U.S. economy, but there is a potential for an economic recession as a result of the hurricane.

Disaster planning is all about compromises.

That's hard to deal with emotionally; it's kind of like the idea of "acceptable losses" in an army battle. On the one hand, we don't want to give up *anything* if disaster strikes. On the other hand, there is a cost of being ready for a disaster, whether or not the disaster ever occurs. Making compromise decisions is tough.

Risk and Hazard aren't the same thing, and our business continuity plans have to take the difference into account.

Risk communication consultant Peter Sandman sums up the risk reaction in an equation: Risk = Hazard + Outrage. The idea is that the perceived riskiness of something is not just based on the probability of the bad thing occurring (what Sandman calls "hazard") but also on the level of outrage that is felt when the bad thing happens. For example, car crashes have higher probability but lower outrage, while plane crashes have lower probability but higher outrage. That's why planes are considered "riskier" than cars by most people. And that's why Hurricane Katrina, which destroyed the city of New Orleans and killed hundreds (maybe thousands) of people, is getting so much press coverage: people are outraged that something like this could happen.

When IT people do business continuity planning, or even when we plan for a new system, we typically include a list of risks in our project plan. But being the analytical people that we are, we don't usually factor in the emotional "outrage" side of the equation. As a result, we focus our attention on the things that are more likely to go wrong, and not on the things that are more likely to get a bad reaction from the public if they go wrong. For example, in the typical systems planning scenario, we spend our time on things like high availability and verifiable financial transactions, and we neglect the relatively low probability ways that bad people can steal data, and the even lower probability ways in which people can be injured or killed. Guess which type of event hurts your company more in the long run.

2026 Commentary

When I write an article about a current event, the article becomes dated very quickly. Twenty years after Katrina, most people have forgotten all about it, even though it killed over 1,300 people, and caused damage estimated to be $125 billion. It's sad that we don't learn from our mistakes. And that's why business continuity planning is so difficult: People don't want disasters to happen, and they don't want to even think about the possibility. But when disasters *do* happen, an emergency plan can make a huge difference.

35. Information Overload: Why You Won't Read This Newsletter

Originally published on OCTOBER 11, 2005 in #CAREERS, #MANAGEMENT

More than half of the subscribers who receive this newsletter won't even open the email. I can understand why: you've only got a limited amount of time, and you have to be selective about how you use it. But let's be honest; are you really being selective? Or are you just randomly reading some things and not others? Being selective implies some logic behind your selection process. Do you have any logic behind your own approach for coping with information overload?

Once upon a time, a century or so ago, there was actually a shortage of information. In small towns, people would even interrogate strangers passing through just to find out what was going on in the world. Those days are long gone, and today we have the reverse problem. There's too much information in the world. Everyone is telling you that they have the solution to your problems, even if they don't know what your problems are. Magazines bombard you with subscription offers, and some of the publications are so specialized that you wonder how they can find more than a few people who are interested in the topic.

Are You Thrashing?

"Thrashing" is the term used to describe a computer operating system which is being overwhelmed by handling an excessive number of user requests. When a computer is thrashing, it's spending too much time on the system overhead of swapping user tasks in and out of memory, and not enough time on getting productive work done for the users. You can thrash too, when you spend too much time handling trivial day-to-day details, and you don't focus on what's truly important.

An Approach for Dealing with Information Overload

1. Pick one or two sources that you'll use as your primary source of news. Subscribe to other sources if you like (many are free to "qualified subscribers"), but just open them up long enough to scan their table of contents to see if anything catches your eye. If you see something interesting, save a copy using whatever tool you prefer, and use keywords to label it for later reading whenever you have a spare moment—maybe on an airplane or while you're in a waiting room somewhere. Then every so often delete the older stuff that you haven't read.

2. Use your own people as information collectors. Have people volunteer to be the focal point for collecting information about certain topics that are important to you and your business, like one person for new software development breakthroughs and another person for new developments in a specific technology. Have the volunteers summarize their findings at a periodic meeting.

3. Don't be a source of information overload for others. I've seen too many managers route or forward information that they don't bother to read themselves. If something isn't worth *your* time, then think carefully about whether it's worth the time of someone else in your organization. You're not helping your employees to focus if you're overwhelming them with things to read. You should filter the information you disseminate based on your strategy and direction. If you have a personal information overload problem, then make sure you solve it. Don't just pass it along to your employees.

4. If anything significant happens in your company, make sure to summarize it at your employee meetings. If your people know

you'll keep them up-to-date, then they won't have to waste as much time keeping their ear to the ground or reading all those internal puff pieces distributed to employees by your company communications department.

5. Remember that focus is the key to success. Focus on the information that you need based on your strategy and direction. Ignore almost everything else. Your knowledge and expertise is useless if you don't apply it to something relevant to your business.
6. Schedule a certain amount of time each month to evaluate your progress toward increased focus. Think about how things are better or worse than last month, and determine why. Take steps to do better the following month.
7. Continue to read articles like this one which provide you with ideas, insight and motivation for improvement. Even better, discuss them with co-workers and share your views.
8. Share this article with your boss, and you'll benefit from the improvements your boss makes in their own information overload.

Conclusion

Don't blame the volume of information you receive—information overload comes from poor self-management and lack of focus. If you know how to swim, you can swim in water of any depth, even in the middle of the ocean. Dealing with information overload works the same way—you only drown if you let the water overwhelm you, but if you stay on top of it, you'll be fine.

2026 Commentary

When I originally wrote this article, magazines like *ComputerWorld* and *BusinessWeek* were my primary sources of information. Then I shifted toward sources like *Twitter*, certain electronic newsletters, and some web sites. Now I use various news aggregators and sources. The technology keeps evolving, but your choice of sources should depend on your profession, your industry, and your interests.

For saving copies of information, I generally rely on the Bear app for key information, and Instapaper for web-based articles I want to read later. But there are a number of good tools available, so pick the ones that work best for you.

36. Creating Wildly Successful Projects

Originally published on NOVEMBER 8, 2005 in #CAREERS, #ALIGNMENT, #LEADERSHIP, #STRATEGY

I saw a shooting star the other day, a meteor streaking across the sky as it burned up in the atmosphere. A lot of IT projects are like that meteor: they briefly get a lot of attention, they brighten the lives of the people who observe them, but then they fade into oblivion when they're done, having no real impact on anything. Only a few meteors survive their trip through the Earth's atmosphere to arrive on the ground as meteorites. And only a few wildly successful IT projects that are done each year make a significant difference to the business.

How do you tell which IT projects are going to make a significant difference? For the answer, you have to look at the business itself. What's holding your company back? What's the biggest limitation to your company's business? Or, to put it another way, what's stopping your business from having ten times the revenue, or ten times the profit, or, if you're a non-profit organization, ten times the effectiveness in accomplishing its mission?

Business is about dealing with limitations, and any business grows until we encounter one of those limitations. It might be a limit on the number of customers we can find and support; it might be a limit on our ability to manufacture products; it might even be a limit on the number of good product ideas we can come up with. It could be a tangible limit, like a limit on the capital we can raise and invest, or a limit on the space our business occupies. Or it might be a mental limit, like a limit on the number of people we can manage, or a limit on the number of new employees we can hire in a certain period of time.

When a business bumps up against one of these limits, its growth stalls. Slow growth may continue, but not at the rate that could be achieved without the limit in place. Various strategies may be employed to circumvent the limit: heading in a different direction, acquiring another company that has found a way around the limit, or even breaking up the organization into pieces so that each piece can attack the limit individually. For example, W. L. Gore and Associates, Inc., creator of Gore-Tex® fabric, doesn't like the idea of creating complex organizations. so it has a large number of plants each made up of a few hundred employees or less, with the work in each plant done using a "lattice" team structure. Gore has grown by adding more plants rather than by making any one plant bigger.

Most significant limitations are considered immovable

You've probably heard about elephants in India who are trained to do heavy lifting and construction work. When the elephants are just babies, they're kept from running away by chaining them to small stakes driven into the ground. The baby elephants struggle against the chains, but they're not strong enough to pull the stakes out of the ground. Gradually, the elephants accept their fate and stop struggling. Later, as the elephants mature and grow stronger, the same size stake is still used, even though the mature elephant could easily pull it out of the ground. But the elephant doesn't realize this; it continues to recognize a limitation that no longer exists.

Businesses are like these elephants. I'm constantly amazed when I talk to executives and business owners and I hear the assumptions they make. They're restricted by limitations that haven't been tested in years: certain kinds of products don't sell, certain sales techniques don't work, customers won't use self-service tools, automation can't be used for certain processes. The executives think they're right, at least until someone (usually a competitor) comes along and proves them wrong.

Remove an important limitation, and you'll have a significant project

The most significant projects performed in a business are the ones that remove a critical business limitation, or drastically reduce the impact of that limitation on the business. Projects which don't address business limitations are fluff: they may make business life better for a few people, but they don't make a significant contribution to the business, and they're not likely to be remembered in the long term.

Unfortunately, most IT projects are fluff, because most IT projects make improvements in relatively trivial business processes. And while there's nothing wrong with such improvements, there's not a lot that's right about them either.

If you want your IT organization to make a real difference, if you want to get the national recognition that you feel you deserve, then you've got to earn that recognition by removing an important limitation in your business. For example, in the mid-80s I worked with a financial services company which performed a very labor-intensive accounting service for its customers. Since labor was the most important limitation to the business, we did a study on how the information workers spent their time. We discovered a result consistent with the 80-20 rule: 80% of the labor time was spent on 20% of the processes. We rank-ordered the processes that took the most labor hours and focused our IT efforts on improving the ones that would make the most difference. Within less than a year we increased labor efficiency by over 200%, enabling the same workers to handle three times the volume. We didn't spend a lot of money, and we didn't use any "bleeding edge" technology; we just applied common IT principles to the processes where they would have the biggest impact.

And we didn't stop there; we continued to focus on the limitations of the business and make progress in making things better. When efficiency was no longer the primary limitation, we focused on the other limitations of the business: quality, keeping the customer informed, providing good financial controls. The journey never ended—it just shifted direction—as the business grew, got past one limitation and then another.

How do you know what significant business limitations are?

Here are some tips:

1. Look at your company from the outside-in. What limitations are imposed on you from the outside? How might those limitations be overcome? What do your customers want that you don't offer? What do your competitors offer that you can't provide?

2. Look at your company financial statements. Where is most of the money going? Is there an opportunity to make drastic cuts in expense by applying technology?

3. Challenge the idea that certain business limitations are immovable. It's important to periodically test "hard and fast" assumptions. Technology evolves and becomes less expensive, and customer circumstances change over time. For example, online access to certain kinds of data was impossible several years ago because of bandwidth limitations and because of people's reluctance to use the Internet for critical tasks. But attitudes change, and now might be the time to do something that couldn't be done a few years ago.

4. Use fill-in-the-blanks statements to help you think of ideas. For example, use the statement "My business would be _____ [better, or faster, or more efficient, or more profitable, or whatever] if only I could _____, or if only we didn't _____." Brainstorm a hundred ideas for each of the blanks, and then pick the best ones to pursue.

5. Although larger meteors are more likely to make it to Earth as meteorites, the reverse is true for projects; larger projects are

more likely to fail. So even when you put together a project to eliminate a huge business limitation, break the project into small chunks which can be accomplished one at a time. The smaller chunks will be easier to fund, and will have a much higher likelihood of success.

Conclusion

There's a tradition of making a wish when you see a shooting star, but it takes more than wishes to create a wildly successful project. If you identify significant business limitations, and overcome them by applying process improvement skills and technology, then you'll be assured that your project will get more attention than the brief light of a meteor streaking across the sky. And succeeding at such a project can make you a star in your own right.

2026 Commentary

In hindsight I wish I hadn't used that elephant story. I hate that the elephants are treated that way. But it does describe the way that a lot of business people think. A slightly more up-to-date metaphor would be the also dated "think outside the box" example. Businesses have built artificial boxes around themselves with limitations to the ways they operate, and they can derive maximum benefit from process changes which remove those limitations.

37. How to Become a CIO

Originally published on DECEMBER 13, 2005 in #CAREERS, #IT

I'm a bit concerned that aspiring CIOs are looking for a "silver bullet," a magic solution that they can easily apply and thereby instantly qualify to be a highly paid CIO. The real world isn't like that. Many years ago I had an employee who wanted to be promoted to a higher level position, and he kept telling me that if I promoted him to the higher level then he would show me that he can perform at that higher level. My answer to him was always the same: show me that you can be a star in your *current* position, and demonstrate some of the skills that go along with the higher level position, and I'll promote you. But the work comes first—not afterward.

The same thing applies to aspiring CIOs. If you demonstrate the capabilities that a CIO needs, and if you get some proven business successes under your belt, then you can qualify for a CIO position in your current company, or you can convince a CIO recruiter that you can do the job elsewhere. But you must have the successes first. That's what recruiters are looking for—proven success stories—not people who have the potential to succeed.

Eight Qualifications for a CIO

Here's my own list of qualifications that you'll need to develop to become a CIO. First, you need the same kind of skills that are required for any senior executive position:

1. **The ability to inspire confidence and trust**. Some call this "poise;" others call it "charisma." It's a quality that's hard to describe, and even more difficult to develop and achieve. It comes from experience, self-confidence, self-reliance, and knowing that you're going to be successful at anything you at-

tempt. And what's more important, you make others know—without you telling them—that you'll be successful at anything you attempt.

2. **The ability to communicate clearly, succinctly, and emotionally to any audience**, whether it's stockholders, investors, your board of directors, or an entry-level clerical person. When I use the word "emotionally" I mean that the communication has to occur at multiple levels. At the surface the communication is intellectual, conveying facts. But at a lower level there's a communication that makes the person want to agree and comply. At this lower level you're doing more than communicating—you're motivating and inspiring.

   You may have noticed that the two skills I've mentioned thus far are the same abilities that politicians need. That's true, but if these are the only skills you have, then you may find yourself described as an "empty suit," and you'll certainly never be successful at an executive position. So we need to add a conscience and a belief system:

3. **Consistent, demonstrated ethical behavior**. You know what's right—for yourself, for your organization, for your company, and for your country and the world—and your actions consistently demonstrate that you want to move things in the right direction.
   With these first three qualifications we now have an ethical politician, but we need an agenda.

4. **Intelligence and wisdom**. Intelligence to comprehend and truly understand what needs to be done. Wisdom to know what shouldn't be done, and to understand the best way to accomplish things.

This fourth qualification has ruled out most politicians, so we're beginning to zero in on good executives.

5. **Good judgment**. Knowing the time and place to say or do what needs to be said or done. Good judgment comes from experience. Unfortunately, experience comes from bad judgment—either your own or the bad judgment of others. The only way to get good judgment is to learn from your mistakes and from the mistakes of others.

6. **Recognition of your own limitations**. This might seem to contradict Skill #1, in which I said that you know you're going to be successful at anything you attempt. But if you truly know your own limitations, then you won't attempt something that's beyond your capabilities. You'll constantly extend your limits, but you know the difference between being aggressive and being foolhardy.

7. **Motivation to be a CIO** in spite of all of the responsibilities and headaches that come with the job. If you don't live and breathe to be a CIO, then you probably won't get there.

8. **A strong interest in applying technology to improve the business**. You probably wondered when I would mention technology. But note the words I used in describing this skill. A successful CIO doesn't have to be a technology expert, and in fact most CIOs *aren't* technology experts. People who are highly motivated to become technology experts typically don't have the other seven skills. Some CIOs may have come from a technology background, but if they were technology experts in their past, they've now left that skill behind them.

How Do You Become a CIO?

First, be realistic. Do you have the eight qualifications I've listed? If not, then can you get there at some point? Some of the qualifications can be developed over time, but I believe that some of these qualifications are a result of who we are—our basic personality type. There are some people who will never be a CIO, and that's not a bad thing. What *is* a bad thing, however, is seeing some of these people deceiving themselves into believing that they can overcome insurmountable obstacles. They obviously don't have skill #6, and it's unfortunate that these people won't find happiness until they recognize their own limitations and focus on things that truly do make them happy.

But let's say for the sake of argument that you do possess all eight of these skills, or you're close to possessing them. What do you do next? Recognize that having the skills themselves doesn't make you a CIO; you have to create *business* successes to be recognized for what you are. And I emphasize the word "business;" technology successes don't get you the recognition until those technology successes translate into specific measurable monetary business success. Use technology to make your company a star (like eBay), or to make your company more competitive (like Walmart). Use your technology to improve productivity (like UPS) or to change an industry (like CheckFree). Technology leaders usually aren't leaders of technology—they're leaders in the use of technology for business purposes.

You don't have to start at the top, and in fact no one does. You can apply CIO skills to your job as an IT project leader or as a department manager, or even as a programmer-analyst. But you have to remember that CIO success comes from business success, and business success comes from thinking from your customers' viewpoint.

How do you become a CIO? It's not a leap—it's just one more step in a continuing career progression of meeting and exceeding business needs.

2026 Commentary

While the bar for becoming a politician seems to have lowered in the last twenty years, I think the list of skills for the CIO is still correct.

38. Interview Performance Doesn't Equate to Job Performance

Originally published on JANUARY 10, 2006 in #CAREERS, #MANAGEMENT

I don't do job interviews well, at least not as an interviewee. But I do a great job when I'm on the interviewer side of the desk. And it's partly because I've learned from my mistakes as an interviewee.

What a lot of people don't seem to realize is that interviewee performance is not at all related to job performance, unless the job for which you're hiring is an actor. That's what being a successful interviewee is all about: acting. But even the acting is tough, since you have to figure out as you go along what part you're supposed to be playing. You might start out thinking that you're playing a confident, well-informed manager type person, and then you realize after a few minutes that this particular interviewer wants you to be a yes-man flatterer. If you're good as an interviewee, you adapt on the spot, just like a chameleon. Good interviewees are hired not because they'll be good at the job, but because they figure out how to push the right buttons for the interviewer. That's too bad.

My own biggest problem as an interviewee is that I care too much about the job. I tend to think two or three steps ahead, and I start solving the problems of being in the job before I ever get the job in the first place. It's like the football player who starts to take off toward the goal line before he has a firm grip on the ball he's just caught. The result isn't pretty.

I've heard of all sorts of trick questions to ask people during interviews, but I've never found much use for them. Yes, you can ask the interviewees why manhole covers are round or how long it would take a coin dropped overboard in the deepest part of the ocean to reach bottom (both questions are supposed to help you understand the way they think). Or you can have applicants drive you to lunch so you can assess their driving style (aggressive or too cautious) or their way of handling incorrect food orders (tolerate it or make an issue). But what's the point? Why is it that you think this type of behavior will directly translate into job performance?

Even worse, I've seen interviewers start participating in the play acting themselves, deliberately being overly aggressive to see how the interviewee reacts, or putting on little skits with multiple interviewees simultaneously. How ridiculous is that!

Motivation

When I'm interviewing candidates for a job, I use a different approach. First, I try to get a feel for the person's motivation, because motivation is what ultimately is going to make the person successful. I like it when I see people who want to work because they're curious or because they sincerely want to solve problems or make things better for others. I avoid people who consider their work "just a job" and give signs that they'll walk away in the middle of solving a critical problem just because it's dinner time (I'm not advocating long hours as a requirement, but I don't like clock watchers). And I quickly rule out people who are in it for the money. They're welcome to use money as a scorecard, but truly successful people don't have money as their primary motivator. If money is your motivator, then how ethical will you be?

How do you assess motivation in an interview? Ask what they enjoy about a job. But don't listen to the words of their answer; listen to the "music" instead. Listen to the emotion (or lack of it) that they have in their voice. Listen for someone who still has the energy and drive to be truly great, and who isn't burned out or jaded. Many people will be reluctant to answer the question honestly—they'll still be acting—so you may have to approach the question indirectly. Ask what they enjoy doing when they're not working.

Ask what caused them to pick their major in school. You'll begin to get a feel for their real motivation in spite of any acting attempts.

Qualifications

The second thing I look for in an interview is the person's qualifications for the position, but I don't do it in the usual way. Most interviewers want proof that the interviewee can do the job, and the more proof the better. If you think this way, then someone who has done the exact same job for twenty years or more is the best candidate. But that's not the one I would choose. I want to hire someone who has the capabilities to do the job but who is still challenged by it, at least a little bit. I prefer someone to be challenged because it means they'll look at the position with a fresh viewpoint and they'll spend time thinking about how to do the job better. If you hire people who have done a similar job for twenty years then there's no doubt how they'll do—they'll do the job exactly the way they've done it for twenty years, whether or not that's the best way to do the job in this particular case. I don't want repetition—I want fresh perspective and high motivation.

I remember a champion downhill ski racer who talked about his winning strategy. The way to win, he said, is to ski almost out of control. You have to ski so close to the edge of your capabilities that if you skied any faster you would fall. You have to push yourself to the edge to win, and, similarly, you have to be working at the limit of your capabilities to really be motivated and do your best in a job. I look for people who want to work at the limit of their capabilities, and I look for a match between their limits and the requirements of the position.

Communication

The third thing I look for in an interviewee is the person's communication ability and style. There are a lot of different characteristics rolled up in this one area, everything from intelligence to listening skills. I try to assess whether I'll be able to carry on an intelligent conversation with this person in the future. Do I have to spell things out, or can the person read between the lines? Will my intent be misinterpreted because the person imposes a different agenda on the words? In IT, this is especially important because we have to communicate constantly with both technical and non-technical people. Requirements are often stated in ways that aren't absolutely clear, and it's up to the individual to sort out the true intent. People who jump too quickly to conclusions based on speculation have no place in my organization.

Communication goes the other way too, and the person should be able to clearly summarize a situation and net it out. This is partly verbal skills and partly reasoning ability, the experience to know what's important to say and what's fluff.

Personality

The interviewee's personality comes across in the interview if you can pierce the actor's veneer. I don't look for any one type of personality—I appreciate diversity. But there are a few personality types that I try to weed out in the interview process. I don't want know-it-alls; they're dangerous and they poison an organization. I don't want people who think it's all about them; they need to be part of a team. I value both introversion and extroversion as long as the personality is a fit for the particular job. I tend to weed out people who constantly talk when they have nothing to say, but that's just my own preference.

Salvaging Bad Interviews

I sometimes run into people who are worse at being an interviewee than I am, but I've learned to deal with the situation. I interviewed one woman who was incredibly nervous and could hardly respond to questions. I spent a few minutes with her, then pointed out the problem and suggested a solution. I said, "Ok, let's forget about this job, but I'll help you practice for your next interview." I told her to start over, just for the practice. She left the room, came back, and we proceeded to do a "practice interview" in which she reverted to her normal self. The interview went well. I didn't hire her, but not because of her bad first interview—she just didn't have the qualifications I was looking for.

Another time I was interviewing to hire a recruiter, and I turned the interview around to get a sense of the recruiter's own interviewing skills. I described a hypothetical job, then switched chairs and had the recruiter interview me for that job. The result told me exactly what I needed to know.

Conclusion

If you hire people for their interview performance, then you deserve the actors you get. Try to see through the performance to get to the underlying qualities of the person. Why do they want the job and what will they do with it? Are they qualified but not jaded? Can they listen and understand? Can they communicate their thoughts clearly to you? Do they have a personality that's appropriate for the position?

An interview is an artificial situation that doesn't at all reflect someone's ability to do a job. Keep that in mind when you're hiring, and remember that interview performance doesn't equate to job performance unless you're hiring actors.

2026 Commentary

The thing I fear most about interviewing in the current business world is that the really good candidates will have been screened out by automated resume processes before they're ever given a chance to interview. I know the world has changed since I did any hiring, but when I was hiring, there were no automated filters. I would typically review 10-20 resumes for every person I interviewed, and I would typically interview 10 people for every person I hired. My recommendation would be to make sure that you interview enough people to find the really good ones. Don't settle for someone who "checks all the boxes" but doesn't look like a future superstar.

39. Top 10 Reasons Why Men and CIOs Don't Ask for Directions

Originally published on FEBRUARY 14, 2006 in #CAREERS, #IT, #LEADERSHIP, #STRATEGY

I don't usually like to talk about stereotypes, but it occurred to me that the stereotype about men not asking for directions applies equally well to CIOs of both genders. So let me talk about why men don't ask for directions, and I think we'll find some lessons that will apply to senior executives as well. Here is the countdown of my Top 10 Reasons:

Reason #10: They think they know where they're going

Obviously if you think you know where you're going then there's no need to ask for directions. But do you really know, or do you just have a general idea? Once you get to the area of your destination, are you going to have to drive around for a while to figure out exactly where your destination is? That costs money and time, and it hurts your image as a leader.

If you're a senior executive, do you know exactly where you want to go, and do you have a plan to get there? If not, then you need help with directions.

Reason #9: They're too proud

This is an ego thing. But intellectually you know that no one is omniscient, so why should it damage your ego to ask for advice? You still retain control, since you can choose to follow the advice or to ignore it, based on the credibility of the advice. Get over your ego, and do the best thing for yourself and your passengers.

Reason #8: They've been brought up to believe that men and senior executives don't ask for directions

This is a self-fulfilling prophecy, but it's why a lot of stereotypes persist long after their original cause has disappeared. It's like one of those traffic jams caused by an accident that occurred three hours previously; cars still slow down at the site of the accident even though the wreckage is long gone. How do we break the stereotype? By ignoring it. Repeat after me, "It's Ok for men and senior executives to ask for directions. It's Ok for men and senior executives to ask for directions. ..."

Reason #7: They believe men/CIOs have a natural sense of direction

There might be some truth to this. Thousands of years ago in the hunter/gatherer days the men had to hunt down game, and many biologists believe there might be a genetic basis for men having a better sense of direction than women. There's certainly a difference between men and women in the way that they read a map and give directions. Men are more likely to hold a map upright (North at the top) and they intuitively know how to make the map directions correspond to physical reality; women are more likely to rotate the map to align it with the direction they're facing. Men are more likely to give directions using distances, street names, highway numbers, compass directions and navigational markers like traffic lights (e.g., go five miles on the highway, and then go west on Main Street); women are more likely to give directions using landmarks (e.g., turn left at the Baptist Church, then right at the grocery store).

For CIOs, it's possible that they have a natural sense of organizational direction based on their many years of work experience. They know "in their gut" when their organization is headed in the right direction, and so they don't feel that they have to ask anyone else. The only problem comes when their gut hasn't been reoriented to face the rapid changes occurring in our world; maybe their gut is pointing them in a direction that's no longer appropriate.

Reason #6: It takes too much time to ask for directions

A cliché comes to mind: "There's never enough time to do it right, but always enough time to do it over." Unfortunately, the cliché is true in a lot of situations. Saying that there's not enough time for something is equivalent to saying that you don't consider it important enough to take precedence over other activities. So if you cite this reason, then what you're really saying is, "I don't think it's important to know where we're going or how to get there."

Reason #5: They can't find someone to provide an answer

Usually this situation occurs when the destination hasn't been defined well enough. If you just know you're looking for something like the "old Moore house," then it will be a miracle if anyone can help you (unless you're in a very small town). Similarly, vague organizational destinations like "Six Sigma quality," "100% Up-Time," or "a more effective organization" are likely to cause confusion and lead to mixed advice.

Reason #4: They know they're going to get conflicting directions

As I said in the explanation for the previous reason, if you define your destination better, then you'll be more likely to get consistent directions. But realistically, there are usually many ways to get to the same destination, and you'll need criteria for picking the best approach. Do you want to get there quickly but with more mileage? Then take the highway or its IT equivalent: spend more money to buy time. Do you want to take the safest possible path? Then consider using public transportation or its IT equivalent: wait for mainstream use of a new technology before employing it yourself.

Reason #3: They trust themselves more than they trust a stranger

Of course you trust yourself more than you trust a stranger—who wouldn't? But that's not the real issue. You don't know how to get to where you want to go, and it's likely that a stranger *does* have that information. So put trust aside as an issue, and accept the fact that you have to talk to other people to obtain additional information. Do you trust yourself enough to be able to differentiate between good advice and bad advice? If you do, then apply that differentiation skill in your discussions with strangers.

Reason #2: They figure, "If I know where I am, then I'm not lost"

Senior executives with accounting backgrounds are particularly fond of this excuse, and they often overemphasize detailed tracking of project expenditures. But you can't drive by looking in the rear-view mirror, and knowing what you've spent doesn't necessarily help you know how much more you have to spend in the future. Are you spending money on the right things, or are you just being economical in spending money on the wrong things? There's a big difference between doing things right and doing the right things.

And [drum roll, please], the number 1 reason why men and CIOs don't ask for directions:

Reason #1: They think, "I can't stop; we're making such good time."

This is my favorite reason because it seems so illogical. In IT we see it as a focus on irrelevant measures. For example, you may make a big deal about your 50% improvement in lines of code per hour or about your 95% server up-time, while ignoring the fact that your software isn't meeting user needs and your servers are always failing during the peak customer service hours. Think about what you're measuring, and make sure that your measurements directly relate to business needs.

Conclusion

Not asking for directions is obviously a stereotype, and some men and CIOs *do* ask for directions. But even if you don't think the stereotype applies to you, please think about the 10 reasons, and see if any of the reasons sound like something you might say. Maybe you'll learn enough about yourself to ask for directions a little more often. And if we all get a better idea of where we're going and how we're going to get there, then maybe we can all reach our destinations a little bit sooner.

2026 Commentary

I have no idea whether the male/female map use stereotype still applies in today's era of using smartphone apps instead of paper maps. In my personal experience, it was true when I wrote the article.

But I suspect that human nature hasn't changed at all, and men probably still don't ask for directions. I wonder if that means that women are more likely to use smartphone GPS apps.

40. The Blind Men and Information Technology

Originally published on MARCH 14, 2006 in #CAREERS, #IT, #ALIGNMENT, #LEADERSHIP, #STRATEGY

A recent article in *ComputerWorld* by Curt Monash entitled "Three Views From the Top of the Software World" reminded me of the poem about the Blind Men and the Elephant. The *ComputerWorld* article talked about different points of view from leading technology vendors. According to the article, Oracle and IBM view IT as data-centric, Microsoft views IT as people-centric, and SAP views IT as business-process-centric. In the Blind Men and the Elephant poem, six blind men approach an elephant from different vantage points and then argue over whether the elephant is like a wall, a spear, a snake, a tree, a fan, or a rope. Of course "each was partly in the right, and all were in the wrong," and so it is with the technology vendors as well.

IBM and Oracle are correct in emphasizing data; of course IT is impossible without a reliable means of storing and retrieving data and information. But IBM and Oracle are historically vendors of infrastructure, and we all know that it takes more than infrastructure for IT to be successful; it takes user-friendly systems and solutions. IBM and Oracle provide useful tools for IT organizations, but only a few tools for business users.

Microsoft is correct in emphasizing people and the tools to help them; people are the reason that systems exist in the first place. But Microsoft has traditionally appealed to individuals and small business, and many Microsoft solutions are difficult to implement and support when you scale them up to handle large multinational, multi-location corporations. Microsoft provides useful tools for business users—not scalable tools for IT organizations.

SAP is correct in emphasizing business processes; efficient processes are at the root of all profitable companies. But the inherent complexity of SAP's solutions has created a whole industry of consultants just to explain how to best use SAP software. SAP provides tailorable information systems for large corporations—not easy-to-implement solutions that are quickly adaptable to change.

In the poem the six blind men argue "loud and long" about the correct viewpoint, and leading technology vendors do the same thing, as if there is only one correct view of the world. If you've fallen under the spell of this vendor propaganda, then you may find yourself arguing that one of the viewpoints is better than another, but here's the reality: information technology doesn't do anyone any good unless it makes business people happier and more productive. Data in itself has no value without a context; tools are no help if they're undependable and hard to use; and business processes don't mean anything unless people understand them and want to follow them. No IT vendor has a complete solution, and frankly that's the way I like it. Competition and different viewpoints encourage new ideas and new approaches for solving problems, and we get progress. Ultimately this makes things better for all of us.

Think about how the role of an IT organization has changed over the years in response to vendor advances. First there were electronic accounting machines and operators. Then there were early computers and assembler language programmers. Next came more advanced computers together with higher-level programming languages. Computers become less expensive and a Data Processing department evolved into a Management Information Systems department, and then into an Information Technology organization. Most of IT today isn't writing code—it's improving the business by utilizing vendor hardware and software. It's systems integration and data integration and project management. Underneath is an incredible infrastructure, especially when compared to the early computers with their card readers and tape drives.

Conclusion

We've come a long way in IT due to vendor progress, but no single vendor has a complete solution, and that's not likely to change. You should focus on the elephant—not the various bits and pieces that the vendors advertise. All of those bits and pieces are important, but only as components of the elephant: only as things that fit together with other components to make the people in your business happier and more productive. No vendor is selling elephants, so you'll have to do the assembly yourself. That's the role of the CIO: figuring out how to assemble the elephant from all of its components, and then leading the assembly process.

2026 Commentary

Vendor orientation has changed over the years, and other major vendors have emerged. But the reality seems to be the same: No single vendor gives you everything you need, but you can assemble pieces from different vendors to do the best thing for the business.

41. You Don't Have to Measure It to Lead It

Originally published on APRIL 11, 2006 in #ALIGNMENT, #LEADERSHIP, #MANAGEMENT, #PROJECTS, #STRATEGY

I don't know about you, but I'm sick of people telling me, "If you can't measure it, you can't manage it." I blame this aphorism for a lot of the wasted measurement effort spent by today's organizations. The problem with this statement is the use of the word "manage." People see the word and assume that it's the *Big M* use of the word: Management, as in "the Management of the Company." But instead the statement uses the *Little M* version of the word: management, as in "management of the department budget."

Don't get the distinction? You're not alone. But think about it this way: management and leadership are not the same thing. Management is the optimization of existing resources. For example, Question: "How will you cope with such a limited budget?" Answer: "We'll manage."

Leadership, on the other hand, is the introduction of change into an organization by defining a clear picture of a different future, motivating people to desire that future, and then showing your followers the first steps in achieving that future (see "1. How to Get People to Change—the Human Side of IT Projects"). Unlike management, leadership doesn't deal with things the way they are —it deals with things the way they ought to be.

Need some examples? You *manage* an existing production line to optimize its efficiency, but you *lead* the creation of a new factory. You *manage* the care and feeding of cattle on a ranch, but you *lead* a cattle drive to move the cattle somewhere else. You *manage* a customer service call center to maximize customer satisfaction and minimize cost, but you *lead* the creation of a web-based self-service solution to serve as an alternative to call support.

In IT organizations, we have the need for both management and leadership activities. The IT infrastructure, including data centers and networks, lends itself to management. Application software development, on the other hand, requires a careful combination of leadership and management.

Almost no job is one hundred percent management or leadership, but as a general rule, higher-level managers and executives are likely to need more leadership and less management. That's because executives typically have a longer-term view, and are focused on changing their organizations to meet future needs.

Misunderstanding of this concept by executives is prevalent because most executives start out as managers, and they have difficulty getting away from their old detailed measurement habits when they rise up through the organization. But think about the situation this way: management is like *pushing* a rope; it works only if you make continuous small adjustments to keep the rope in line. Leadership is like *pulling* a rope; the parts of the rope that are out of line will automatically align themselves with the direction of motion. When you push a rope you have to focus on the rope; when you pull a rope you focus on the direction in which it's being pulled.

So let me go back to that measure-it/manage-it aphorism again. It is certainly true that you can't optimize the use of existing resources ("manage it") if you don't have a good understanding of how the existing resources are being used ("measure it"). But how many times have you heard the measure-it/manage-it statement when in fact it's leadership that's needed—not management? You don't need that same detailed information about existing resources if you intend to lead an organization in a whole different direction. In other words, you don't have to measure it in order to lead it.

I'm not saying management is unnecessary—I'm just saying it's totally different from leadership. For every organizational change that is led, there are many little changes that have to be managed. But don't confuse the two. You can't be an effective leader if you spend too much time measuring things.

2026 Commentary

I still like that rope example: pushing the rope versus pulling the rope. I think it makes this clearer.

42. Are You Wasting Your Resources on "Honey Projects"?

Originally published on MAY 9, 2006 in #ACQUISITIONS, #LEADERSHIP, #MANAGEMENT, #PROJECTS, #STRATEGY

When I was a child I learned a funny nonsense rhyme:

I eat my peas with honey.

I've done so all my life.

It makes the peas taste funny

But it keeps them on the knife.

The logic of the rhyme argues that:

1. I eat peas with a knife.
2. But when I try to eat peas with a knife, they fall off.
3. So I should eat the peas with honey so they'll stick to the knife and not fall off.

I coined the phrase "Honey Project" based on the rhyme; a Honey Project is one that makes a bad solution easier to tolerate. When faced with a Honey Project, you should seriously consider whether the project is worthwhile. After all, wouldn't it make more sense to replace the bad solution with a better one instead of compounding the problem by dressing up the bad solution?

Here are some examples of Honey Projects:

An Attempted Central Customer Service System

A friend of mine told me about a customer service system that his company tried to put together. They wanted a single system that a customer service rep (CSR) could use to tap into the data of his company's twenty or thirty different business divisions. The IT people working on the project made the incorrect assumption that a copy of the data had to reside in a central database (eating peas with a knife), so they spent tons of time and money trying to work out the systems and processes required to take daily copies of the data from all over the world. Ultimately the project collapsed due to the number of data translations required, the politics of moving data across multiple IT organizations, and the realization by the users that the data provided in the CSR system wouldn't be real-time anyway.

In hindsight, the project team should have considered a layered approach, with a single CSR system providing a common interface that goes out to other existing customer service systems under the covers to retrieve data as it is needed. And they should have solved the problem in phases, adding a few business divisions at a time in order to build political support for the central solution. But the project failed because the project team jumped to a bad solution before fully understanding the problem.

Inappropriate Product Acquisitions

I did some work with a corporation that has a history of inappropriate acquisitions. Although it's a technology corporation, the non-technology-savvy marketing executives have repeatedly discovered small companies with supposedly complementary products (eating peas with a knife). Then the corporation spends big bucks on "integrating" (Honey Projects) the small companies with the corporate products, in most cases spending more than it would have cost to develop an equivalent product internally, and in many cases abandoning the new products after the unsuccessful integration effort.

Doing the right technology due diligence should have prevented some of these small companies from ever being acquired. In this case the corporation asked the right questions during technology due diligence, but the corporate executives didn't like the answers so they ignored the due diligence results and acquired the companies anyway. Ultimately the due diligence proved correct, and the Honey Projects failed to make up for the shortcomings found in the due diligence. It's hard to dress up a bad investment.

Microsoft Virus Protection

Microsoft Windows has long been criticized for its susceptibility to computer viruses (eating peas with a knife). Microsoft's answer to the problem is to introduce its own virus protection product, Microsoft Windows OneCare Live, rather than to fix the vulnerabilities in the Windows product. Thus OneCare Live will become the "honey" that helps to cover up the problem that Microsoft itself created.

Virus vulnerability is a difficult problem for Microsoft to solve. Historically Microsoft has had mediocre initial releases for each of its products, but it has invested large amounts of money on product improvements for later product versions. The computer virus susceptibility in Microsoft Windows dates back to the initial versions of Windows, is embedded deep in the architecture of the product itself, and will require huge amounts of effort to correct. It's likely that Windows users will live with the Honey Project approach from Microsoft for some time to come, probably until a solid competitor to Windows is available.

Conclusion

The "Peas with Honey" rhyme is about an attempt to salvage a bad decision, and the lesson from the rhyme is to be very careful of your decisions before you start building on them. Too often we make this mistake in IT: we don't fully understand what's needed, we put a lot of effort into building an inappropriate or incomplete solution, and then we waste effort on Honey Projects to make the bad solution more palatable.

Take a long hard look at the projects you've got going on now. How many of them are Honey Projects? What can be done to solve the real problems and put your resources to better use? How can you change your processes to prevent this kind of problem in the future?

Honey Projects happen to all IT groups, but if you're going to do a Honey Project, then make sure you're going into it with your eyes wide open. Try to minimize wasted effort on Honey Projects by taking the time to make the right decisions in the first place. But if that doesn't work, then try making the right decisions the second time around.

2026 Commentary

My comments about Microsoft Windows are dated, and I haven't used Microsoft Windows in over twenty years, so it's up to the reader to assess whether the comments still apply.

Honey projects still continue, of course, and they're not restricted to IT. You see them in construction when a workaround is implemented to lessen the impact of an architectural design problem. You see them in manufacturing when someone comes up with a temporary fix to a production defect. When I was in school, we used to call these Honey Projects a "kludge," which Wikipedia defines as "a workaround or makeshift solution that is clumsy, inelegant, inefficient, difficult to extend, and hard to maintain. Its only benefit is that it rapidly solves an important problem using available resources."

43. Heroes Don't Scale

Originally published on JUNE 13, 2006 in #LEADERSHIP, #MANAGEMENT, #PROJECTS, #STRATEGY

"Hero" is one of those positive words that gives us mental images of rescuing children from burning buildings or saving troops from certain death. We admire, praise and imitate heroes; they set the standard for bravery and going "beyond the call of duty."

But despite the personal admiration associated with heroism, there's a dark side to some heroism stories. If we look past the glory of the heroic act itself, we sometimes wonder whether the need for the heroic act could have been avoided. Why didn't the child leave the burning building earlier? How did the surrounded troops get in that position in the first place?

Our asking these questions doesn't make the heroes any less heroic; these people still performed a selfless act that deserves our admiration. But it's important to think about heroic situations in order to reduce the need for heroism in the future. There's only a limited supply of heroes in the world, and it would be a shame to waste them on unnecessary heroism.

I once worked with a business which overemphasized heroism. A consultant once told them, "Your company is the best crisis management organization I've ever seen, but unfortunately you treat everything as a crisis." The observation was absolutely correct; the culture of the organization was to spend most of the time dealing with exceptions to the rule, but without establishing very many rules to begin with. There was minimal emphasis on planning, but a huge emphasis on being ready for the unexpected. It was a service business, and the guidelines for the service were minimal: just do anything the customer wants, and hope that the business can make a profit by doing so.

Contrast this with NASA, the U.S. space program, where the emphasis is on anticipating any possible thing that could go wrong and doing endless simulations to prepare astronauts to deal with those problems. Space walks are practiced underwater to simulate weightlessness. The astronauts are even subjected to real weightlessness in a diving airplane (the so-called "Vomit Comet") so that they are prepared for the physical sensations they'll get in space.

Of course, in spite of all of the NASA preparation, things happen that aren't anticipated. The Apollo 13 moon mission was a great example; we almost lost the astronauts altogether but for the heroism of our people in space and our engineers on the ground. Preparation, practice and established processes don't handle all of our problems, but they position us well, and they prepare us better for dealing with the occasional situation where true heroism is required.

Why Talk about Heroes?

You may ask, "Why is all of this talk about heroes relevant to me?" Here's why: People in our organizations act the way they do because of the way we motivate and reward them. It's ok to reward an occasional hero, but it's critically important to equally reward the people who do things on a day-to-day basis which make it unnecessary for heroism to occur.

If you only reward heroism and crisis handling, and you ignore crisis prevention, then you run a serious risk of turning your organization into a crisis handling machine. Such an organization can operate successfully for years, but it can't scale; it's very difficult for the organization to grow and handle bigger challenges because it's just barely meeting the challenges that exist. An organization that depends on the heroism of individuals won't survive in the long term. People burn out when they have to be heroic on a day-to-day basis, and after a while they get frustrated and they begin to realize, "Why do I have to do this time after time?" An organization of heroes requires a constant supply of new heroes, and unless you have some political basis for their motivation (e.g., the security of the country), then you'll eventually run out of recruits.

What's worse, even if you develop a heroic organization that works well with 100 heroes, you'll find it very difficult to increase its capacity to handle higher volumes. Can you find 200 heroes? 300 heroes? 1,000 heroes? When you have an organization that's dependent on the heroism of individuals, you're artificially limiting its capacity for growth. One person acting beyond the expectations of any individual is a hero. But to expect heroic behavior on a daily basis is unfair to the people and the organization. Heroism is, by definition, an exceptional behavior. But the necessity of heroism can be avoided in most cases by establishing solid rules and processes that govern day-to-day activity.

Conclusion

Reward your heroes, but recognize them as exceptions, and establish processes that make day-to-day heroism unnecessary. Then reward the people who create the processes and who keep these day-to-day processes running. Remember that heroes don't scale, but that you *can* scale well-defined processes. The capacity for growth is a necessity in any successful organization, and you can only achieve growth when you're not dependent on heroism.

2026 Commentary

The title for this article came from a comment made in a presentation at the April, 2006 SIM Atlanta meeting by Dr. Guido Sacchi, CIO and Executive Director of Shared Services at CompuCredit. My thanks to Dr. Sacchi.

44. 5 Reasons Why IT People Love Lists

Originally published on JULY 11, 2006 in #CAREERS, #IT, #MANAGEMENT, #STRATEGY

Ok, admit it. You're reading this because you want to know the 5 reasons. That means that you're probably like most IT people and you love lists yourself. Of course it's not just IT people who love lists; it's almost anyone who falls into the Myers-Briggs category called "Judging" which describes people who need structure and need to have things decided. Using the Myers-Briggs categorization, most IT people are ISTJ, INTJ or occasionally ESTJ or ENTJ.

I'm a list person myself. You can identify true list people by the way they use a To Do list. When true list people accomplish something that's not on the list, they add it to the list just so that they can have the satisfaction of crossing it out. Does that sound familiar?

In thinking about lists the other day, I decided to make a list of the reasons why people love lists. Here they are:

1. There are two kinds of people in the world: those who classify people and things into categories, and those who don't. Lists are for the first kind of person; lists are used to categorize.

2. Lists are logical and make it seem easier to cope with what seems to be an inherently illogical world. Lists are used to create a micro-world that's somehow more satisfying than the real world. It allows us to avoid confusion by restricting our focus to the things on our list.

3. Lists make it appear that there are a finite number of choices for something, rather than the infinite number of choices that are actually available when you consider alternative ways to solve your problems. If you have a list then you can eliminate choices one by one until there's only one choice left. That makes us feel like we've done our homework, although perhaps

if we had really done our homework then we would have put more types of choices on the list in the first place.

4. Lists define the "box" that people don't think outside of (remember the popular business admonition, "think outside the box"). Once you make a list you seldom consider other factors that aren't on the list. That's a good thing if you want to limit your thinking to "the box," but a bad thing if you're in a situation where your thinking shouldn't be limited.
5. Lists give us a sense of precision even if there's no accuracy. [see my sidebar comments below for more on the confusion between accuracy and precision]

Why This List?

So far this article has been a bit tongue in cheek, but let's apply the concept of lists to the real world of business and IT. There is a tendency among analytical people—IT people included—to boil issues down to numbers. That works in some cases. When we're trying to design widgets to do 77 operations per second, creating a widget that does 86 operations per second is generally an improvement.

But when it comes to issues that involve people, or issues that affect groups of people, we need to be careful about the way that we overanalyze the situation. When trying to decide whether to close down an office, there are non-monetary costs that are not given enough weight in the decision. When trying to hire a new worker there are intangible factors that are much more important than a list of items on a resume. When trying to change a business there are social and humanistic issues that are often left unresolved.

There's an old saying, "when all you have is a hammer, everything starts to look like a nail." A list is like a hammer, and it shouldn't be the only tool in your tool chest. If you're trying to make a decision and your list tells you that the decision ought to go a certain way, double-check that decision with your gut. How do you feel about that solution? Does it make you uncomfortable? Does it make your stomach tighten up? Does it seem like something isn't right? Then maybe it's the wrong decision and you're being deceived by your list. I'm not talking about something supernatural here; your gut is telling you that you've left important aspects of the decision off your list, and that you ought to take those aspects into account. Listen to your gut; it's the voice of the experience that you've gained over the years.

Pick A or B

As a manager I've often had employees come to me for help with a difficult decision. They'll list the reasons why they ought to pick alternative A, and then give me another list of reasons why they ought to pick alternative B. They'll say they can't decide.

I'll tell them, "Pick B." Then I'll ask them, "How do you feel about that decision?" Sometimes they'll say, "I'm relieved, I felt like B was better, but I couldn't figure out why." Sometimes they'll immediately argue with me, saying, "But I feel like A would be a better choice."

I didn't pick B because I thought it was better; I picked B arbitrarily to get a reaction. Most people know intuitively what decision they should make; they just can't rationalize the decision using a list. By arbitrarily picking one of the alternatives, I bypassed the list process and got to their sense of what ought to be done. After they understood their reaction to the arbitrary choice then they could make the real decision.

Conclusion

Be very careful in your use of lists. Use them to help you with problems that truly have a finite number of possibilities and a finite number of factors that can be translated into numbers. Don't use them for decisions that depend on ethics, human values or world issues. And to expand your box a little bit, always add an item to each list: "other things not on this list."

Trust your gut; the experience you've gained is more valuable than you think. When something *feels* wrong, try to figure out why. Then make the best decision based on a combination of logic *and* your experience.

Lists can be valuable, but they're only one tool in your toolbox. If you rely on them too much, then you won't be taking advantage of your experience, and you won't be making the best decisions for your business and for your life.

Exercises for the Reader

1. If you're a true list person, try to make a list of the situations where a list shouldn't be used.
2. Make a list of your friends and coworkers who like lists, then show this article to them.

Sidebar: Confusing Accuracy and Precision

People tend to confuse accuracy and precision. An accurate number is one that corresponds to reality and has been proven in unbiased testing. A precise number is one that's more exact or has more decimal places, whether or not it's accurate.

Advertising people in particular know that it sounds more reassuring to say that "4 out of 5 dentists recommend" a certain toothpaste or that a certain soap is "99 and 44/100% pure." Putting a precise number on something makes it sound more accurate and official, even if the number is invented out of thin air.

There's an art to putting numbers in ads; certain numbers are better than others. Round numbers like 90% sound like they've been made up, numbers like 97% sound more reliable, and adding a decimal place or two like 97.49% makes the number sound more scientific. Is there a real scientific basis for a number like 97.49%? Maybe, but it might just be a made-up number that's deliberately more precise because people think that a more precise number is more accurate.

I can say things like "here are the five reasons why people love lists" even though someone else might include seventy-six reasons and have no overlapping items with my list. The number 5 is precise and so the list of 5 reasons is precise as well. In being precise, the list fulfills the reader's need for structure and it provides a satisfying sense of closure. But unfortunately the list may not be accurate at all. Think about that when you put together your own lists.

2026 Commentary

For more on the confusion between accuracy and precision, see the article, “114. How to Create Misleading Statistics in 6 Easy Steps.”

45. 10 Rules for IT Job Success

Originally published on AUGUST 8, 2006 in #CAREERS, #IT

I got a call last month from a newsletter reader in India who wanted help in making a career decision. He was a bit vague about the details, but it seems that he impetuously quit his previous job over a difference of opinion with his manager. Now he wasn't sure whether to try to patch things up or to move on to another company.

I won't go into the details of the advice I gave the reader, but the phone call prompted me to consider what general career advice I would offer to lower- and mid-level people in IT. I came up with the following set of 10 rules:

1. You don't become an executive by being a better employee.

As Robert Kiyosaki pointed out in his excellent book *Before You Quit Your Job*, people usually gravitate toward either being entrepreneurs or being employees. The education system in the United States (and probably elsewhere as well) is focused on training people to become employees. It teaches them to follow rules and procedures, to maximize their use of existing resources, and to follow the orders of the people in charge. Entrepreneurs, on the other hand, learn through trial and error to obtain the resources they need and to make something out of nothing. Employees want the security of a full-time job; entrepreneurs want the freedom to accomplish their dreams.

I point out the difference because lower- and mid-level people in IT are very definitely employees, but the higher you go in the executive ranks the more likely you are to find an entrepreneurial spirit. This creates a natural barrier which prevents lower-level people from ever becoming executives. Only by leaving behind the very essence of why you became an employee can you ever rise to the executive level.

2. There are no right answers to any question—just opinion guided by ethics, legality, experience and tradition.

This of course contradicts what you're taught in school, but see rule #1. Even the answers to basic questions like "How much is 2 + 2?" are based on years of tradition that developed our modern number system. In a different context, for example in a chemical reaction, 2 + 2 might not equal 4 at all.

If someone tells you that there's only one way to do something, what the person really means is that they always do it that way. It doesn't mean there isn't a better way, and it doesn't mean you can't improve on things.

3. Pick your battles.

Of course, life is full of people telling you that there's only one way to do things. And although there are probably better ways to do almost everything, it often isn't worth the time or investment to figure out a better way to do most of them. The key is to prioritize: to focus on improving the things that are most important to you or that are most important to your business.

4. The people who make the fewest mistakes are the ones who have the most experience. But the people with the most experience are the ones who have made (or observed) the most mistakes in the past; making mistakes is how they got the experience in the first place.

The only way to learn is to make mistakes; babies don't walk the first time they try, but they keep trying until they stay upright. If you do your job by copying exactly what someone else does, then you might do the job as well but you won't know why. You won't be able to respond to changing conditions by altering what you do, because you won't understand the implications of doing things a different way.

Overprotective parents often correct their children's homework because they don't want their children to "fail." But the lesson they teach their children is that it's unacceptable to fail, while the lesson they ought to be teaching their children is that it's ok to fail as long as you learn from your mistakes.

Try new things, and try to do old things new ways. Some new approaches won't work, but you'll learn why and you'll be better for the knowledge.

5. The more credibility you have, the more people will listen to you.

If you're perceived as an expert, then people will pay attention to what you say. This seems obvious. But it's the route to this credibility that confuses most people. See the next rule.

6. Build credibility by being successful at what you do—not by telling other people they're wrong.

When you first start your career, just do the best you can at the work you're assigned. Treat each assignment as a challenge to exceed (not just meet) your goals, and gradually you'll build your credibility.

But be very careful to keep your opinions to yourself about topics for which you haven't yet established credibility. It's very easy to get a reputation as a troublemaker when the number of your critical comments exceeds your demonstrated capability.

7. Titles convey temporary credibility, but the credibility expires if it isn't proven.

New managers or people who are new to senior technical positions often expect that they get credibility just because of their title. While it's true that new people in titled positions are given the benefit of the doubt for a month or so, all of the credibility associated with the position will disappear if you don't show evidence that the credibility is warranted. You don't get credibility because of a title—you get a title because of your credibility.

8. The lower you are in an organization, the more you have to do what you're told.

There are four scenarios for doing what you're told:

1. If you do what you're told and things go well, then you'll be perceived as right.

2. If you do what you're told (and it's well documented) and things go badly, then your boss is wrong. But you're also wrong if you didn't point out the problem earlier.

3. If you don't do what you're told and things go well, then you might be right or you might be lucky, but you've got to be careful in either case. Was this a battle worth fighting?

4. If you don't do what you're told and things go badly, then you'll be perceived as wrong whether or not you really are.

Try to learn from every assignment you're given, even if you don't happen to agree with the way you're supposed to do the work. Figure out why the work is being assigned that way; there's always a reason even if it's not a very good one.

9. You can learn something every day if you look for it.

You might learn something you should do, or you might learn something you shouldn't do. Either way it's progress.

10. Enjoy the journey.

Sure it might be great to be an executive someday. But enjoy what you're doing right now. Life is about more than anticipation—it's about the learning experience you go through to get to your destination. To me, that learning experience has always been the fun part.

2026 Commentary

Although the title is "10 Rules for IT Job Success," I actually think the rules are applicable to *any* job.

46. Achievement is Not the Absence of Failure

Originally published on SEPTEMBER 12, 2006 in #CAREERS, #LEADERSHIP, #MANAGEMENT

There are some jobs where achievement is the absence—or maybe the avoidance—of failure. Driving a bus is one of those jobs; if you make it through the day without an accident, without hurting or annoying anyone, and without falling behind your schedule, then you're successful. There are other jobs where carrying out your duties and avoiding failure is more of a challenge; defending an army outpost against an enemy is an example.

In the IT world, much of our success with IT infrastructure is measured this way. Systems and network availability is determined by the absence of failure. IT security is similarly measured by the lack of security breaches.

In some companies (a declining number) where IT is not a critical part of the company's products or services, the IT organization is invisible. In those companies, in fact, you can say that IT is measured by its invisibility. If business people can do their jobs without an IT problem, then IT is successful. This kind of thinking is at the heart of Nicholas Carr's May, 2003 *Harvard Business Review* article, "IT Doesn't Matter." Carr's main point in the article is that IT is increasingly being judged by its reliability and cost—not by its breakthrough contributions to the business.

I won't go into an argument against some of Carr's thoughts here (for more on that subject, see “5. IT Doesn’t Matter?”). But I want to point out that even when overall IT success is judged by the absence of failure, the success of individuals within the IT organization is still judged by their unique contributions—not their absence of failure. If you're personally responsible for IT infrastructure, then you'll be personally measured by *improvements* in infrastructure reliability and cost—not merely by a continued level of constant reliability and cost. Senior managers and executives are expected to make things better: to *achieve* something. If things around you stay the way they are then you will be regarded as a failure even while your organization may be regarded as a success.

There is only one exception to this rule, and that's the case of an organization continuing at the same level of reliability and cost in the face of a dramatic increase in the issues facing the organization. If you're defending an army outpost and you continue to hold off the enemy while the attacking force doubles, then surely you must be a success, and it's the same way in business. But there's a catch: the people measuring you must *recognize* the increase in the attacking force. That's more difficult than you might think, since many IT organizations are not well understood by the business itself. How does the business know that you're in a struggle to hold the line on reliability and cost in the face of an increasing number of problems? As an example, how many IT security management teams have been regarded as failures due to extraordinary increases in hacker attempts in the last few years?

The key to being recognized as an achiever is to clearly demonstrate your achievement, and to show that things are much better because you're there and they would be much worse without your contribution. You have to prove your contribution to the business. And then when you're done proving, go right back and prove some more, because you won't be rewarded again for yesterday's achievement.

Conclusion

Yes, in some jobs, achievement *is* the absence of failure. But if you want to be truly successful in life you have to do more. It's not enough to avoid failure—you have to make things better.

2026 Commentary

No notes.

47. IT Lessons from a Manufacturing Shop Floor

Originally published on OCTOBER 10, 2006 in #IT, #MANAGE-MENT, #PROJECTS

When there's too much work to do, most people try to multi-task and get it all done simultaneously. But the reality is that multi-tasking often hurts your productivity, and it drastically increases your stress level. Let me tell you about a parallel situation in manufacturing that illustrates my point.

I once worked with a manufacturing organization that specialized in making custom products using a variety of machine tools. Their shop floor consisted of a large number of "stations," with each station corresponding to a particular tool: everything from lathes to punch presses to grinders to cutting machines. Each station had one or more workers, as well as a rack that contained a "queue" of work to be performed. Most custom products required more than one operation at more than one station, so when each product was scheduled, a "routing" was created that showed the route that the component material would take from station to station until the work was completed.

This kind of manufacturing organization used to be pretty commonplace, and you'll still see a lot of small businesses run this way. Work is initiated when a customer places an order, and when the order is scheduled to begin, the raw material is taken from inventory and issued to the first work station listed in the routing. Then as each operation is completed, the work-in-process material is moved to the next work station where it sits in queue until the next operation begins. When the last operation is completed, the finished product is moved to an area where it is packaged for shipping to the customer.

This all sounds fairly straightforward, and in fact most manufacturing operations like this operate very well as long as they're operating below full capacity. But as orders grow, it becomes more and more difficult to avoid scheduling problems. Workers at each station have a choice on what job to do next, and they often choose the easiest job even though it might not be the one that's most important to the business.

When manufacturing capacity gets scarce and the queues build up, a lot of game playing begins. First there is an attempt to label jobs by their priority. This is confusing, however, because a job's priority may change from hour to hour based on how much time is left before the finished product due date. So shops begin to get more creative, attaching red flags to the jobs that are in danger of being late. But in an overcapacity situation there will be a sea of red flags, and the poor workers don't have a good idea which job to work on next. Some shops get even more creative at this point; I heard about one manufacturing company that attached red helium balloons to important jobs.

When flags and balloons fail, a typical manufacturing shop will hire expediters whose job it is to follow certain jobs through the shop floor and make sure that they're given higher priority. But when there's not enough capacity there are always some jobs that will be late, and so multiple expediters are hired. Of course the workers get caught in the middle: which expediter do they listen to when there are conflicting views? In some companies there have been fist fights between expediters, each of whom has been tasked with ensuring the on-time delivery of their assigned orders.

Does any of this sound familiar? In an IT environment we have the same sort of problems. We have multiple project priorities with limited capacity. We have people with different skill sets (analysts, programmers, database designers, QA, infrastructure, etc.) and our projects often require routing work from person to person and skill set to skill set. Most of the people have a queue of work to do, and they often get confused about what work to do first. When in doubt they tend to do the work that's the most fun or the easiest, even though that particular task may not be the most important overall.

In IT we have project managers who, like expediters, are responsible for moving a particular project along. And resources are often shared among projects, so the project managers have to fight each other (at least figuratively) on a day-to-day basis to ensure that their own project keeps moving.

You can see the parallels between the manufacturing shop and the IT organization, but why am I telling you this? Because most manufacturing shops have solved the capacity problem, and I think we can learn from their solutions. A lot of time and research has been spent on creating various techniques, algorithms and software to help manufacturing shops optimize their use of capacity. This area of Production Activity Control has been studied extensively by various professors and consultants because there's obvious money to be made in optimizing the use of high-priced machinery.

There are a number of ways in which the manufacturing experience can be applied to IT, but let me tell you the most important. And it doesn't require you to buy software or invest in high-priced consultants.

The most important thing found in manufacturing to optimize the shop floor is to keep work off the shop floor until it's really ready to be started. That sounds very simplistic and even counterintuitive, but you would be surprised how effective it is. If the amount of work sitting in queue in front of a workstation is limited to the things that are supposed to be done in the near future, then the workers have little choice but to do that work. All of the confusion goes away. The need for expediters goes away. All of the red flags and balloons go away. The simple rule is that if it's on the shop floor then it's important. The workers do the important stuff first, then go back to the scheduler to get the next piece of work assigned.

Japanese manufacturers have gone so far as to formalize the process used in higher-volume manufacturing by implementing something they call "kanban." In the kanban process the work is pulled through manufacturing instead of being pushed. The IT equivalent would be to assign new work to an individual only when the individual has completed their previous assigned work.

Think about it. A lot of the multitasking that we claim to do isn't really multitasking at all. We're not actively working on many of the tasks that we're assigned; we're just juggling them until we have time to actually put some effort into them. Until a task is actually started, it's just occupying our thoughts and giving us stress. That's fine if you're an executive or manager (no, it isn't really fine, but that's a subject for another day), but it's not something we want to pass on to our employees. Shop floors that are overwhelmed by partially finished work are drowning for lack of clarity and proper prioritization. Does that sound like your IT employees? Or are you focusing their efforts by giving them only the work they have time to do? By giving them the important work and holding back the rest, you'll be implementing the most important principle that has been discovered in manufacturing shop floor research: people do what's important if they know what's important and if they're not distracted by unimportant demands.

Conclusion

It seems like a long way from a manufacturing shop floor to IT, but the management processes are remarkably similar. The key to getting work done productively is to focus effort on doing the important things one at a time. Don't distract your employees by assigning them more work than they can possibly do. It makes life more stressful for the employees, and you're abdicating your responsibility as a manager because your employees are setting the priorities—not you.

2026 Commentary

I wonder how many people will have trouble relating to the manufacturing example used in this article. I hope the message is clear even if you don't have any manufacturing experience.

48. The Politics of Information Technology

Originally published on NOVEMBER 14, 2006 in #CAREERS, #IT, #ALIGNMENT, #LEADERSHIP, #MANAGEMENT, #PROJECTS, #STRATEGY

I'm writing this on November 7, 2006. That's election day in the United States. During the last few months we've been besieged with television and radio advertising for candidates, and even recorded messages sent to our telephones. Now it's time for all of the campaigning to end as we go to the polls to vote.

It's unfortunate that most of us will vote based on perception rather than reality. We don't really know whether these candidates for office will do a good job; we only know what they've told us or what their opponents have told us. And even more unfortunately, these costly campaigns (television and radio ads aren't cheap) have been financed by people and organizations who have a stake in the election, people who in many cases are more interested in their own personal agendas than in the choice of the best candidate.

The whole process is so vague that we're really voting on the perception of a perception: our perception of advertising that's been designed to create a certain perception of its own.

That's what politics is all about: creating a perception of success that leads to career advancement. And like it or not, politics is a part of everyday business life just as much as it is a part of government. An Information Technology organization isn't immune to politics, although many of the analytically oriented people who belong to IT organizations believe that to be the case. Wherever there are people there will be politics, and learning to cope with politics is a necessity for any manager.

Answering a Question on Politics

In government organizations we see politics in its purest form. I was the keynote speaker last year for the University System of Georgia's annual computing conference, a gathering of IT people who support the colleges and universities in the state. After the keynote I did a question and answer session, and one of the questions got to the heart of the political situation. The question was, "How do we in IT achieve the longer-term objective of providing optimal information technology for our customers when the heads of our customer organizations change from election to election?" The questioner cited the tendency for elected officials to come to their new jobs with specific agenda items to be worked: agenda items that were at the heart of their election campaign. The result for IT is that there is a huge directional discontinuity from the term of one elected official to the term of the next elected official, and programs that are started under one administration are often abandoned under the next.

I answered the question two ways. First, I mentioned one of my standard rules for successful projects (see more in Chapter 6 of my book *Boiling the IT Frog*): projects should never have a duration longer than the remaining job tenure of the sponsoring executive. If the sponsoring executive is an elected official, then all projects should be scheduled to be completed before the next election. Most projects should be even shorter in duration, or should at least be broken up into sub-projects which provide customer benefits within the shorter timeframe of each sub-project.

Second, I dealt with an obvious objection to the first part of my answer: What about necessary IT projects like a major update in infrastructure that can't be done in less time than the tenure of a sponsoring executive? Or what about projects which are obviously necessary from an IT viewpoint, but which aren't directly tied to the agenda of the elected official? In these cases you have to use the same methodology employed by sailboats which need to sail against the wind. You have to tack back and forth, sailing alternately in directions other than your ultimate direction, but ultimately moving closer and closer to your objective. In Chapter 16 of my book *Boiling the IT Frog* I describe the process this way:

Rather than to go after these objectives directly, you might have to define some intermediate steps along the way, and tie the accomplishment of these intermediate steps to other more business-critical projects that they support. In effect, you are "tacking" the long-term project back and forth to take advantage of the winds: the current business projects that support your intermediate steps. And just as a sailing ship has to sail a longer distance to reach its destination because of the tacking, you'll find that sometimes the achievement of a long-term IT objective will involve some steps that might seem a little bit off the route. But that's the price to pay for achieving the best compromise between business objectives and technology objectives. And the ultimate measure is whether you can achieve both sets of objectives simultaneously.

Political People

So far I've talked about "real" politics: elected officials and their impact on IT. But the other use of the word politics applies to anyone who puts their own self-interest before the interests of a company or organization. This kind of person, often described as "very political," is "all about me." They are only interested in personal advancement and glory, and anything else is just a way to achieve that advancement and glory.

If you're personally more focused on doing the best thing for the company, then it's difficult to work for a political person like this. You have to be very careful in conversations and presentations to make sure that anything you want to do is presented in such a way as to maximize the personal benefit to the person without seeming too obvious. It's almost as if everything has to be thought through on two levels: at the first level you justify a project or an action by showing how it's the right thing to do for your company. On the second level you justify a project or action by showing—possibly in a veiled way—that it's the right thing to do to advance the career of your boss.

This same course of action applies if the "very political" person is a customer or a user of your systems. Everything has to be approached at two levels, almost as if you're keeping two sets of books.

Conclusion

Politics are a fact of life in a world of human beings. Politics in government cause you to take special steps to deal with a boss or customer who has a very specific timeframe in the job. Political people may not have quite the short-term focus of real politicians, but they have the same focus on self-advancement that politicians do. As a result, all communications between you and the political people have to be approached on two levels: one level for the company's best interests and another subtextual level for the individuals' best interests.

For people who claim to "avoid politics," I just have one piece of advice: people are going to be political whether you acknowledge it or not. It's best to develop an approach where you *always* try to present things in a way that appeals to both the company's best interests and the personal best interests of the individuals. For the non-political people it won't matter, but it will be critical for the politicians.

2026 Commentary

The title could have just as easily been "The Politics of Business" or even "The Politics of Life." All of these issues come up whether you're in IT or not. And the answers are the same for non-IT organizations.

49. The Best IT Organization in the Country?

Originally published on DECEMBER 12, 2006 in #IT, #ALIGNMENT

Recently a newsletter reader told me that his CEO asked the question, "Is our IT organization the best in the country?" The reader wanted to know how it's possible to "benchmark yourself against other IT organizations so to be in a position to answer such a question."

First Answer

I think there are two answers to the reader's question. The first answer is that the CEO needs some help in repositioning their thinking about IT. The CEO is viewing IT success as separate from the success of the overall business, and that view is a common warning sign of misalignment between IT and the business. IT is a part of the business—often a key part—and IT should be measured by its contributions to the business. If it's a service business, then how does IT contribute to the processes used for delivery of the service to customers? How does IT improve those processes, providing more service at higher quality and lower cost? If it's a product business, then how does IT improve the development and production of products and then facilitate the sales and distribution of those products? You get the idea.

My point is that it's a trap to think about IT separately from the business. It reminds me of the old medical cliché, "The operation was a success but the patient died." Clearly, a medical operation can't really be a success if the patient dies, and similarly an IT organization can't be successful if it isn't contributing to the overall success of the business.

The tendency to measure process success independent of business success is not unique to IT. The United States has an award, the Malcolm Baldrige National Quality Award, that recognizes American businesses which meet specific criteria for process improvement. But what's particularly interesting about the Baldrige award is that most of the business winners aren't leaders among their business competitors. I've always had the view that companies who spend the time to qualify for the Baldrige award are spending their time on the wrong things—they're focusing on their own processes and not on the way that those processes contribute to business success.

Second Answer

A second answer to the question is that there are two ways to measure an organization: you can measure it on *what* it's doing or you can measure the organization on *how* it's doing those things. Another way to say this is that you can measure whether the organization is doing the right things, or you can measure whether the organization is doing things right. Of course the ideal is that the organization is "doing right things right"—that the organization is doing the right things *and* it is doing them in the right way.

When I talk about an IT organization's contribution to the business I'm mostly focused on *what* the organization is doing. Are the IT projects contributing to the bottom line? Is the IT infrastructure resilient and secure enough to make the business customers happy? Where deficiencies in systems or infrastructure are noted, is the IT organization prioritizing its work to solve the problems and create the most benefit in the quickest possible way?

When you compare two IT organizations in two different companies—even in the same industry—you can't easily compare what they're doing, since the needs of the two businesses are likely to be substantially different and the two businesses are likely to be at two different points in their evolutionary path. It's impossible to say whether one IT organization is better than another on this basis, except by asking the executives in the two companies whether they feel that the IT organizations are successful.

It's like comparing the engines in two dissimilar race cars. Which engine is better? The one that makes the race car win the race, regardless of horsepower, compression ratio, or any other internal measure of the engine. Each engine has to be matched to the needs of its race car; you can't measure the success of an engine without considering the needs of the car.

Inside-Out Measurement: Measuring How

I've been talking about outside-in measurements: measures of the success of something by looking at the effect it has on other more important things. There is some legitimacy in doing an inside-out measurement as well, but only as a secondary, less important measurement. For an engine, an inside-out measurement would be something like its horsepower-to-weight ratio, and these ratios can be compared between two engines. But again I want to emphasize that what's truly important is whether the car wins the race—not its engine horsepower-to-weight ratio.

So with that caveat then yes, it's possible to benchmark two IT organizations and measure their efficiency at doing things right—to measure *how* IT is doing things. You see this all the time in IT organizations because (a) the measurements are relatively easy to do, and (b) the IT people are very analytical and so they like to talk about these efficiency measurements. Common doing-things-right measurements include bugs per line of code, lines of code per person per week, up-time percentage of various hardware (A very deceiving number. Would you want to fly on a plane that has 99% up-time?), average phone wait time for support calls, average response time of a transaction (usually averaging across both peak and non-peak periods so that the average is nonsensical). I could go on and on, and many IT organizations have indeed gone on and on, with hundreds of these measurements. If you think about it, it's almost certain that a creative analytical person can find some measurement in which their organization appears better than someone else. But the key question is, "Does the measurement have any correlation at all with business success and user satisfaction?"

I don't mean to come across as totally negative on the idea of benchmarking and measurement. Some inside-out measurements truly *are* meaningful, and for those measurements, it's a great idea to benchmark against other companies (or other organizations within your own company) to try to learn from the best practices of others. A measurement of unnecessary customer call-in hold time (for example, caused by the customer service rep waiting on a computer response) can be useful, and an improvement in the measurement directly correlates with business success. Similarly, the amount of time spent by an average salesperson filling out an order processing screen can be measured and improved through various processes, and the improvement can directly contribute to the productivity of the salespeople.

What you need to remember about inside-out measurements is that (a) they only matter if improvement in the measurement will lead to a tangible improvement in the business, and (b) it's the processes that you're comparing when you benchmark against another company—not the measurements themselves. It's also important to keep the number of inside-out measurements small; too many measurements will prevent your employees from focusing on the process improvements that really matter to the business.

Conclusion

Do you have the best IT organization in the country? I don't know, and you probably won't know either. But the more important questions are,

- "Is my IT organization holding back the company, or is it contributing to the betterment of the business?" and

- "If my IT organization is contributing to the betterment of the business, then is it doing all that it can, and is it doing so in the most effective and efficient way possible?"

If you can answer that last question in the affirmative, then your IT organization is successful, and you can focus on the bigger question, "Do you have the best *company* in the country?" Ask your CEO to answer *that* question.

2026 Commentary

In hindsight, there is another way to view the CEO question, "Is our IT organization the best in the country?" The CEO is really telling you, “I don’t know how to measure IT, so I don’t know if you’re doing a good job. Help me by giving me some criteria to evaluate your performance.” Your best response to this statement is to give your CEO some concrete measures of how IT is contributing to business success. And take the opportunity to give the CEO a list of other things that the IT organization could be doing that would make an even bigger impact on the business.

50. How to Reduce Risk in IT Projects

Originally published on DECEMBER 15, 2006 in #IT, #MANAGEMENT, #PROJECTS

If you only consider one type of risk, then you're headed toward project failure.

Most software projects are unsuccessful. According to studies by the Standish Group, in 2002 only 34% of Information Technology (IT) projects were completed on time and within budget, 15% of software projects were abandoned altogether, and $55 billion was wasted on abandoned projects and cost overruns. Those statistics are bad enough, but in fact they represent an improvement over earlier study results. In 1994 the Standish Group found that only 16% of IT projects were completed on time and within budget, and a huge 31% of software projects were abandoned, for a total waste of $140 billion.

Obviously, IT projects are risky, but in this article I'll tell you how you can reduce that risk. I'll start by using a simple example of a project. Then I'll apply the ideas from the simple project to more complicated IT projects.

The Sample Project

For a sample project I'm going to use a project that has nothing to do with Information Technology. People have different expectations for IT projects, and I want you to first think about projects without considering IT. In our sample project – deliberately made as simple as possible – our objective is to carry ten 20-pound cannon balls across a plank from the roof of one building to the roof of another.

Let's start by defining what we mean by risk. What kinds of risks are there in this simple project? Before you read any further, stop for a minute and think about the risks that you would expect in such a project.

Here are three examples of some of the risks associated with this project:

- Accidentally dropping the cannon balls off the side of the building
- Falling off the plank
- Getting the cannon balls to the other side but failing to meet some other business requirement

What other risks did you come up with? As you read further, see if you can figure out what type of risk they are.

The First Type of Risk

The first example—accidentally dropping the cannon balls off the side of the building—is a type of risk I'll call "the uncertainty of good things not happening." This type of risk is directly related to the achievement of project objectives; if you drop cannon balls off the side of the building then the cannon balls don't make it to the roof of the second building. To reduce this type of risk,

- Increase the certainty (the probability) of the good things happening, or
- Make the good things better.

One way in which we can increase the certainty of good things happening is to use a simpler process, like a plank with a v-shaped groove in the middle so the cannon balls can be rolled across, and you don't have to carry them.

One way in which we can make the good things better is to build a permanent process to move the cannon balls if this turns out to be an ongoing need.

The 2nd Type of Risk

Falling off the plank is a second kind of risk, one I'll call "the uncertainty of bad things happening." This is the risk of things that interfere with project success, but which aren't directly related to the achievement of project objectives (my falling off the plank doesn't necessarily affect the cannon balls). It's more difficult to anticipate this kind of risk, but it's important to try. You should brainstorm possible things that could go wrong, and after you generate a long list of possibilities, go back through your list and figure out which bad things are more likely.

To reduce this second type of risk,

- Reduce the uncertainty (the probability) of the bad things happening, or
- Make the bad things not so bad.

In our sample project, here are some ways in which we can reduce the uncertainty of bad things happening:

- Widen the plank
- Install side rails on the plank
- Pick a different building that's closer (shorter plank)
- Wear shoes with good traction

And here are some ways in which we can make the bad things not so bad:

- Pick two buildings that are shorter so the fall isn't as bad
- Install a safety net under the plank
- Wear a safety harness anchored to the roof

The 3rd Type of Risk

The third kind of risk – getting the cannon balls to the other side but failing to meet some other business requirement – is the risk of not doing the best thing for the business. This risk is often forgotten in project planning, but it can be the death of any project. To reduce this type of risk,

- Find out what the best thing for the business is, and then do it, or
- Make the best things for the business even better.

Here are some ways in which we can find out what the best thing for the business is, and then do it:

- Get a better understanding of the underlying business requirement to ensure that the solution is a good fit: ask why they need cannon balls anyway
- Find out how long your company will be using cannons, and tailor your project solution accordingly

One way to make the best things for the business even better is to convince your company management that missiles are a better solution than cannons.

Other Common Project Issues that Increase Project Risk

You can also increase the risk in a project by doing stupid things. Here are some examples:

- Unnecessary enhancements. For example, you must juggle the cannon balls as you carry them across the plank.
- Unnecessary constraints. For example, you must move the cannon balls in the dark of night.
- Scaling up the project beyond the solution. For example, it's not just 10 cannon balls – it's 10 million cannon balls.

These examples may be overly simplistic, but this type of behavior is very common. And it's difficult to recognize unless you're looking for it throughout the project.

The Hardest Risks

The hardest risks to deal with are the ones that are embarrassing or politically incorrect, because you don't want to acknowledge publicly that you're thinking these things—even though they could sink your project. For example, "I'm afraid of heights," or "some of the project team members are going to leave," or "the project sponsor is unsupportive."

Applying What We've Learned to IT

OK, so how many of you have had to move cannon balls across a plank lately? Not many, so let's apply this to the real world of Information Technology (IT). In the following tables, I'll give you some real-world equivalents for each of the risks I mentioned with the sample project.

Increase the certainty of good things

Sample project	**Any project**
Use a simpler process, like a plank with a v- shaped groove	Clarify the project objective, and don't let the objective specify the solution

In our sample project, the word "carry" was part of the objective, but the real objective was just to get the cannon balls from one place to another. Often we'll find that the wording of an objective unnecessarily focuses us on a particular solution. But don't assume that this solution has to be used. Try to uncover the real objective, and redefine the project as required.

Make the good things better

Sample project	Any project
Build a permanent process to move the cannon balls	Consider the context into which the project will be implemented. How can you make things better for the business in the long run?

Often there is a rush to do a project, and in the panic of getting started no one thinks about the longer-term implications of the objective. Some projects have "throwaway" solutions; the result won't be used too many times, and a short-term least-effort approach is appropriate. Other projects may sound like they're solving a small, short-term problem but when you dig deeper you discover that there is a long-term need for a better, more sustainable solution. It's important for you to figure out which type of project you're dealing with and to choose your solution accordingly.

Reduce the uncertainty of bad things

Sample Project	**Any Project**
Widen the plank	Make allowances for error in your plan
Install side rails on the plank	Use proven techniques; have fault- tolerant equipment
Pick a different building that's closer (shorter plank)	Break the project up into a series of smaller deliverables
Wear good shoes	Use proven tools

It's always safer to do a project in a series of smaller verifiable steps than to attempt to use a "big bang" approach in which there are large amounts of effort and/or time expended without a verifiable result.

It's also important to limit the number of unknowns in a project. Having inexperienced people use new unproven tools to achieve a radical objective is too risky for most projects.

Make the bad things not so bad

Sample Project	Any Project
Pick two buildings that are shorter so the fall isn't as bad	Try unproven technology on a less important project
Install a safety net under the plank	Have contingency plans to mitigate the risk
Wear a safety harness anchored to the roof	Have "trigger events" that will cause contingency plans to be invoked

Many project managers tend to regard risk as black or white—it will happen or it won't. But risk is a grey area, and even if something bad happens, steps can be taken to minimize the impact of the bad thing on the project. Risk mitigation is a subtle art that's a key part of successful project management.

Find out what the best thing for the business is, and then do it

Sample Project	Any Project
Get a better understanding of the underlying business requirement to ensure that the solution is a good fit. Ask why they need cannon balls.	Make sure the project is truly what the business needs. Are there better things to spend the resource on?
Find out how long your company will be using cannons, and tailor your project solution accordingly	Adjust the scope and quality of your solution to match the real need

This is one of the toughest things for a project manager to do: question the need for the project. But even a "successful" project is a failure if the project shouldn't have been done in the first place, so it's important that someone ask the tough questions. Think about things from a high-level business perspective. Is this the project that should be done right now? Is this the best thing for the business?

Make the best things for the business even better

Sample Project	Any Project
Convince your company management that missiles are a better solution than cannons	Look at things from the point of view of benefits instead of features. Does this project offer the most benefit to your company?

Often a project is chosen based on a set of business assumptions that may be incomplete, out-of-date, or biased by the personal views of the people championing the project. Think about the business assumptions and how the success of the project will be impacted if they change. Should the nature of the project be revised to anticipate those possible changes? Or should the project be changed to a "throwaway" project, with everyone knowing that its benefits are only temporary?

Other common project issues that increase project risk

Sample Project	**Any Project**
Unnecessary enhancements: juggle	Keep the project focused on the real objective. Push back at requirements inflation.
Unnecessary constraints: do it in the dark	Ask why. If there is a real business problem, then solve it.
Scale up the project beyond the solution: 10 million cannon balls	Watch out for unspoken future requirements that will invalidate your project solution.

Remember: Requirements aren't "gathered"—they're negotiated. There's nothing forcing you to say yes to project requirements that will increase your risk without providing any substantial business benefit.

Similarly, it's important to question any volume or quantity assumptions that will have a significant impact on the project solution. It's better to plan for possible volume increases (or decreases) now than to try to react to them later.

Hardest Risks to Deal with

Sample Project	Any Project
I'm afraid of heights	Explicitly deal with emotional reasons why this project could fail (e.g., lack of user commitment, strong sense of "not invented here")
Some of the project team members are going to leave	Turnover will happen in most projects. Have contingency plans for key resources
The project sponsor is unsupportive	Reconsider whether the project ought to go forward if it doesn't have the right support. If necessary, "fire" your project sponsor to ensure success

They're called "hardest risks" for a reason. It's very difficult to bring these kinds of risks up in a discussion with executives. But whether or not they're discussed openly, the risks exist, and some plan should be made to deal with them.

Conclusion

By using a simple, non-IT sample project, I've shown you the three kinds of project risk to anticipate and reduce. Unfortunately, most projects only address the first type: the uncertainty of good things not happening. As a result, these projects are more likely to fail because of the unaddressed risk of bad things happening or the risk of not doing the best thing for the business. Add in unnecessary enhancements, unnecessary constraints, and occasionally scaling up a project beyond the solution, and it's a wonder that any projects ever finish on schedule and under budget. And when you consider that most projects fail to deal with embarrassing or politically incorrect risks, it's a miracle that any project ever completes at all.

Don't let your project fall victim to uncertainty. Anticipate all types of risk, and come up with appropriate contingency plans and mitigation strategies. Your project will be far more successful, and you'll be the star of your organization.

2026 Commentary

The cannonball example is a bit farfetched, but I wanted to use an example that was outlandish but relatable. When I did this presentation for a group of project managers, it stimulated a lot of thought and discussion. There is so much emphasis on the detailed part of project estimation and scheduling, it's rare for people to talk about these high-level issues that are actually more likely to cause project failure.

For additional perspective, see "15. The Opposite of 'Project Risk' isn't 'No Risk.'"

51. 8 Attributes of an Ideal Boss

Originally published on JANUARY 9, 2007 in #CAREERS, #MANAGEMENT

The first part of January is time for the annual introspection exercise known as resolution making. Many of us will take a hard look at ourselves and try to focus on ways in which we can improve our lives. Some of us will take a particularly hard look at our jobs: Are we happy in our current job? Should we look elsewhere for a different job?

One of the key factors in our job happiness or unhappiness is our relationship with our boss—the person to whom we report in our company's organization chart. A good boss can make a bad job tolerable or even enjoyable, and a bad boss can turn an otherwise gratifying occupation into a miserable experience.

Take a minute or two to think about your own boss. How do they rate in the following categories?

1. Understanding

There are a lot of different aspects of understanding. First off, does your boss relate to you as a person or are you just a cog in the machine? Boss-to-employee personal relationships are tough to do successfully. In an ideal relationship, the boss values you for your contributions to the business but recognizes that you are an individual and so you have individualized needs. An ideal boss understands your skills and capabilities, but also understands your individual goals and objectives—both in your career and in your personal life.

A second aspect of this factor is the boss's understanding of where your organization fits into the "big picture" of the company, how it's supposed to contribute to the business, and how to make that happen. Good bosses also have an understanding of their own strengths and weaknesses, and they know how to create an organization of individuals that is stronger than any one individual.

Another aspect of understanding is the boss's understanding of the work you do. It's not unusual for a boss to be managing a business function that they have never done themselves, nor do they have any idea how the work is performed. It takes a very special boss to be successful in managing such an environment.

2. Flexibility

The worst bosses are like those cheap little battery-operated toy cars you buy for your kids: they only go one direction, and if something gets in their path they just bang up against the obstacle until the obstruction moves (if it ever does). People like this are very difficult to work for. They don't listen to your objections or ideas for improvement; they just continue to try to get all of their employees to do things the "right" way (i.e., the way the boss has always done it).

A good boss is flexible enough to understand that there is more than one route to success. Good bosses know the difference between giving you an assignment and telling you how to carry it out. They'll help you when you need assistance, but they won't force you to follow their way of doing things.

The flexibility also extends to personal considerations. Flexible bosses know that the most important thing about your work is that the job gets done. They know that your prompt arrival at the official start of the business day isn't that big a deal as long as it doesn't interfere with your results, nor does it matter what you wear as long as it isn't disruptive. Flexible bosses understand that personal issues sometimes require you to take unscheduled time off from work, although they expect that you'll be mature enough to find someone else to take over any pressing responsibilities.

3. An Insulator

I once heard of a manager who described his job this way, "I absorb uncertainty." What he meant was that he acts as an insulator between his own boss and the people who work for him. When higher management seems to be vacillating on the best approach to take on a given subject, a good boss will insulate their employees from the confusion, keeping the employees headed in a consistent direction until there's reason to change that direction. Similarly a good boss will absorb uncertainty from their employees, passing upward in the organization a consistent unified view even if the employees are not in total agreement.

4. A Conductor

Unlike electrical devices, a boss has to be a conductor as well as an insulator. This is the idea of conductor in the transmission sense—someone who passes messages downwards from upper management and upwards from employees. The real skill in this area is knowing when to conduct and when to insulate. Judgment is obviously required, as is the skill to summarize information and clarify things that aren't clear to begin with.

Of course a good boss is also a conductor in the musical sense of the word; someone who optimizes the performances of individuals to produce a result that is much bigger and better than the sum of the individual contributions. This skill starts with understanding the capabilities of the individuals, and includes the ability to fit the individual skills, aptitudes and motivations together into something that creates a performance worthy of praise.

5. Clear about expectations for what you should do

I've worked for bosses who are never specific or complete in their assignments. They neglect to reveal important information—

like the fact that the result of a big project has already been promised to customers for next month. I think these bosses are ashamed of the situation, but rather than to face the problem head-on, they figure they will "just see how you do" and then maybe a miracle will happen.

Good bosses are very clear about expectations: both their own and those that have been imposed upon them from higher up the organization chart. Good bosses are honest about what has been requested, and they're willing to discuss requirements openly and negotiate them where possible.

6. Helps you prioritize

When requirements are clearly overwhelming, and there's more to do than you can possibly handle, a good boss doesn't just say "do it all." Instead, they tell you what's important and what can wait, thereby giving you the flexibility to focus on the most important things without undue stress.

7. Listens

Underlying many of the other attributes I've already mentioned is a listening attitude. Listening is more than just hearing what you say; it's understanding both the text and the subtext of your message, empathizing with your point of view, and responding in a way that shows you that the boss cares about your situation.

8. Clear about what you have to do to get to your next career goal

Last but not least, a good boss should understand where you want to go with your career, and should help you take steps to move in that direction. For example, if you're a technician and you want to move into management, then a good boss will help you recognize the skills that you'll need and will help tailor some of your assignments to build those skills.

Career planning in business is usually associated with performance reviews, but the American annual performance review is mostly a joke, albeit one with a huge impact on your salary. A useful performance review should happen whenever it's appropriate—after a project is completed, after you've done something particularly good, after you've done something particularly bad, but probably not on a certain calendar date. Useful performance reviews hold up two yardsticks to your work: one that measures you against the expectations of your current job and another one that measures you against your personal expectations for career growth. The measuring is done jointly by your boss and yourself, and it's important that there be a two-way communication: you have to understand how your boss feels and your boss has to understand how you feel.

Conclusion

I've never had an ideal boss, although I've had a few who were good in most of the categories. Think about how your own boss measures up to the criteria I've defined here. Then consider four more things:

1. Are you being fair in your measurement, or are you refusing to communicate with your boss in some of these areas? Are *you* part of the problem?

2. How would you measure up as an employee using the same eight criteria in reverse (i.e., are *you* understanding, flexible, etc.?)

3. How do you think your boss would rate your boss's boss? There's a top-down effect here; if your boss's boss doesn't treat your boss fairly then it makes it harder for your boss to treat *you* fairly.

4. If you're a boss yourself, then how would your own employees rate *you*?

Finally, if you're feeling brave or just confident, show this article to your boss and your employees to start an interesting discussion.

2026 Commentary

I still agree with the list of attributes, but I'll admit that there aren't many bosses who act like this. That's a shame. And if your boss doesn't measure up, see "110. How to Deal with a Bad Boss—3 Approaches."

52. 5 Approaches to Software Strategy

Originally published on FEBRUARY 13, 2007 in #IT, #STRATEGY

I recently visited a potential client company who wants help in setting strategy for its licensed software products. In the last few years I've mostly helped companies with IT strategy, so I had to think back to my product development days and consider the differences between IT strategy and software product strategy. And in doing so, I realized that there are lessons from software product strategy that can be applied for use in IT organizations as well.

From my own experience, software companies (and maybe non-software companies as well) set their product strategy using one of five different approaches:

1. Internal Vision
2. Surveys and Check Lists
3. Customer Requirements
4. Personas and Scenarios
5. Some combination of the above

1. Internal Vision

Steve Jobs, founder of Apple, is an example. Jobs himself created the initial vision of what Apple products should be, and so the products were an implementation of his vision. It's like this in some other companies as well, or at least it starts out this way. Typically it's the founder of the company who sets the initial product strategy, and then the entire company rises or falls by that vision. If the vision is one that customers want, then the company succeeds. If the vision is out of sync with customer wishes and the visionary doesn't adapt the vision, then the company fails miserably.

The visionary in this approach doesn't have to be the founder of the company, or even part of the company when it starts. Sometimes an individual makes it their personal mission to create the vision and strategy for a product—even if the individual didn't create the product—and the product strategy is then set based on that individual vision.

The biggest advantage of this approach is that the product strategy is consistent, and the products typically appear to be well thought-out and well engineered for their purpose. But when the product vision becomes too big for one person to control, other approaches have to be considered.

2. Surveys and Check Lists

In some ways this is the exact opposite of the Internal Vision approach. Companies who use this approach don't have an "inner compass" that shows them the way—or they choose to ignore their inner compass. Instead, they rely on "scientific" methods to determine what the product strategy should be. Surveys and focus groups identify the desired characteristics of a successful product. Check lists for product features are created by consultants and third-party product reviewers. Then the product strategy is created by the company to maximize the product score on these checklists, while often paying little or no attention to overall product consistency and usability.

The result is a product that scores well on the checklists but usually fails to be more than a mediocre performer in meeting customer needs. Thus this approach is a good way to avoid product failure, but not a particularly good way to achieve product success. Software defined this way often has feature bloat (an unreasonable swelling of the software to accommodate too many check list items), and it becomes non-intuitive and difficult to use. This is particularly true when the software is cobbled together from the source code of multiple acquired products. Some of the Microsoft products fall into this category.

3. Customer Requirements

This is like the Surveys and Check Lists approach but it's targeted at specific customers instead of relying on a sample of potential customers. The concept is simple: If you want to know what customers want, just ask them, then do what they say. Of course there are a few problems with this approach. For example, if you ask ditch diggers what they need to do a better job, they are likely to say a bigger shovel if they have never seen a backhoe. Most customers can only tell you about minor improvements that are needed in a product because they're too close to the existing model of product use. It takes a different perspective to step back and look at the bigger problem, then recommend an improvement that radically alters the processes being used. Customers are not likely to have that perspective, so the Customer Requirements approach typically only gives you incremental product improvement.

Customers also tend to have conflicting requirements. The tradeoffs among the various conflicting needs are not easily made without a detailed understanding of how the customer actually uses the product (not typically obtained using the Requirements approach), but it's these tradeoffs which ultimately cause the product to fail or succeed.

4. Personas and Scenarios

In his wonderful book, *The Inmates are Running the Asylum*, Alan Cooper describes a method that goes beyond customer requirements to focus on the process itself. Cooper asks you to create a few imaginary users, each of whom has specific goals for using the software. For payroll software, for example, you might create Joe Employee (who just wants to get his paycheck on time and with minimal hassle), Susie Payroll Clerk (who wants to minimize her own weekly effort while simultaneously minimizing the number of people calling her about problems with their paychecks), and Barbara HR Manager (who needs summary reports to satisfy company executives and to meet various regulatory demands). Now Cooper proposes that you imagine typical scenarios that each persona might follow in the use of the product. For Susie Payroll Clerk, those scenarios might include the weekly time sheet process, data entry, checking for errors, and distributing paychecks. Design your software to maximize the satisfaction of each persona in each scenario, and you've gone a long way toward providing the ideal software product.

As Cooper points out, "To create a product that must satisfy a broad audience of users, logic will tell you to make it as broad in its functionality as possible to accommodate the most people. *Logic is wrong*. You will have far greater success by designing for one single person [the persona]."

5. Some Combination of the Above

I'm always amazed at the way various companies mix and match these approaches. Sometimes the combination creates a better product, but more often it just raises internal conflicts among the designers. If you have to use multiple approaches, then make sure you pick one of the approaches as primary, and then use the other approaches to fine-tune the primary approach or to sanity-check the design. For example, you might choose Personas and Scenarios as primary, but still do focus groups to help you improve your definition of the personas and scenarios that you use.

Conclusion—Getting Back to IT

"Ok," you may be saying, "what does this have to do with IT? My IT organization doesn't have any products, and I know exactly who is going to be using my software. Is any of this relevant to me?"

Yes it is. Consider that:

- The people who use your software will eventually leave their jobs and be replaced by others who have different opinions about how to do their jobs.
- Organizations change, and you may find that your software will be used by people in newly acquired divisions, in new locations, or even in different countries.
- Job definitions change, and the work that's currently being done by several people may someday be done by one person. Alternatively, the work being done by one person may be split apart to be done by multiple people.
- Transaction volumes change, and a system that is adequate for today's volume may be overwhelmed by higher demands in the future, or may be too cumbersome to satisfy a future decrease in volume.
- Technology changes, and the user interface method used today may be replaced by a different method in the future.
- The IT organization itself will change, and the people working on a system today will be replaced by different people down the road.

In light of all of this change, it's not enough to just modify the system based on what your users tell you. Whether you realize it or not, you need a product strategy for each system even if the system is only used inside your company.

So think about the five approaches. Which approach do you want to use to define the product strategy (or "system strategy" or "software strategy" if you prefer one of those phrases) for each system you have? How will you use that approach to optimize the user effectiveness of each system within the budget limitations you're given? And how will you use that approach to justify an increased budget for those systems that need more work, and a decreased budget for those systems that are good enough?

Product strategy isn't just for software you sell; you really ought to have a product strategy for every one of your systems.

2026 Commentary

In today's world where business software is often purchased instead of written from scratch, you still need a product strategy for each system. Many vendors offer multiple versions of their products, ranging from basic function to advanced integration with companion products. It's important to decide what function you're going to buy from which vendors, and how you're going to interconnect all of your purchased products with each other and with whatever custom software you have in your business.

53. How to Fail as a CIO

Originally published on MARCH 13, 2007 in #CAREERS, #IT, #ALIGNMENT, #LEADERSHIP, #STRATEGY

Success in any job is measured by the alignment of expectations and performance. If the company expects "X" and you deliver "Y" then you fail, no matter whether or not "X" is achievable and no matter whether or not "Y" is actually better for the company. If it's impossible to deliver "X" in the desired time frame and with the resources you've been given, then job success depends on your being able to credibly convince executives in your company that their expectations are unrealistic. And if "X" is inappropriate—something that's not really where the company should be putting its energy—then your job success depends on you determining the more important objectives and then convincing executives that another focus is better.

A good CIO knows that there are two important components of success in an executive position, and *both* components are required for success. On the one hand is what most people think the job is: making the things happen that you committed to do. But on the other hand is a more subtle aspect of the job: determining the right thing to do for the company and then convincing other executives in the company that the right thing is what needs to be done. A CIO who neglects the second component of success doesn't deserve to be called a CIO; they're not a leader—they're just a manager.

How most companies find a CIO

Finding the right CIO for a company requires three things:

1. You have to know what's really needed in the company.
2. You have to get executives in the company to agree on those objectives.

3. You have to find someone as CIO who has the experience and talent to be able to achieve those objectives.

The typical CIO's job is more difficult because most CIO searches do these three things in the wrong sequence. The search focuses on finding someone who can achieve the objectives that the company has defined, even though in most cases the company really doesn't know what the right CIO objectives should be. As a result, newly hired CIOs often go into their new jobs focused on the wrong things—the things they were told to do by the hiring executives. Instead, a new CIO should focus on doing three things simultaneously:

- Make sure that existing bad things don't get any worse.
- Work on the key areas that were identified by the hiring executives.
- Learn as much as possible about what the company *really* needs, and then convince the business executives that priorities ought to be changed.

Of course, once in a while the key areas identified by the hiring executives really *are* what the company really needs, and then the second and third items in the priority list become the same. It's great when that happens, but I haven't seen it happen often.

The CIO Prospect Game

There's a game that CIO prospects have to play during the interview process. The game is to try to sell yourself for the job while simultaneously trying to get a handle on how far out of whack the perception of IT is in the hiring company. You want a job, of course, so you have to be nice. But you also want to avoid walking into a situation where failure is inevitable.

A CIO friend of mine who's now in the job market told me about one such situation. The interviewing executive gave him a whole list of complaints about the company's current IT, but when my friend asked about the company's commitment to try to improve things, he was told that the company wants a total revamp of IT without investing any additional money or business executive time. My friend declined to go further in the interview process, but I know that eventually that company will find a CIO candidate who's not as picky. My guess is that the CIO position will be open again within a year if the new CIO can't change the minds of the company's executives.

How to Fail as a CIO

The awful truth about being a CIO is that there are a lot more ways to fail than to succeed:

1. You can fail to deliver key projects.
2. You can be ambushed by deficiencies in your infrastructure (e.g., customer credit card information leaks out).
3. You can be overwhelmed by thousands of small things that add up (e.g., massive support problems due to poor control over PCs).
4. You can waste millions of dollars on tools and infrastructure because unreasonable user demands force you to support too many conflicting platforms.
5. You can focus your organization's efforts on the wrong things (usually discovered when your company's competition comes out with something that *you* could have done).
6. You can even do everything else right but fail because you don't hit it off with the other executives.

But let's be clear about all of these failures. They *all* come down to the statement I made earlier in this article: If the company expects "X" and you deliver "Y" then you fail. Sure it's important for you to be able to deliver, but in my experience ninety percent or more of CIO failures come from not managing the expectations of your business executives. Make sure that if "Y" is going to be delivered, then "Y" is what's expected. Managing expectations is the way to avoid CIO failure.

2026 Commentary

What I didn't mention in the article is that different business executives in a company often have very different expectations. Before taking a CIO position, you need to be clear on whose expectations you'll be expected to meet. Sometimes the answer will be "everyone's expectations." If that's the case, then insist on one specific focal point for measurement—maybe the CEO, or maybe someone else. Alternatively, insist on specific measurable goals and objectives. Otherwise the probability of your success is very low.

54. To Succeed in Corporate IT, Play Hard to Get

Originally published on APRIL 10, 2007 in #IT, #ALIGNMENT, #PROJECTS

According to popular fiction, "playing hard to get" is a strategy sometimes used by women to snare a man. It makes the assumption that men want something more when they can't have it, so if a woman acts like she's not interested in a man, it makes the man more interested in her.

I have no personal opinion about whether or not this works for women, or whether the strategy is even used in real life. But I do know that this approach works for corporate IT.

Small companies have IT organizations, and large companies often have multiple IT organizations: some that are decentralized to support the various divisions, factories or offices, and another centralized "Corporate IT" organization which performs various IT duties for the overall corporation. I've worked in corporate IT organizations, I've talked to people who have worked in corporate IT organizations in other companies, and I know that most corporate IT organizations all over the world share one particular problem: corporate IT is regarded by everyone outside of the corporate headquarters as a huge pain in the neck.

Many years ago, when I worked in corporate IT for Boston-area based Digital Equipment Corporation (DEC), the people in the various manufacturing plants had a name for people from corporate IT. They called us "seagulls" because we would fly in from the coast, eat their food, poop on them, and then fly away. Seagulls were unwanted creatures; corporate IT just got in the way of things that were going on locally, and the corporate people always wanted the local people to do something that they didn't want to do.

So here I was in this corporate IT organization, working at the mandate of corporate executives to develop a Material Requirements Planning (MRP) system to be used by the 29 DEC plants. How, I wondered, do I sell such a system to these plants? How can a seagull be successful?

The answer, as you may have guessed from the title of the article, is to play hard to get. We could not be successful if we forced the manufacturing plants to use a corporate system. They would go along begrudgingly, but they wouldn't commit to success, and in fact they would probably try their hardest to prove that the corporate system was no good. We had to take a different approach.

A few of the manufacturing plants were seriously interested in getting an MRP system, so we started with them. We worked directly with those plants to develop something that would meet their needs, but we tried our best to generalize the solution so that it would be usable in other plants as well. We didn't force these initial plants to use *our* solution; instead, we volunteered our expertise to build and implement *their* solution.

We told the managers in the initial manufacturing plants that there are four keys to MRP success:

1. Changing their processes to include the use of the new system. The system has to be built into their day-to-day activities.

2. An emphasis on data integrity. You can't make critical decisions using a system that you don't trust, so it's paramount that the data in the system be trustworthy.

3. A new approach to measuring people. Employees have to be rewarded or punished based on their use of the system. For example, a stockroom clerk cannot be considered successful if they do things the old way and don't use the system. The measurement criteria for each job have to be revised to include some sort of measurement related to the use of the system and the importance of data quality. Failure to use the new system correctly has to be grounds for termination.

4. The software. When we listed this item, we deliberately listed it last, and we would usually say "oh, by the way" when we described the item. Software wasn't the key to MRP success—it was an "oh by the way." The software doesn't do the work—the people do. That message absolutely has to get across: giving someone a tool doesn't do anything for the business unless the tool is incorporated into daily life.

The first couple of plants were a success, and the manufacturing managers in those plants saw benefits: lower inventory, fewer out-of-stock situations, and smoother production. We could have immediately started blasting the software out to the other plants, but we didn't. We played hard to get, and we waited until other plants came to us, asking for the software. But even then, we didn't just give the software to them. We continued to play hard to get, telling the new plants about the four keys to MRP success. We told them that we would only give them the software if the plant managers signed a pledge that they would commit to all four keys. Once they signed the pledge, then we would do training in the new plant, explain the processes that are required by the software, and help the plant people change their processes and measurements. Software would be installed only after the people in the plant were ready, and by then they were eager to start using it.

More DEC plants signed the pledge and implemented the software. Not every plant did, but all of the plants were measured by corporate executives in the same way, so the ones using MRP typically came out ahead on most measurements. That motivated the remaining plants to either implement the software or to come up with an alternate way to achieve the goals.

What a difference—the corporate IT organization went from being a flock of seagulls to being a respected consultant and vendor. And all because we played hard to get.

Conclusion

Is your corporate IT organization treated like seagulls? It's usually because you act like you work for some sort of bureaucratic government agency where you can force people to do things your way, instead of acting as a partner to your customers. When you play hard to get, you have to create a demand for your services and then graciously satisfy that demand. It's quite a change from the way most corporate organizations operate, but it's a whole lot more successful.

2026 Commentary

The term Material Requirements Planning (MRP) has been replaced with a more comprehensive solution called Enterprise Resource Planning (ERP), but the concept is the same. And I would recommend this approach to any IT organization that's attempting to implement new software to any part of a company—large or small. An IT organization can't force anyone to use their software—it has to be something that the business users *want* to use.

55. Everything You Ever Wanted to Know about Firing People

Originally published on MAY 8, 2007 in #CAREERS, #MANAGEMENT

A few weeks ago *BusinessWeek* had an interesting article on firing people, pointing out the legal risks associated with terminating employees. Here are the things you need to know about firing that *BusinessWeek* didn't mention.

5 Kinds of Firings

Let's start at the beginning. In my experience, there are five kinds of firings:

1. Firing someone due to poor performance. As the *BusinessWeek* article points out, this one requires all sorts of documented conversations with the employee explaining how the employee is failing.
2. Firing someone due to "gross misconduct." This is more obvious and doesn't require much in the way of conversation: it's usually something like catching the employee stealing from you.
3. The scapegoat firing. This is like a combination of #1 and #2 except it's usually done in reaction to extreme negative press or publicity about something going on in the business. You fire a scapegoat and hope that this will defuse the problem. Sometimes the press does the documentation of poor performance for you.
4. The "you don't fit my direction and goals" firing. This one is so soft that most people don't know they're being fired; it's more a polite suggestion that the employee would do better elsewhere.

5. The workforce reduction or "reduction in force" (RIF). Although this isn't firing per se, it accomplishes the same thing. Most work force reductions give managers an easy way to eliminate poor performers without having to go through all of the documentation headaches of #1. But then again, in some countries there are stringent legal requirements if you reduce "redundant" employees this way.

Firing is Like a Breakup, Sort of

Firings are sort of like breaking up with your girl/boyfriend; you were in a long-term relationship and now you're not. The logic behind the decision is similar to the logic you go through in breaking up in a romantic relationship. You think to yourself, "Do the long-term pluses outweigh the long-term minuses?" If not, then you think about breaking up.

Most people who face a firing decision think that the decision is more clear-cut, but it's not—not even in cases where the employee is stealing from the company (I've heard of companies who tolerate low levels of theft as sort of an off-the-books compensation program.) No employee is entirely bad, just like no girl/boyfriend is entirely bad. There was a reason the relationship started in the first place, and some of the positive aspects of that reason still exist. Inertia plays a role in both firing and breakup decisions as well; it's a pain to have to start all over, and that pain justifies a certain amount of tolerance of the person's negative traits. Personal situations further complicate both firing and breakup decisions; if the person is going through some tough times then you feel like you ought to cut him/her a little slack, at least for a while.

Of course there are also differences between a firing and a breakup:

- Documentation isn't required for a breakup.
- There's less opportunity to sue in a breakup (but palimony suits have certainly made it possible).

- You can say "It's not you, it's me" in a breakup and people will sometimes believe it.
- Monogamous relationships force you to break up with one person in order to be with another "better" person, while most businesses can hire additional employees to fulfill the needs left unanswered by a slacking employee.

How to Fire

I won't describe the long drawn-out procedure that your HR department undoubtedly requires. The main rule of thumb is to document everything, even conversations with the employee that don't otherwise have any written record. The document for each incident should note the date and time of the event, an overview of what happened, a recap of what you told the employee, and a summary of how the employee responded. If specific agreements were made (e.g., the project will be done by x date), then include that in the document as well.

For even better documentation, summarize the event in an email to the employee and ask the employee to reply if the employee sees any discrepancies. Of course the email doesn't replace your own personal record of the event, which you need because the email won't record things like the mood of the employee, the level of animosity or apathy shown, or other intangibles which should be recorded.

There is a tendency among many companies—particularly large companies—to use regular "performance appraisals" as a substitute for this documentation. I'm not a strong advocate of periodic performance appraisals, and I particularly hate the ones that are done once a year. I certainly recognize the need for feedback to the employee—I just don't like the idea of waiting until an arbitrary date to provide that feedback.

In my opinion, it's better to deliver feedback at the time of various milestones in the employee's performance, for example when a project is completed (or abandoned), or when the employee has just made a contribution to business success. It's a long-established fact in the performance improvement field that feedback must come right on the heels of the action that deserves feedback. To wait until later is unfair to both you and the employee.

Annual performance appraisals have been instituted as a poor bureaucratic substitute for the feedback that a good manager should provide. They're better than no record at all, but it makes more sense for each manager to keep a file folder or word processing document for each employee and to record significant events and interactions that way.

Firing: Stage 1

Ok, so regardless of how you've documented the shortcomings of the employee, you've come to a decision that the employee should no longer work for you. What do you do? My suggestion is to try a few things before you resort to firing. First, it ought to already be obvious to the employee that you're unhappy with the employee's performance; if not, then why have you been keeping it to yourself? The most reasonable thing to do as a first step would be to have a heart-to-heart talk with the employee. Explain to the employee that you don't think they are a good fit for the current position, and offer to help the employee find something elsewhere. I mean this sincerely, and so should you. I've never seen an employee who wasn't good at something—it just might not be the thing you need. So if you can see the employee's true strengths then you ought to help steer the employee toward a job (inside or outside your company) that builds on those strengths.

Some employees will immediately panic when confronted with this talk, but it's really in your best interests—and the best interests of the employee—to push through the panic and try to get the employee to recognize the mismatch between the employee's strengths and the job the employee is in. Some employees will promise to "turn over a new leaf," "buckle down," or some other appropriate cliché that signifies the employee's interest in rectifying the performance problem. That might work if the problem is motivation (and even then it might only be temporary), but in most cases there's nothing that the employee can do to make things better; a mismatch is a mismatch no matter how motivated the employee is.

I've had about an 80% success rate in getting poor performing employees to move to another job; the key is to put a deadline on the move so the employee doesn't put it off. For the remaining 20% of problem situations I've had to go to the next stage.

Firing: Stage 2

At this point you've got an employee who hasn't made any headway in finding a better fitting position, even though you've followed up regularly with the employee to get progress reports, and you've done everything you can to help the employee find a different place to work. It's now time to get hard-nosed. This is when you impose what many companies call a "performance plan." If you're unfamiliar with the term, I'll explain it briefly here.

A performance plan is essentially a contract between the employee and yourself (representing the company) that lays out very clear, measurable criteria for success, and defines a very clear penalty for non-performance. Essentially the document says, "Do this or be fired." The specifics of the performance plan will vary from person to person and from job to job.

For example, I had a customer support person who had clear responsibilities to be available to take calls during certain hours, but who was always an hour or so late in arriving. The performance plan called for him to check in with me when he arrived every day so I could be sure he was actually at work. This is a simple example; most performance plans are much more complicated, with deliverables tied to completion of certain project deliverables by certain dates at a certain quality level. The key to a successful performance plan is to have a plan with no loopholes; the success criteria have to be very tightly defined so that there's no doubt whether or not the employee is successful.

One of three things will happen to an employee doing a performance plan:

1. The employee will successfully complete the plan and (at least temporarily) be considered a good employee, or
2. The employee will redouble their efforts to find a position with a better fit and will finally move to such a position, or
3. The employee will fail at the performance plan and, according to the terms of the plan, will be fired.

One of the worst things that can happen is for the employee to work hard at successfully completing the plan but then fall back into old ways and go back to being a problem employee again. If this happens then all you can do is to put the employee on a plan again, although this time the plan can be much shorter. One way or the other the employee will either do the job or will leave.

How Not to Fire Someone

While I'm on the subject of firing, let me give you an example of how not to fire someone. I once worked with a company which was doing a round of lay-offs. The cuts were substantial, and it wasn't at all clear who would stay and who would go. While senior managers called employees into one-on-one meetings to give them the news, other employees milled around discussing the situation. Then one of the waiting employees discovered that an over-eager security manager had already disabled the card keys of the employees who were going to be terminated. Immediately everyone went to the front door and swiped their card keys through the reader. What a way to find out you're losing your job!

If You're Fired

As difficult as it is to fire someone, it's much harder to be on the other side of the desk and find out you're losing your job. If it happens to you, remember that:

- The fact that you're a bad fit for your job doesn't mean that you're a bad person. It just means that you haven't yet found the ideal job for you. Don't take it personally; just try to learn from the experience.
- Losing a job is just like losing a loved one, and you'll go through the five stages of grief: denial, anger, bargaining, depression and acceptance. Let yourself grieve, try to avoid any sort of retaliation or lashing out (it will come back to bite you later), and give yourself a little bit of time before you start looking for another job. Interviews done right after being terminated won't be a fair reflection of your talents and personality.
- Don't take the bad news out on your family and friends. Let them help you through the grieving process, and be kind to them even if you're angry.

Are You Likely to Be Fired?

First off, recognize that if you're higher up in the organization, if you make a lot of money, or if you have a very visible position associated with the success or failure of a key project, then you have a higher probability of being fired. The downside of high levels of success is correspondingly high expectations, and if you fail to live up to those expectations even for a moment then your job is too important for you to stay in it as a non-performer. You seldom see people in CxO positions who have been there for very many years; any lapse in performance will lead to being fired.

For those of you in lower level positions, here are some warning signs to watch out for. If you see any of these things happening then seriously evaluate your job alternatives.

- A change in your boss's attitude toward you, or a change in the level and type of communication between the two of you.
- You miss clearly defined goals and objectives.
- You start to tell lies to cover up problems.
- Your office mates stop talking when you approach them, or they talk about you behind your back.
- You get in public disagreements with your boss.
- The press calls for you to be fired (rare, but more likely for government jobs).

Here's the bottom line on being fired: do a good job, make sure you know who's calling the shots and stay in their good graces, and have honest communication with your boss about your performance. Don't worry about being fired; focus on doing the best job you can.

Specific Firing Issues for IT Organizations

Most issues relating to firing apply equally well to all types of organizations, but IT organizations have a few specific concerns. First, due to the technical skills and training required for IT jobs, IT jobs are less likely to be interchangeable; it's not always so easy to just move someone else into a job. That makes the firing decision a more difficult one. You still have to ask yourself "Do the long-term pluses outweigh the long-term minuses?" but you also have to consider how long it will take to find a trained person or to train someone new. This is a particularly bad problem for small organizations in which there is little duplication of skills. For example, firing a person who provides support for a certain system may require you to leave the system unsupported while you're finding a replacement and bringing the person up to speed.

If you find yourself in this situation, talk to the business users who are getting support from the soon-to-be-fired person. You'll find that if you get their support for the decision then often they're willing to tolerate a lower level of support during the transition to the new support person. In fact, you may find that they have had concerns about the support person's fit for the job anyway but didn't know how to broach the subject with you. Whatever you do, don't blindside a support person's customers by firing the person without their prior knowledge; the customers will see the action as a slap in the face.

Conclusion

"Firing" has a very negative connotation; the word "fire" seems to imply a violent act done in anger. But although the result of a firing is sudden, it's only the culmination of events that have taken a long time to build to that sudden conclusion. The key to success in firing is to recognize it as an act of desperation: something you do when everything else has failed.

You can avoid firing by hiring the right people, motivating them well, and then giving them frequent feedback on how they're doing. If you find that a person is not right for a job—because the job has changed, because the person has changed, or because the person didn't live up to expectations—your best course of action is to try to move the person into a job that's a better fit. That better fitting job might be inside your organization, somewhere else inside your company, or even somewhere outside your company. If you can't help the person yourself, then at least have the frank discussion required to communicate the issues to the person.

No one wants to stay in a job that's not a good fit, so in most cases you can help both the employee and yourself by moving the person elsewhere. If all else fails, then fire, but only as a last resort. A fundamental part of being a manager is looking out for your employees; firing should mean that you've exhausted all other possibilities.

2026 Commentary

The whole idea of firing sounds brutal, and it's difficult to write an objective article to talk about it. However, if you're a manager, then sooner or later, no matter how good you are at hiring, you're going to run into a situation where you have a poor-performing employee. And you're going to have to face the possibility of letting them go. I'll go back to one of the sentences in the article, "The key to success in firing is to recognize it as an act of desperation: something you do when everything else has failed."

56. How to Organize IT

Originally published on JUNE 12, 2007 in #IT, #ALIGNMENT, #LEADERSHIP, #MANAGEMENT, #STRATEGY

I'm frequently asked the question, "how should IT be organized?" Let me start by saying there is no right answer, at least no answer that's right for all situations. There are a lot of different aspects of the IT organization issue, and I address some of them in this article.

Where should IT report?

Where IT reports in a company is directly related to how IT is viewed in that company. If IT is viewed as absolutely key to the future success of the business, then IT will usually report to the CEO or President. If you're in a manufacturing company and IT is critical to the management of the manufacturing and manufacturing engineering processes, then IT might report to the executive in charge of manufacturing or operations. If IT is viewed as a way to automate accounting systems and manage PC purchases, then IT will typically report to the CFO or even the Accounting Manager.

According to a joke I once heard, you can always tell when the head of IT reports to a financial department because the company knows to the penny how much it's losing. The idea behind the joke is actually pretty accurate: a company which focuses its IT efforts on accounting won't be putting technology to use in creating and optimizing revenue or profit, but it will be able to track its losses accurately.

Where does IT report in your own company or organization? If you're in IT and you're not happy with the answer, then you'll have to convince your company's executives that IT can serve a bigger role. Just bear in mind that the bigger role may require a more experienced senior executive as head of IT. If you're in charge of IT and your own personal limitations are holding the IT organization back in its reporting relationship, then you'll need to gain some experience and training, or you'll be left behind by your own organization.

Should IT be Centralized or Decentralized?

I talked about this question in the article, "31. The Information Technology Merry-Go-Round." Most companies go through cycles of centralization and decentralization unless they arrive at a compromise that seems to satisfy everyone. Centralization offers economies of scale, and decentralization offers better responsiveness to user needs. If you're lucky, you'll figure out how to centralize the aspects of IT that need economies of scale (e.g., purchasing or standards) and decentralize the parts of IT that need more user responsiveness.

Should IT reporting be split? Matrixed?

Some companies don't have a single IT organization; instead they have divisional IT or departmental IT. They may also have different parts of IT reporting to various functional managers, for example infrastructure management and system maintenance to someone like a COO (Chief Operating Officer) and new projects reporting to the various departments sponsoring the projects.

Matrix management offers a way to split the IT organization while keeping it together at the same time. Under matrix management, each IT manager has two bosses: one functional boss in a user department who provides functional leadership and direction, and another technical boss in the IT organization who provides technical leadership and direction. Picture a matrix as a rectangular table of numbers with rows and columns, only in this case the numbers represent the IT managers, with a manager's leadership provided from both the row (functional) direction as well as the column (technical) direction. The functional leader makes sure that the manager is doing the right things for the company, and the technical leader makes sure that the manager is doing things right (i.e., doing them using company standard technology, tools and techniques).

Although matrix management offers a theoretical way to balance functional and technical needs, many companies have trouble making matrix management work. When conflicts arise between the two leaders, it's not always easy to figure out the most reasonable solution, and often one of the sides of the matrix becomes much stronger than the other. When matrix management works, it works well. But when it doesn't work, it confuses the employees who are caught in the middle between two opposing forces.

If the IT reporting relationship is split, what should be the basis of the split? User oriented? Technology oriented? Project oriented?

I didn't mention reporting basis in my "31. The Information Technology Merry-Go-Round" article, but this goes through cycles as well. There are good reasons to organize a department in a certain way. A user-oriented department is more attuned and responsive to user needs. A technology-oriented department is more effective in implementing a new technology (e.g., RFID) that requires focus on issues with the technology itself. A project-oriented department is best for organizing a group of people brought together solely for the purpose of building and implementing a major new project.

All of these types of organizations make sense in specific cases, and that's the best way to use these organization types: for specific cases. Compare this type of organization to the group of people who create a motion picture. The organization doesn't exist before the movie, the organization doesn't exist after the movie is done, it just exists for the duration of the production process, and it's optimized specifically for that movie. Similarly, you can put together temporary teams to accomplish specific objectives for your company, and then dissolve the teams (moving the people to other teams) when the objective is accomplished.

Although temporary teams work great, I do not favor the idea of permanently splitting an IT organization along user or technology lines. When you make such teams permanent, the organization begins to suboptimize itself around the lines of the split, and the organization responds poorly to overall change. So my preference is to create teams for specific purposes but to leave the teams in place only long enough to accomplish their objectives.

What should your IT department look like?

You know the answer: It depends. You already have an IT organization (even if it's just a few people), so you have a starting point for making improvements. There's no "right" organization, so you choose the best compromise between economies of scale and user responsiveness, focusing your employees on the things that matter most. That best compromise is likely to change over time as the players change and as the needs of your business change. The key is to listen to feedback about the performance of various parts of your organization, and to make your organization better as a result of that feedback.

In my own experience I've found that some of the more routine aspects of IT like infrastructure management and support tend to benefit from being pulled together across departments. And if you want your various systems to work together and share data or objects, then standards for data and software should be managed—or at least coordinated—centrally as well.

Projects, on the other hand, tend to be primarily associated with specific sets of users, so you need to do what it takes to build a strong relationship between the users and the project team. That way the project gives the users what they need, and the users feel confident during the project.

Although having projects report to the user organization will achieve this result, it tends to deemphasize the technical side of a project, and you can end up with a project result that may not fit well with other systems. There are alternative ways to build a strong relationship between project teams and users without changing the organization structure; sometimes you can build the relationship just by physically moving the team members into the area near their users, or by having the IT people on the team spend more informal time with the users.

The key is to end up with an organizational structure that allows project functional requirements to be determined in cooperation with the people who will use the results of the project, while project technical standards remain in the hands of a (usually central) group that looks out for the technical needs of the overall company.

The Fundamental Rule of Organizations

People who aren't satisfied with the service they're getting like to change the reporting relationship of the service organization so that the service organization now reports to them. In this way, they feel more in control of the service. On the other hand, people who *are* satisfied with the service they're getting usually don't care where the service organization reports.

If you keep your customers satisfied, then you can organize any way you like. Just pick an approach that seems to work, and adjust it as required by changing conditions and projects.

2026 Commentary

The reference to RFID as a new technology is pretty dated, since that technology is well understood these days. A technology like Artificial Intelligence would be a better example in today's environment.

57. You've Got to Specialize

Originally published on JULY 10, 2007 in #CAREERS

If I've learned anything in life, it's that being "well-rounded" is a sure path to mediocrity. You're much more likely to achieve career success if you get really good in one area than if you achieve moderate mastery in a number of different areas. That's true even for managers and executives. You may not think of executives as specialized, but in fact executives often focus on specific types of corporate actions like start-ups, turnarounds, mergers or quality improvement, and they're often brought in specifically to achieve results in their areas of expertise. In fact, executives often specialize more than the rest of us because they can afford to delegate almost everything but their specialty.

We're taught to be generalists

We're not given very good training in achieving specialization. Our schools in the U.S. try to make us well-rounded by forcing us to learn all of the basic subjects. This makes sense in early education—everyone needs to know how to read and do basic math—but the emphasis on generalization continues throughout high school and college. Most schools consider a successful student to be one who does fairly well in *all* subjects rather than the one who does exceptionally well in one subject. Yet it's doing exceptionally well in one subject that will serve you better in the long run.

When we join the corporate world, the performance appraisal process continues to try to make us well-rounded. Most appraisal systems evaluate us in a wide variety of categories, and put emphasis on the categories we don't do very well. Our corporate training then attempts to improve our abilities in these weaker categories, and so we typically spend little or no training time making us stronger in the areas in which we're already strong.

Why do we think generality is good?

There's a precedence for this behavior in other parts of our lives. Take the human body as an example. We want everything in our body functioning at a normal level because all of the parts of the body work together to keep us alive and healthy. It doesn't matter if we have the best brain or liver or spleen if the other parts of our bodies aren't around to keep our body running. It's the whole body that matters, and we need to have overall health.

Human beauty is another example of well-roundedness. Studies have shown that people's faces are perceived as beautiful when the proportions of the face are closest to the overall *average* proportions for the type of people we see every day (which is why definitions of beauty vary in different parts of the world). If your nose is longer or shorter than average, then you're perceived as less beautiful. If the distance between your eyes is more or less than average then you're considered less beautiful. Scientists explain that this perception of average as beautiful is programmed into us to help us select a healthy mate, since facial proportions radically different from average may be a sign of "bad genes."

Based on these two examples you can see how we might begin to think that well-roundedness is good. It *is* important to be well-rounded if you want to be healthy and beautiful, but a well-rounded life with nothing to distinguish you from others is the very definition of mediocrity. To succeed in life you need to stand out from others, and the only way you'll do that is to find the one or two things you're really good at (which are usually the things you most enjoy doing), and develop your skills until you become outstanding at those one or two things.

How do you know what to specialize in?

This is the big question, kind of like the question, "What do you want to be when you grow up?" For some people, the decision is easy; they naturally gravitate to their strength at an early age. For others it's more difficult; they can do a lot of different things and it's hard to pick one of them to be a specialty.

Here's an exercise adapted from the wonderful book *The Three Boxes of Life and How to Get Out of Them: An Introduction to Life/Work Planning* by Richard N Bolles, who also wrote the job-hunting book, *What Color is Your Parachute?*. Make a list of the 10 things you've done in your life that you're most proud of, whether at work, in school or in your personal life. Make sure that these are things you've *done*—not just objects you've owned or titles you've held. For each item on the list, write a short paragraph that explains the item in a little more detail, telling what you did and why you're proud of it.

Now make a second list of the 10 things you've done in your life that you most enjoyed. Again, write a short paragraph for each item that gives more details about the enjoyment you experienced.

When you have completed the lists, go back through the two lists and try to find one or two common threads that run through multiple items. Are there similarities in the situations, in the things you did, or in the processes you followed? How did you feel when you were doing these things? How were these things different from other things you've done?

Using exercises like this (see more exercises in the *The Three Boxes of Life and How to Get Out of Them* book), you can begin to identify your specialty. In general you're looking for a specialty that:

1. Makes you happy when you do it
2. Makes you stand out from the crowd because you're so good at it

3. Is specific enough to make you attractive as an employee or service provider
4. Is not so specific that there's very little market for the service
5. Is based on a process rather than a product (Don't tie your specialization to expertise in a specific product from a specific vendor, or you may find yourself obsolete in a few years)

Here are some examples of IT specialization:

- Building data centers, consolidating them, or optimizing their performance
- Designing human-engineered information systems for mobile service workers
- Planning and managing software development projects that implement a technology for the first time
- Finding and fixing software or system problems that other people have been unable to solve
- Designing e-commerce web sites for teenagers
- Putting together a highly-motivated team of competent software developers
- Understanding the needs of a particular kind of business and determining the best way to utilize IT talent to contribute to the business

A Personal Example

I went through the list-making exercise many years ago, and I was initially confused. There were lots of different things on the list that didn't have any obvious similarity. Then I thought about it some more, and I discovered that my list consisted mostly of things that I had learned and taught to others. I discovered that my special skill is in simplifying complex things and then leading organizations to make changes as a result.

Over the years I've applied this skill in corporate due diligence by helping businesses understand what's really going on in the IT organization of a company they want to acquire. I've applied the skill to simplify and revamp the processes of a financial services business. And I've applied the skill in numerous other situations ranging from product development and systems design to the creation of IT strategy. In each case the specialization was the same—simplifying complex things—but I was able to apply that specialization to a wide variety of cases.

A warning: my skill is an example of specialization that cuts across traditional organizational lines—it doesn't correspond to any typical job title. Some specialization is like that, but you'll find that it's more difficult to sell to employers. Buyers of a skill usually translate the need for the skill into the need for a person in a conventional job category. If your skill isn't associated with a common job category, then you'll have to work extra hard to find jobs.

Next Steps

Once you've picked your specialty, you should take steps to make it your own. Learn more about it, find role models that you can emulate, and find ways to improve on the processes that your role models use. Go beyond all of the established published literature on the subject and blaze new trails by inventing your own processes. If you like to write, then publish articles on your subject to advertise your expertise. Speak at conferences to share your learning, or if speaking or writing isn't your style then partner with someone else who can write or speak on your behalf (vendors love to do white papers on companies and people who use their products). In whatever way you choose, make your specialty known to the outside world.

While you're focusing on your specialty, however, you need to always keep an eye on the demand for your talents. If you see the demand shifting then consider altering the focus of your specialty to move it in a different direction. You can still do things you like —you just might have to reposition your talents in the eyes of your customers and employers.

Conclusion

There are lots of mediocre people in businesses all over the world. If you want to be successful then you have to stand out, and the best way to stand out is to excel in an area of specialization. Pick an area of specialization that you're good at and that you enjoy, and focus your efforts on becoming the best in that area. You'll differentiate yourself from all those mediocre people, your career will take off, and you'll be happier as well.

2026 Commentary

I still stand by my advice, but I realize that current hiring methods make it much more difficult for anyone with specialty talents to be hired by today’s companies. In some cases, that’s not a problem, because certain specialties are more entrepreneurial in nature. For those specialties, you’re likely to be more successful starting your own business, maybe starting with an apprenticeship in the field in which you’re interested.

For other situations, if you need to start your career in a larger business, then you might have to compromise on applying for jobs that are related, but not exactly what you want. Then, once you’re hired, you can use the experience you gain to enlarge the position, or to branch out into other areas that are closer to your ultimate specialization.

58. Harwell's Unfortunate Laws of Human Organizational Behavior

Originally created on APRIL 1, 2004, but not published until AUGUST 14, 2007 in #CAREERS, #LEADERSHIP, #MANAGEMENT

[Note: I wrote this on one of my more cynical days when my tongue seemed to be stuck in my cheek. For a newsletter article offering advice for new managers on how to minimize the effect of these laws on your own organization, see the next article.]

In the spirit of the Peter Principle and Parkinson's Law, I've created my own set of laws that show how managers behave in organizations unless you explicitly take steps to prevent the behavior. The laws are deliberately extreme to make a point: human motivation is a funny thing, and you have to be very careful to align motivation and reward with the behavior you want. Here are the laws:

1. People always try to use their experience even if it doesn't apply to the current situation.

2. If someone is brought in as the new manager of an organization, then the new manager will feel obligated to change the organization. Otherwise, what value is the manager contributing?

3. If a manager doesn't know how to improve an organization, then they will change it to look like the last organization that they remember.

4. Either things in an organization will get better or things will get worse. If things get better, then the manager will be rewarded, regardless of whether the manager is responsible for the improvement. If things get worse, then the manager will move on to another organization to try again.

5. Sometimes things in an organization get a little better before they get a lot worse. If the manager is lucky, then they will be

promoted to a better job before things start to get worse. Of course, then the cycle will start all over again.

6. Rarely, a new manager will actually ask the employees what should be improved. If the employees understand how their organization contributes to the company's overall performance, then they will undoubtedly provide a better answer than the manager's experience.
7. Sometimes the manager will hire a consultant. The consultant will then ask the employees what should be improved, and the result will be put into a PowerPoint® presentation, then ignored by the manager.
8. Good organizations will continue to get better until someone in higher management decides they're not good enough. Then they'll get worse.
9. Bad organizations will continue to get worse until a new manager tries an idea which should have been obvious from the beginning.
10. If a company contains multiple similar organizations, then they will always be compared, but the true differences will not be discovered. The good organizations will get worse as a result of the comparison, and the bad organizations will get worse as well.

2026 Perspective

All true. It's human nature. Now read the next article.

59. Advice for New Managers on How to Avoid Harwell's Laws

Originally published on AUGUST 14, 2007 in #CAREERS, #LEADERSHIP, #MANAGEMENT

In 2004 I wrote a tongue-in-cheek essay called "Harwell's Unfortunate Laws of Human Organizational Behavior." I put it on my web site, but I just sent the link to a few close friends. Frankly, I thought the content of the essay was too different from my normal, more up-beat type of article. But I was looking at the "Harwell's Laws" essay a few weeks ago, and it struck me that there's a positive statement that can be made from every negative. So I've decided to repeat the 10 laws here, together with the advice that I would offer based on each of the laws.

Law #1: People always try to use their experience even if it doesn't apply to the current situation.

See my article, "25. Don't Get Stuck in a Learning Stage." It describes how people who are new to a situation—for example, first-time managers—don't know what they don't know. As a result, they make the incorrect assumption that their prior experience must apply to the current situation, even when it doesn't. There are ways to avoid this problem, but they chiefly depend on having a good mentor who can help you differentiate between things you know and things you don't. It also takes an open mind, which of course is difficult to achieve in itself, especially under the pressure of a new management job.

Law #2: If someone is brought in as the new manager of an organization, then the new manager will feel obligated to change the organization. Otherwise, what value is the manager contributing?

This again gets back to the "don't know what they don't know" situation. If you ever jumped rope as a child with other children holding the rope, then you know that you have to get into the rhythm of the rope before you jump in; otherwise you'll just get all tangled up. The same thing applies to a new manager of an organization: you have to get a sense of the "rhythm" of the organization before you understand it enough to make changes. Yes, you want to make the organization better, but it helps to understand the strengths and weaknesses of the organization—and of the people in the organization—before you start making improvements.

To use a different analogy, if you're going to tear down some of the walls in a building, it helps to know which ones are structural supports before you start demolition. Otherwise the whole building may come down around you. Some of the processes, attitudes, people and relationships in an organization are just as much structural supports as a supporting wall. Change them only after you have put alternative supports in place.

Law #3: If a manager doesn't know how to improve an organization, then they will change it to look like the last organization that they remember.

In Chapter 3 of my book, *Boiling the IT Frog*, I say that "It's human nature to want to minimize change, and if [you're] used to working with a product, and if [you've] been successful with the product in the past, then [you're] going to want to work with it again." The same thing applies to processes: you want to use the same processes that you've used in the past because you're used to them. What you might forget, however, is that there's just one of *you*, and there are *lots* of employees who are already used to doing things the way they're doing them now. So if you're *sure* that changing things will provide a benefit, then go ahead. But don't make changes just because *you're* used to things working a certain way; you'll waste your time, the time of your employees, and the limited credibility you get as a new manager.

Law #4: Either things in an organization will get better or things will get worse. If things get better, then the manager will be rewarded, regardless of whether the manager is responsible for the improvement. If things get worse, then the manager will move on to another organization to try again.

If this is true in your company then it's not the fault of the new manager—it's the fault of the manager's manager. Don't reward people based on organizational improvement without understanding whether the person being rewarded actually caused the improvement. You might have to do a little digging to find out, but it will be time well spent.

And if the reward you're giving the "successful" manager is a promotion, then you need to consider the fit of the potential manager for the new position into which they are being promoted. Managing engineers doesn't quality someone to manage customer service people; managing clerical people doesn't qualify someone to manage engineers. Look at the specific needs of the new management role, and then take a hard look at whether those needs can be met by the prospective manager. If there's a gap, then are you prepared to help the new manager attain the skills necessary to close the gap?

Law #5: Sometimes things in an organization get a little better before they get a lot worse. If the manager is lucky, then they will be promoted to a better job before things start to get worse. Of course, then the cycle will start all over again.

See my comments on Law #4. This is what happens when you promote based on organizational success without determining whether the manager of the organization had anything to do with that success. It's also what happens when you declare victory prematurely; it sometimes takes a while to determine whether an initially positive change will continue to be positive.

Law #6: Rarely, a new manager will actually ask the employees what should be improved. If the employees understand how their organization contributes to the company's overall performance, then they will undoubtedly provide a better answer than the manager's experience.

There are two statements included in this law. First, there's the statement that employees often know what can be improved in an organization. This is almost always true, and I talk about it in my article, "19. Hidden Consultants within your Organization." Second, there's the statement that new managers will rarely ask employees what should be improved, and I explain why in that same newsletter article. I won't repeat these lessons in this month's newsletter article, but here's the bottom line: listen to your employees, but put their advice into perspective.

Law #7: Sometimes the manager will hire a consultant. The consultant will then ask the employees what should be improved, and the result will be put into a PowerPoint® presentation, then ignored by the manager.

Smart consultants know how valuable a resource the employees of an organization are. The consultants do a lot of listening, and then they filter the things they've heard based on their prior experience (see Law #1). Nevertheless, much of the employee advice will find its way into the consultant's recommendation, and thus into a PowerPoint presentation to management. Alas, the manager will then proceed to filter the consultant's recommendation based on their own experience, thus gutting the key aspects of the consultant recommendation.

Realistically—and very unfortunately—most consultants are hired not to provide new recommendations, but to legitimize the actions that the manager already wants to take. That's why the consultant (and employee) recommendations are so often ignored. But it doesn't have to be this way, and as a new manager you are in a position to actually listen to employee and consultant recommendations. If you use those recommendations, then you'll be pleasantly surprised at how much the recommendations can help, and you'll be amazed at the morale boost you'll get from employees who know that you listen.

Law #8: Good organizations will continue to get better until someone in higher management decides they're not good enough. Then they'll get worse.

When you're on a path toward improvement, it's easy to get impatient and try to speed things up. But there are many hazards in doing so, not the least of which is that you often don't really understand why things are getting better. You might think that the improvement is due to a particular change you've made, when in fact the improvement might be caused by other factors entirely, like improved interaction among employees or a better sense of pride in the work. If you pick the wrong factor and attempt to speed it up, then you're likely to sabotage the progress you've made. Patience is rare in today's organizations, but it's extremely important when coping with complex organizational change.

Law #9: Bad organizations will continue to get worse until a new manager tries an idea which should have been obvious from the beginning.

Actually, there is no such thing as a "bad organization"—there are only organizations which aren't focused on the right things. Most people in an organization know what the right things are (for example, customer satisfaction or truly enjoying the work), but most managers are distracted from the right things by instructions from higher up in the chain of command. When a manager finally sees the light and realizes (maybe by talking to employees or customers) what the right things are, and when the manager redirects the organization to do those right things, then the organization will be able to turn its performance around.

Law #10: If a company contains multiple similar organizations, then they will always be compared, but the true differences will not be discovered. The good organizations will get worse as a result of the comparison, and the bad organizations will get worse as well.

It's human nature to compare things, and in business we like to compare things using numbers. "XYZ organization is more profitable, more productive, [pick your own numeric superlative] than ABC organization." We think we know what we're talking about, but we have to be very careful when we try to explain *why* one organization has a better numeric score than another. That's the first mistake we typically make: we jump to the conclusion that the difference between XYZ and ABC can be attributed to a certain factor. Then we make a second mistake: we assume that there is a cause and effect relationship between that factor and the numeric performance. As a result we try to change this factor in organization ABC and assume that the numeric performance of ABC will improve.

Now the people in organization XYZ begin to believe that this magical factor is the cause of their success. They begin to put an inappropriate emphasis on the factor in the hope of making things even better, and their performance begins to degrade. Maybe they were doing things "right" before, but now they're caught up in the psychological game of thinking they know how to improve, and because they're focusing on improvement instead of on the "right things" (see my comments on Law #9), XYZ's performance gets worse.

Meanwhile, the people in organization ABC try to improve their performance by changing the identified factor. But however they were doing before, their new emphasis on this factor is likely to distract them from the "right things" and the result is likely to be poorer performance for ABC.

You might be able to make temporary improvements in an organization by focusing on some factor that's not directly related to the end goal of the organization, but improving an organization is like improving your golf or tennis game: you have to integrate all of the factors into a single focus on the result, and not be distracted by a short-term focus on contributing factors like your backswing, your stance or your follow-through.

Conclusion

I wrote the ten laws to point out some patterns in the way managers and organizations relate. The laws are funny, but in a Dilbert sort of way; there's an edge to the humor because we all see situations in our own lives where these laws hold true. Humor is good because it helps us address problem areas without being embarrassed; we can learn from humor by trying to avoid situations where our actions would be seen as a joke. Laugh all you want at my 10 Unfortunate Laws—just try to make sure the joke is not on you.

2026 Commentary

A corollary of Law #10 is that once a company starts comparing organizations, then the number of things being measured will increase. Soon there will be so many different things being compared than no one will know what to focus on. As a result, none of the measurements will actually be used.

See references to the concept of focus in a number of the other articles in this book, but particularly in "113. There are Only Two Reasons for Strategy."

60. How Do You Talk to a CIO?

Originally published on SEPTEMBER 11, 2007 in #IT, #ALIGNMENT, #LEADERSHIP

Last week I was interviewed by a writer from a popular business magazine. He wanted my views on the question, "How do you talk to a CTO or CIO?" It's an interesting question, and what makes it most interesting to me is that I've never heard anyone ask the question about any other senior executive position in a company. I don't hear anyone ask the question, "How do you talk to a CFO?" or "How do you talk to a COO?" I don't even hear anyone ask the question, "How do you talk to an HR organization?" So why do you think the technology organization is singled out for this type of question?

The answer, unfortunately, is that there's a layer of mystery around the IT organization; IT is less understood than any other business organization. In my book, *Boiling the IT Frog*, I compare IT people to doctors, who share some of the same problems of having to communicate complex technical information. But as I point out in the book, good doctors will get around the communication problems by focusing on the patient implications of the information: How will a particular treatment impact the patient? How long will treatment take? What are the risks? How will the treatment affect the patient's quality of life in the long term? What kind of success rate does such a treatment have? These are all questions that skirt around the "how does it work" technical core of the issue at hand, but most doctors don't feel the need to go into the details of how a particular treatment works, since the description of the patient impact is more relevant and more easily understood by the patient.

IT issues are just like the issues of medical treatment. Every business is made up of hundreds of processes: some which use computers and some which don't. Any IT project will affect some of those processes—some will be improved and some will be negatively impacted. The negative impacts can range from rigidity (making a process more difficult to change in the future) to cost (making a process more expensive, hopefully with a counter-balancing reduction in cost or an improvement in revenue or profitability elsewhere in the business). So just as a medical procedure has an impact on the body's processes, an IT project will have an impact on business processes.

A good IT executive will discuss the pluses and minuses of IT projects with business executives in much the same way that a doctor will discuss various treatment options with a patient. It's a high-level discussion that focuses on positive and negative outcomes, project durations, risks and success rates. At the end of the discussion the business executives should feel like they've been presented with enough information to make choices, and that's all that the discussion is intended to do. A good IT executive won't attempt to explain technical issues unless the business executives ask, just as a doctor won't go into the medical details unnecessarily. If a project requires an unproven technology, then that's important to know, but only as it impacts the risk of the project and the probability of success.

Bad IT executives talk to business executives in great detail about the technology that will be used, why it was chosen, and why it's the best technology for the project. These bad IT executives don't seem to realize that (a) the business executives don't care, and (b) the business executives just get more confused and uncomfortable when they are provided all of this useless (in their opinion) information. It's bad IT executives who make the business executives ask, "How do you talk to a CTO or CIO?" But the real answer is to replace your IT executive with someone for whom the question doesn't have to be asked.

The Flip Side

Now you may ask, "Aren't business executives sometimes at fault as well?" Absolutely. Just as there are bad patients who try to tell surgeons how to operate, there are bad business executives who try to force specific solutions on an IT organization rather than keep the discussion at the level of benefits and costs. In medicine we see drug manufacturers doing an end run around doctors by advertising specific drugs directly to the patient. And in the IT industry we see technology vendors using the same end-run approach to try to sell hardware and software solutions directly to business executives.

But when a patient who is sold on a new drug goes to their doctor, it's the doctor's responsibility to determine whether or not the drug is the best treatment for the patient's condition. And the same thing applies when a business executive who has been sold by a technology vendor goes to their IT organization to request the technology. It's the IT executive's responsibility to look at the real business problem that the business executive is trying to solve, and to recommend a solution that is optimized for the specific business need. That might be the technology that the business executive wants, but there may be an alternative technology which better meets the real business need, has lower risk, or fits better with the overall IT architecture of the business.

If there is in fact a better alternative and the IT executive fails to convince the business executive, then one of two things is going on:

1. The IT executive doesn't have enough trust and support from the CEO to stand up for what's right, or
2. The IT executive doesn't have the communication and political skills required for the senior IT position.

In either case the IT executive—and in fact the entire IT organization—is headed toward failure, just as a doctor will fail who prescribes drugs just because a patient wants them. An IT organization can survive a few of these failures, but if there are any more than a few, then the quality and reputation of the IT organization will sink rapidly as the number of unconnected and uncontrolled systems grows within the business. When non-technology people control technology choices, chaos will ensue.

2026 Commentary

This is still true. I've seen a number of CIOs fail due to not having the communication skills to build trust with their fellow executives.

61. 10 Reasons You Need an IT Architect

Originally published on OCTOBER 9, 2007 in #CAREERS, #IT, #ALIGNMENT, #STRATEGY

Every IT organization needs skilled architects who are experienced in designing systems, databases, networks and user interfaces. In this article I'll give you 10 reasons.

1. Architecture is critically important to IT.

Let's start with a definition. In my book, *Boiling the IT Frog*, I explain the IT use of the word "architecture" this way:

For a building like a house or a skyscraper, the word "architecture" refers to the way that building materials are assembled to form a structure. Similarly, the architecture of an information system refers to the way that software and hardware components are assembled to form the overall system. Just as building architecture has styles, IT architecture has different styles for system construction. Just as building architecture has building codes and best practices to ensure the safety and longevity of a building, IT architecture has system standards and best practices to ensure the security and longevity of a system. And just as building architecture has approaches which make the maintenance and alteration of a building easier and less expensive, IT architecture has approaches which make the maintenance and alteration of software easier and less expensive.

The book then talks about architecture's critical importance:

The architecture part of the IT strategy is often omitted, but this omission is a huge mistake. Would you want someone to build a new city without a plan for where the streets should go or how the buildings should look? An IT strategy without an architecture definition is similar—systems would be developed without regard for how they will fit together or even how they should be built to minimize ongoing cost. In an IT organization without an architecture, it's not uncommon to do the equivalent of adding two additional floors to a building which has a foundation that wasn't designed to support the extra floors.

2. The IT architect focuses on how to do things, while the CIO focuses on working with the business to determine what to do.

As I said in my book, the most common IT mistakes are doing the wrong things—not doing things wrong. The CIO has to concentrate on figuring out *what* to do, but someone else has to figure out *how* to do those things. That's the architect's responsibility.

3. In companies with centralized IT, the IT architect must manage the architecture approach being used by each IT project.

One of the biggest advantages of centralized IT is the efficiency gained by using common approaches and standards. The architect makes that efficiency possible.

4. In companies with decentralized IT, the IT architect must coordinate the architecture approaches being used by the individual decentralized IT organizations.

Even in decentralized IT, there is a need to share data and information across the various parts of the company, to present a unified view of the company to customers and employees, and to maximize the sharing of effort being expended by each decentralized IT organization. Decentralization doesn't imply anarchy, and even in companies which have decentralized total control of IT, someone should facilitate sharing of architectural information across those decentralized IT organizations.

5. In a company which uses off-the-shelf packaged software like ERP or CRM, or software-as-a-service applications like those from Salesforce.com, the IT architect has to figure out a way to sew together all of the different disparate systems so that they can talk to each other.

In an ideal world (I wish I saw this more), the architect would also be involved in the decision to buy a particular off-the-shelf packaged software solution, so that the solution could be evaluated based on the difficulty of integrating it with the company's existing systems.

6. In a company environment where business people make systems decisions based on articles in airline magazines, the IT architect has to clean up the mess made by the poor system choices.

This is sort of like the Winston "the Wolf" Wolfe character played by Harvey Keitel in the movie *Pulp Fiction*—someone who comes in after major mistakes have been made and tries to sanitize the situation to prevent future issues. Only in this case the mess has to do with software that doesn't play well with others, packages which require nonstandard hardware and operating systems, and the need to customize software to present a consistent user interface to customers and employees.

7. In an international company, the IT architect has to deal with the software implications of internationalization.

The architect must figure out how the company's systems are going to deal with multiple currencies, multiple languages, multiple laws and legal systems, and multiple cultures. The architect must also determine how data will be converted between various combinations of systems. Internationalization is one of the most difficult challenges facing an architect, kind of like building a skyscraper that has to change appearance and behavior based on who's looking at it.

8. The IT architect has to separate truth from fiction in the latest round of buzzword architectures (e.g., Agile, RAD, RUP, SOA).

In this role the architect has to be the gatekeeper and voice of reality: Does a particular architecture or software development approach get tested or used in your company? Which buzzwords should be ignored? Which improvement claims are real?

9. The IT architect has to figure out how to phase in and integrate the latest technology that your company chooses to adopt.

These are technologies like RFID, VPN, security tokens, encrypted disk drives, smartphones and PDAs. Are they viable? What pilot tests have to be done before they're used in production? How will they fit with current technology?

10. The IT architect has to anticipate change.

The only constant in IT is change. Business needs will change, customers will change, technologies will change, legal requirements will change. The biggest challenge for an architect is not to pick the best architecture for *current* needs—the biggest challenge is to pick the best architecture for constantly changing future needs, and to revise that definition of "best" on an ongoing basis.

Conclusion

In the Q&A session after a recent webinar, I was asked to describe the working relationship between the CIO and the architect. I cited reason #2 above, and said that if the CIO is focused on *what* the business needs, then the Chief Architect is in a sense serving the role of the "Chief How Officer." That in a nutshell is what the architect does, and that's the difference between a CIO and an architect: the CIO determines *what* to do, and the architect determines *how* to do it.

If you have a CIO but not an architect, then the CIO will be less effective in setting business direction for IT because the CIO will be distracted by *how* issues. It is also quite likely that a CIO in such a situation will either be a good CIO or a good architect but not both. That's because the personalities and skill sets of the two roles are so different.

The bottom line is that both a CIO and an architect are needed, and they both make a major contribution to IT strategy—each in their own way. Good IT architecture is critical to successful IT, and you can't have good architecture without good architects.

[Note: I've used the singular form of the word "architect" in this article, but obviously in larger companies there can and should be many architects. In such an organization there is a need for a Chief Architect who provides overall architectural direction. This person is also sometimes called the Chief Technology Officer

(CTO) to distinguish the job from the CIO who is more business focused.]

2026 Commentary

I mention smartphones in the article, but I don't specifically address companies who have a bring-your-own-device philosophy, where many different types of smartphones and other devices are in use. In such a company, the architect has to define and coordinate minimum standards for those devices, and ensure that proper security concerns are addressed in the way that business data is handled on those devices.

And in point #6 I mention airline magazines, but do those even exist anymore? These days a business person is more likely to hear of new products from social media.

62. Thrasher's Hierarchy of Business IT Needs

Originally published on NOVEMBER 13, 2007 in #IT, #ALIGNMENT, #LEADERSHIP, #MANAGEMENT, #STRATEGY

I recently attended an IT panel discussion where one of the attendees asked the panel what IT initiatives are strategic to the business. There was a brief debate about what "strategic" means, and I came to the conclusion that it's impossible to tell whether or not a specific IT initiative is strategic to a particular business without understanding what's going on at that company.

In thinking about this further, I remembered Maslow's Hierarchy of Needs, and I decided that there must be an equivalent Hierarchy of Business IT Needs that determine an IT organization's strategic focus at any point in time. This article describes such a hierarchy.

First, A Refresher: Maslow's Hierarchy of [Human] Needs

If you ever took a psychology course, then you probably ran across Maslow's Hierarchy of Needs. The concept is that people focus on the important things that we lack, but once certain levels of these things are achieved, then our focus shifts. For example, if we don't have food and water then that's our focus. Once we have food and water and everything else we need for our bodies, then we begin to think about our next level of concern: safety. Once our safety needs are taken care of, then we focus on love, belonging and social needs. Satisfy those needs and we focus on self-esteem. And so on and so on. Here's a diagram that shows Maslow's Hierarchy of Needs represented as a pyramid, with the more primitive needs at the bottom.

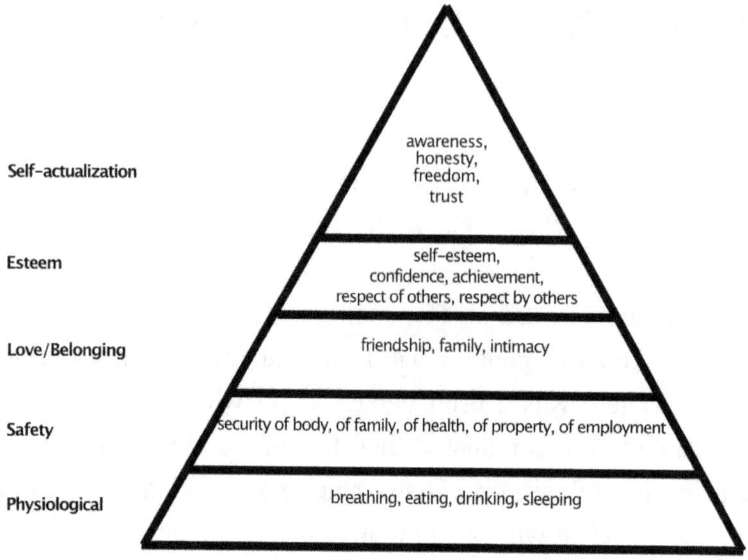

Maslow's Hierarchy of [Human] Needs

Thrasher's Hierarchy of Business IT Needs

Now think about a business instead of a human being. In a sense, the business has the same sort of hierarchy, and depending on the maturity of IT use within the business, the "strategic" IT needs of the business will vary. Here's a diagram that shows my Hierarchy of Business IT Needs.

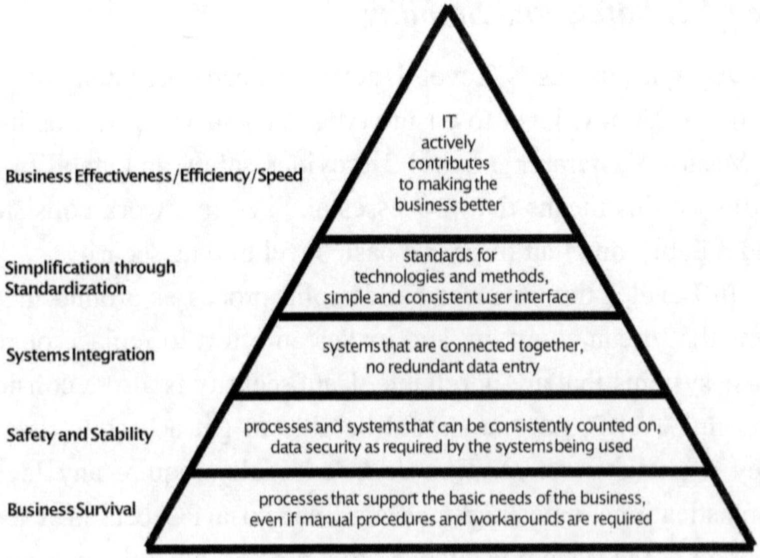

Thrasher's Hierarchy of Business IT Needs

Level 1: Business Survival

At the lowest level, there is a need for the business itself to survive. So at the lowest level of business IT need, there is a requirement for the processes in a business to work well enough to support the basic needs of that business. I'm using the word "processes" because technology is not required to fulfill this level of business need; the business may be able to use manual processes that don't require technology at all. A business achieves its Level 1 needs when processes are working well enough that the business can deal with all of the basic requirements of being in business. This doesn't mean that everything works well—it just means that there are manual workarounds and fallback plans in place to deal with any system issues, and manual processes in place to replace IT systems that are missing or don't work.

Level 2: Safety and Stability

After a business's Level 1 survival needs are met, the focus shifts to the next level to ensure *continued* survival, just as it does in Maslow's Hierarchy. Level 2 provides safety and stability for a business; this means that processes and systems work consistently and reliably, and that there is a basic level of data security.

In Level 2 the business builds solid processes around the systems that it can count on, and makes an effort to replace or repair those systems that are unreliable. Data security is also a consideration in Level 2; no one should be able to get access to data that they shouldn't look at. This doesn't initially require any level of sophistication—data might be confined to a PC behind a locked door, or if the system is attached to a network then the network offers at least rudimentary protection against intrusion. Then as systems grow in importance and complexity, data security should increase to meet the expanding need.

Level 3: Systems Integration

Once a business has systems that are good enough to ensure continued survival, the focus shifts toward reducing unnecessary work caused by systems duplication and overlap. Up until this point, business IT effort has been applied to build or acquire individual systems. Now there's a realization that systems must be integrated to provide the business with accurate, up-to-date and consistent data and information. In Level 3, IT resource is used to connect different systems together, move data among various databases, and eliminate redundant data entry.

Level 4: Simplification through Standardization

After the systems integration efforts of Level 3, the business begins to recognize that not all of the systems duplication comes from data overlaps. Many systems are too complicated, and there is often inconsistency in user interfaces, leading to unnecessary confusion and wasted effort. The same thing is true within the IT organization itself, where there is wasted effort because of the need to maintain and support many different technology platforms.

In Level 4, there is a move toward simplification. This is accomplished by implementing standards for technologies and methods, and using those standards to reduce the number of different software and hardware platforms used by the business, while providing a user interface that is more consistent and much simpler. Standards and simplification also lead to better controlled and tighter data security.

Level 5: Business Effectiveness/Efficiency/Speed

Once Level 4 is achieved, the focus goes beyond optimization of existing processes and systems, and the IT organization actively contributes to making the business better in new ways that are tied to overall business objectives. In Levels 3 and 4, we optimized the systems we built in Levels 1 and 2. Now in Level 5, technology is applied to make the business more effective, more efficient and faster.

Must Levels Be Achieved One at a Time?

A typical business starts at Level 1 and moves upwards from level to level after the needs of the current level are satisfied. Once a higher level is achieved, the needs of a lower level are no longer prioritized and are often forgotten. But if the business regresses—finds that the needs of a lower level are no longer being met—then the business will temporarily reprioritize to focus attention on the unfulfilled needs of a lower level (a data security breach is an example).

A business doesn't *have* to accomplish everything required by one level before doing any work on a higher level. This is the same as in Maslow's Hierarchy, in which human beings are free to make improvements at higher levels while they're still working on fulfilling the needs of lower levels.

But as with Maslow's Hierarchy, there is less payback from work done at higher levels until you have fulfilled all of the requirements from the preceding level. For example, security is nice, and you can lock your doors at night, but if you don't have enough to eat then your focus is going to be primarily on getting food; it doesn't do you any good to starve behind a locked door. The same thing is true for the Hierarchy of Business IT Needs: it doesn't do you a lot of good to focus on IT's strategic contribution to new revenue if your basic systems aren't meeting the needs of the business. Your credibility will be questioned, and you'll probably be building those new systems on a shaky foundation.

A Typical Progression

If a business follows the typical level to level progression then the work at Level 2 will include rework of some of the things done in Level 1, Level 3 will include rework of some Level 2 work, and Level 4 will have rework of some of the work done in Level 3. This is unfortunate, but it reflects the difference in priorities at the different levels. In theory you might be able to convince the business that all of your systems ought to be standardized and integrated from the start. But reality says that you're not likely to be able to deliver a standardized and integrated solution in the timeframe required by your business, nor are you likely to be able to convince anyone that you should do so.

Once a business reaches Level 5 and IT goes into new areas, each of these new areas will tend to regress a bit in the Hierarchy. Initial focus in a new area may be on a standalone system. After the standalone system works, then there will be an effort to integrate the standalone system with other IT systems. Eventually standards and simplification will be applied in the new area. This cycle will be repeated again and again as IT broadens its focus on new and different benefits to the business.

More than One Hierarchy at a Time

In large companies it's possible for different parts of a company—for example, different divisions—to be at different levels in the Hierarchy, just as different members of a family may be at different levels on the Maslow Hierarchy. In such a case, each part of the business should be addressed separately at its own level, with perhaps an overall focus on bringing the entire company up to certain standards.

However, too strong a corporate emphasis on standards—a Level 4 concern—will cause a conflict with a division which is still struggling to meet lower level needs. It's important for each party in such a dispute to understand the motivation of the other party, and to reach a compromise which helps the division go through the lower levels more quickly. Otherwise a premature emphasis on standards for the division will cause the division IT organization to be perceived as failing by the business when it's caught in the middle between corporate and divisional needs.

Conclusion

The word "strategic" is thrown around a lot in IT; everyone wants to be strategic or to be doing strategic things. But it's important to recognize that what's strategic to one business is not necessarily what's strategic to another business because they may be at different levels on the Hierarchy of Business IT Needs. And even for a single business, what's strategic one year probably won't be strategic the next year when some of those strategic goals have been accomplished.

Just as human beings have a hierarchy of needs that causes us to focus on different needs at different points in our lives, so too does a business have a hierarchy of needs that causes the business to have different IT needs at different points in time. Success in IT comes from recognizing the level of your business in the hierarchy, and focusing on the things that are most important to your business in its current hierarchy level. Being strategic means doing the best thing for your business in its *current* state. Remember that, and it will be much easier for you to focus on strategic goals for IT.

Notes

1. Right after I wrote this newsletter, I happened to read an article in the Fall, 2007 issue of the *MIT Sloan Management Review* (SMR) entitled, "Avoiding the Alignment Trap in IT." The "Alignment Trap" mentioned in the title of the SMR article refers to the problem that businesses have when an ineffective IT organization tries to tackle the most important business priorities of a business. The article recommends that IT organizations focus on their own effectiveness before going after the toughest business problems. Sound familiar? Using the terminology of my Hierarchy of Business IT Needs, businesses with an Alignment Trap are going after Level 5 priorities before they've mastered Levels 3 and 4.

2. Some of you may be comparing the Hierarchy to the Capability Maturity Model (CMM) that was used a few years ago to describe the levels of maturity of software development organizations. But the only thing that the Hierarchy of Business IT Needs and CMM have in common is the number 5: each has five levels. CMM refers to measurement of a software development organization *from the inside out*, and the Hierarchy of Business IT Needs refers to an assessment of Business IT *from the outside in* (i.e., from the *business* viewpoint). The Hierarchy of Business IT Needs is intended to explain why different people in different businesses

disagree on what's strategic for IT at any point in time—not to comment on the way that software is being developed.

2026 Commentary

You may wonder how Artificial Intelligence affects the Hierarchy or its use. At the moment, AI may offer a way to accelerate certain improvements in software or processes, but it doesn't fundamentally alter the idea of the Hierarchy. For example, you could use AI to help find data security breaches or process issues. You could also use AI to accelerate data integration of different systems. You could possibly even use AI to identify opportunities for IT to help the business. But while AI might help you progress up the Hierarchy faster, the concept of the Hierarchy remains the same.

63. How to Stop a Runaway Project

Originally published on DECEMBER 11, 2007 in #MANAGEMENT, #PROJECTS

What do you think when you hear the phrase "runaway project"? When I hear the phrase, it reminds me of movies I've watched where there's a runaway car hurtling down a steep mountain road with no brakes, barely hanging on around each curve. Or maybe you think of a runaway train, flying down the track toward imminent collision or derailment. In either case the idea of a "runaway" refers to something that's out of control and getting more out of control by the minute. Unless a runaway is stopped, disaster always ensues.

This article looks at a runaway project, drawing parallels to the runaway car example. Let's start with the cause: what's the project equivalent of losing your brakes? Car brakes in our mountain road situation help us keep our velocity below the point where we lose control over the car. Of course the maximum speed at which we can maintain control depends on the driver and on the vehicle. An experienced race car driver in a small car that grips the road can stay in control at a higher speed down a mountain road than the average person driving a minivan overloaded with baggage.

The same situation exists with projects. An experienced project manager can move a project a lot faster than someone who is inexperienced; and projects that are well-defined with no obvious technical hurdles are likely to move a lot smoother than projects with lots of murky ill-defined objectives. So right from the start you know that a huge, unwieldy project with an inexperienced project manager has to go slower to stay "in control."

Brake Failure

The project equivalent to brake failure comes from:

1. Not stopping to make key decisions
2. Letting key decisions be made for you without consideration of project implications
3. Making decisions based on goals that are unrelated to project success, and
4. Emphasizing project velocity over project control

When you're driving a car down a mountain, you need to slow down as you approach a tight curve. The faster you approach the curve, the more likely you are to hit a guard rail.

The same thing happens in projects. There are points in the project where you need to slow down and make sure that you're making the right decision about the next step in the project. If political or deadline pressures cause you to make a decision too quickly—without proper consideration of alternatives—then you're going to crash into a project guard rail and swerve down the mountain road out of control. You might be able to regain control, but now your momentum has increased, and the likelihood rises of a second crash into a project guard rail. Crash into a few of these guard rails and you begin to panic, causing you to make worse decisions. Bad decisions proliferate, and the project becomes a runaway.

Gravity = Complexity

Now we all understand that gravity is the ultimate cause of a runaway car (you never see a runaway car going *up* a mountain), but what is the project equivalent of gravity? What is the force that makes the project worse and worse, that makes the project take longer and longer until the project ends with cancellation, a major organizational restructuring, a corporate write-off, or even a corporate bankruptcy?

The project equivalent to gravity is complexity. Just as gravity makes everything fall toward the center of the earth, there's a natural tendency in human organizations to make things more complex than they have to be.

A key element in successful project management is taking a complex objective and breaking it down into a series of simple tasks that can be accomplished with existing resources in a reasonable amount of time. But with the pressure and attention focused on a big project, not all project managers are able to get far enough along in the simplification process to stay ahead of project effort, and if project effort gets ahead of your ability to manage it then the complexity gets worse.

The more complexity you have, the more decisions you'll have to make. The project manager gets even further behind in the simplification process. The complexity grows and feeds on itself, building the same momentum of a failing project that you would get from gravity pulling a runaway car down a mountain.

In a runaway project the number of issues will continuously rise, and revised estimates are always too low no matter how many project obstacles are overcome. In essence the runaway project passes the "tipping point" of complexity, and the complexity now feeds on itself to the point where no one seems to be able to get it under control.

How Do You Stop a Runaway Project?

Just as with a runaway car, the first step is to recognize that the situation requires drastic action. Runaway projects continue because everyone refuses to compromise, and so the complexity of objectives gets worse and worse. You can't stop a runaway project by doing the same things you've been doing so far. The *only* way to bring a runaway project to a reasonable outcome is to make a radical change that drastically reduces the complexity. There is no single approach to doing so, but here are some possible ways to accomplish this goal:

- Make a drastic cut to the functionality required by the project, possibly by postponing achievement of some of the objectives
- Break the project into multiple independent pieces
- Take a multi-organization or multi-person responsibility and give it to one organization or one person who can make decisions to reduce complexity
- Take multi-location responsibility and give it to one location

There are other ways to do this as well; what we're trying to do is to get the project past a complexity impasse so that we can restart the project with simpler objectives.

Inappropriate Approaches to Stopping a Runaway Project

You'd be surprised how many people refuse to address the complexity problems of a runaway project. A much more common approach—which doesn't work—is to add more resource to the project. But with more resource comes even more complexity, and a typical runaway project placed in this situation just accelerates more rapidly out of control. Using this approach is like adding more people to a runaway car—it just adds more momentum and therefore more acceleration, and it increases the casualty count when the car finally crashes.

Another common approach to stopping a runaway project is to change project managers. While it's true that a more experienced project manager might be able to cope with more complexity than an inexperienced one, the typical runaway project has more complexity than any human being can handle. Even an experienced race car driver can't drive a car down a mountain road at 500 miles per hour. If the new project manager doesn't take radical steps to decrease complexity then the project is still likely to crash.

How Do You Prevent a Runaway Project in the First Place?

This is a much easier problem to solve. To prevent a runaway car you just have to make sure that your car is in good condition, that it's not overloaded, and that the driver of the car is experienced enough to handle the road ahead. The same rules apply in preventing runaway projects:

1. Avoid huge projects. If at all possible, break big projects into a smaller, less complex series of deliverables.
2. Focus control on one organization and one person. If work has to be done in multiple organizations or in multiple locations, then (a) define clear objectives and deliverables for each organization or location, and (b) make sure that there is still a single organization and person defined as a focal point for decisions that span multiple organizations and locations.
3. Avoid project dependence on a new unproven technology. If you have to use a new unproven technology, then do a smaller pilot project first. If this isn't possible (and I can't think of too many situations where it isn't) then identify a fallback plan in case the new technology doesn't work, and define clear trigger points that specify when the fallback plan will be invoked. These trigger points are important because there's a natural human tendency to keep postponing the use of a fallback plan, hoping that things will somehow work out. A well-defined rule for a trigger point makes the decision clear-cut and you won't waste time.
4. Minimize the number of uncertainties in the project. For uncertainties which remain, do the same thing I recommended for an unproven technology: create a contingency plan and define clear trigger points that specify when contingency plans will be invoked.

5. Minimize dependence on factors outside the control of the project. They virtually guarantee that the project will run late and over budget, and can cause the project complexity to grow to the point of a runaway project.
6. Minimize the unknowns. At the very least, you need to know (a) what has to be done, (b) who is responsible for getting the work done, and (c) who signs off on the successful completion of the project. If these three things are not clarified before the project starts, then you might as well start down the mountain road without any brakes at all.
7. Avoid "over-using your brakes." Most brakes fail in a mountain road situation because they're used too heavily at the top of the mountain and so they overheat and stop functioning. You can do the same thing in a big project if you exert too much control early and then have the control taken away from you in retaliation. Like brakes, project control is something that should be used judiciously and with some finesse and delicacy.

Conclusion

There are probably a few of you reading this article right now who are in the middle of a runaway project, and I wish you luck in applying some of the techniques described in this article. What most people don't realize is how emotionally charged the issues are. There's unbelievable pressure from everyone to make the project succeed, yet that pressure just makes matters worse. Even if everyone agrees that complexity needs to be reduced, there will be a huge debate about *how* the complexity will be reduced, since no one wants to give up on their own objectives for the project. Often it takes an impartial outsider to break the logjam and figure out the best way to reduce complexity.

For those of you who have never been involved in a runaway project, my advice is to be careful—you're not immune. Follow the recommendations in the article for preventing runaway projects, and you may be able to avoid them in the future. The advice is good for all projects so you don't need to worry about when to apply the recommendations—just apply them all the time and your projects are more likely to be successful and less likely to be runaways.

2026 Commentary

I wouldn't consider Artificial Intelligence to be a "silver bullet" to stop a runaway project. Although it's possible that AI might help in some areas of IT, it introduces uncertainty associated with unproven technology, thus adding to the complexity. To use the out-of-control-car analogy, it's a little like speeding up the car but loosening the steering wheel. This would probably hasten a crash.

Also be aware of the "sunk costs fallacy." This is when your business leaders say things like, "We can't stop the project now—we've invested too much in it." It's almost like saying, "I know we're out of control, but go faster," and it almost guarantees a project failure. If the project is out of control, then complexity has to be reduced. No other solution will work, no matter how much money and time has been spent.

For another viewpoint on runaway projects which looks at the problem from a communications point of view, see "96. Irreconcilable Differences and Runaway Projects."

64. 8 Ways to Find IT Talent for an Undesirable Location

Originally published on JANUARY 8, 2008 in #IT, #CAREERS, #LEADERSHIP, #MANAGEMENT

Last month I did some work for a company that has manufacturing plants in a number of rural locations. The IT managers who work in those locations complained about how difficult it is to hire IT people, partly because of the locations and partly because of the image of manufacturing being a dirty industry. Here are some ideas that may help.

Idea 1: Rethink the type of person you hire

There's a general rule for hiring customer service people that you look for someone who can relate well to the public, and then you train the person in the area of expertise for which they'll do customer support. This is because it's much harder to do the reverse: hire someone with the expertise and then teach them to relate to the public.

Consider a similar approach when you're hiring for an isolated or rural area. Find the kind of person who will want to live in that area and who is motivated to stay. Look for certain aptitudes which make the person trainable in the area of expertise you need, and then invest in training the person to suit your needs. When you define your candidate requirements this way, you might even discover that the ideal candidate is already in the area or even working in your company.

Idea 2: Think in terms of a career path—not just a job

Few people want to do the same thing every day for the rest of their lives. Most want to continue to learn and grow, and in many cases to attain positions of increasing responsibility. When you look at your people needs, think of how a person will grow and advance in each position. Try to define a progression of jobs, starting with an entry-level position and advancing to high-level, higher-paid positions. The entire progression may not reside in the IT organization—maybe you start the progression with part-time participation from a factory worker, or maybe the higher-end of the progression leads to supervisory positions in non-IT departments. But once the progression is defined—maybe with variations due to the interests of the individuals in the positions—you can sell the *progression* to prospective job candidates instead of just selling the job. This approach makes it much easier to hire people. They'll feel comfortable that they're not moving into a dead-end job, and they're likely to stay with your company for many years.

Idea 3: Rotate people through an undesirable location as part of their career path

Look at the bigger picture. If you really can't find someone to work permanently in an undesirable location, then make a temporary assignment at the location be part of a career path—maybe for a year or two. There's probably some great experience to be obtained by working there temporarily. And who knows? Some of them may decide they want to stay.

Idea 4: Start an intern program to attract candidates while they're still in school

Team up with local schools—colleges, vocational technical school, or even high schools—to attract candidates early. The schools are happy to help advertise intern positions and in most cases they'll even help you select the most appropriate candidates. It's a win-win for both you and the candidate: they get a way to help pay for their education, and you get a part-time or seasonal person who you can groom for a future full-time position.

Idea 5: Make your company a better place to work

Of course no one wants to stay in a company that treats its employees poorly, has terrible working conditions, and in general shows that it doesn't value its employees. If this sounds like your company then you've got an up-hill battle to fight (and frankly I wonder why you're still working there yourself). Employee surveys in all industries have shown that salary is not at the top of most people's list of the most important things about their job. Instead, the most important thing is usually a feeling that the employees are contributing to something bigger, and that the employees are respected and valued for being themselves. Make sure that you and your organization are making your company a place where employees will want to work.

Idea 6: Use part-time or shared resource

Don't assume that every one of your needs has to be fulfilled by a full-time person in your IT department. If you have a part-time need then see if there's someone in a non-IT job who might be able to do the work. Or maybe you could go in with someone else in your company who has a part-time need, and find a person who can work a full-time job made up of part-time work for each of you.

If part-time resource with the right skills and aptitudes isn't available inside your company then look outside. Are there other non-competing businesses in your area that might have a similar part-time need? If so, then see if you can find someone who is willing to do part-time work for each of the businesses. You could think of this as outsourcing, but on a lower budget.

Idea 7: Outsource

Full-blown outsourcing is also a possibility. Find well-defined tasks that don't have to be done locally; payroll and web site maintenance are two examples. Outsource these tasks to companies or individuals who work outside the area, and deal with them primarily by telephone and email. Or if your company has IT people in other locations, then consider outsourcing some of the work to people in those other locations.

Idea 8: Use standard products and technologies

I've seen companies that build proprietary systems around older or seldom-used technology (like the PICK operating system or dBase) and then wonder why they can't attract people to work on them. You're much better off using standard off-the-shelf software whenever possible and then building any proprietary software using currently popular products. There will be more people available who know the technology, you'll find it much easier to obtain training for new people, and outsourcing will be much easier if you decide to go in that direction.

Similarly, it's important to use good software design principles when building your software. For example, a layered approach which separates presentation, business logic and database will make it much easier to modify or adapt software for future needs, and it will be easier for multiple people (even in different locations) to work on the software without stepping on each others' toes.

Conclusion

Companies in rural or undesirable locations have a slightly different kind of problem from those of us who live in metropolitan areas, but that doesn't mean that IT needs must go unfulfilled. There are plenty of people in rural areas and quite a few people in urban areas who would love to move to a location with a different lifestyle and possibly a better quality of life.

The secret to recruiting for an undesirable area is to figure out what type of person would enjoy living there, and then try to find such a person who has the aptitude to do the job with some training. But that's not all: you have to make the job attractive as a good career move so that people will want to stay.

Even with this approach, you have to be realistic about what needs to be done locally and what can be done elsewhere instead. The Internet and long-distance computing have made it possible to do IT work remotely that until recently would require an onsite presence.

The best approach to providing IT support for an undesirable location is a combination of tactics: reduce the job to the essentials, use standard technology, and find good people who actually enjoy working there. After all, what's undesirable to you may be extremely attractive to someone else.

2026 Commentary

Advances in communication technology since 2008, including FaceTime, Slack, and Zoom, have made remote work even easier. And I didn’t mention cost of living in the original article, but if the cost of living is lower in the rural location, then that’s another advantage to use in convincing someone to work there.

65. Information Technology is like the Stock Market

Originally published on FEBRUARY 12, 2008 in #CAREERS, #IT, #ALIGNMENT, #LEADERSHIP

Success in IT is a lot like success in the stock market. People who don't understand the stock market sometimes think that there's a "right" price for a stock based on some elaborate and mystical formula. Similarly, people who don't understand IT sometimes think there's an objective way to measure an IT organization that will give the same result no matter who does the measuring. Both of these opinions are incorrect. Here is the truth about the stock market and IT:

1. Success depends more on perception than reality.

A stock's price is the result of perception. Investors have an opinion on a stock based on expectations of what the stock will deliver over time. People who think that the stock is undervalued tend to buy the stock. People who think a stock is overvalued tend to sell the stock. It's like a tug of war, with buyers and sellers pulling on opposite ends of a rope, and with the stock price fluctuating back and forth like the flag that marks the middle of the rope.

A company keeps its stock price under control by setting proper expectations. If performance matches expectations then the stock price stays steady (except for overall stock market influences—more on that later). If things go well, then the company raises its expectations based on increased performance, and the stock price rises to match these raised expectations. It's very important for performance and expectations to match. Even if the company is successful, the stock will go down if the market expected better performance. So the rules are: Keep meeting expectations and the stock price will go up. Fail to meet an expectation, and the stock price will go down.

IT organizations are measured exactly the same way. An IT organization's success is the result of perception, with overall success the result of various opinions on how IT is doing. Some opinions (like the CEO's) matter more than others—just as the opinions of large investors have more of an effect on a stock price.

Like a CEO trying to keep the company's stock price under control, a CIO keeps the IT reputation under control by setting proper expectations and then living up to those expectations. A good IT organization will raise the level of expectations as it increases its capabilities. The rules for CIOs are: Keep meeting expectations and the IT organization will be viewed as successful. Fail to meet an expectation, and your reputation will decline.

CIOs who don't understand these rules are sometimes surprised when their organizations are perceived as failing. They disagree with the perception, pointing to "objective" measures of their success. But they miss the point: it's not the objective measures that matter—it's how IT performs against expectations.

2. Consistency is more important for trust than an occasional outstanding accomplishment.

When a company fails to meet expectations, stock investors begin to lose trust in the company. Each additional failure loses more trust and the stock price falls.

The same thing happens with an IT organization. Consistently meeting expectations is the only way to build the trust that is absolutely critical to success. It's better to set lower expectations and consistently meet them than it is to set higher expectations and fail to deliver. When you have the trust of your IT customers then you can get the support you need for increased resources to do more. But without trust you won't be able to get those resources because your customers won't trust you to use the resources appropriately.

3. Your success is linked to the success of others.

Performance of a stock is impacted by the performance of the economy as a whole. If the economy is bad, then investors expect that sales will decline in certain businesses. This expectation of falling sales will impact stock prices.

The real test of a company's strength is how it prepares for and deals with adversity. The stock price of a company that handles itself well in a bad economy can actually go up as investors shift their money into the good stock from other stocks that aren't doing so well.

IT performance is linked to the success of the rest of the company. It's easier for an IT organization to do well when everything goes right. As it is for stocks, the real test of an IT organization's strength is how it prepares for and deals with adversity. How do you keep a project on schedule and roll out a system when you're faced with labor cutbacks? How do you keep your company's data secure when you have to live with severe budget constraints? You have to make sure everyone knows that the situation is a challenge but that you're doing the best that can be done. Show your company that even in the face of adversity your organization is still a good investment for the company's money. Prove your ability to handle adversity, and your organization will do well, even as less trusted organizations do poorly.

4. Success seems to be tied to a few charismatic individuals.

It's usually the CEOs who get the glory in the stock market, the ones who "single-handedly" turn a company around. It's the same thing for CIOs, although the glory is on a smaller scale. But we all know who does the real work—it's the lowly employees who actually put in the hours to grind out the product or to deliver excellent service. So why do the CEOs and CIOs get the glory? It's because they set the direction, they determine the focus, and they set the standard for attitude, philosophy, and ethical behavior. Senior leaders are the ones who determine what commitments should be made, and they are the focal point for setting expectations with investors and customers. Yes, the "real work" is done in the trenches, but without a focal point those workers wouldn't be able to focus on the work that's most important.

5. To be considered a superstar, you need to "buy low."

In the stock market, the investor superstars are the ones who invest heavily in a promising company and then ride the stock price up to riches. The key is to recognize the potential in a company before everyone else does. The CEO superstars, on the other hand, are the ones who recognize the potential of a new market or idea, and then figure out how to turn that new market or idea into a money maker.

In IT, the CIO superstars are the ones who take an inadequate or failing IT organization and turn it around to make it a key contributor to business success. "Buying low" in this case means to take a CIO job that no one else wants. It means recognizing the potential in an organization and turning a failing IT organization into a successful one.

Although it may be satisfying to take over an IT organization that's already successful, you can't really be a superstar unless you make order of magnitude improvements in the organization. It's possible to take a moderately successful IT organization and make it wildly successful, but it's very difficult to be a new CIO taking responsibility for an IT organization that's *already* wildly successful. That's like buying a stock when it's already at its peak price. Can it go even higher? Only time will tell.

2026 Commentary

The superstar idea in #5 is true for lower-level IT people as well, and it works for non-IT areas of a business. You get the biggest rewards for coming into a job in a problem area and turning it around. Then, if you want to continue as a superstar, go to another position in another problem area (maybe in a different company or a different part of the same company), and turn that new area around too.

66. How to Help Your Help Desk

Originally published on MARCH 11, 2008 in #MANAGEMENT

Almost all IT organizations have a Help Desk, and yet it always amazes me how many of those organizations think that the Help Desk has only one purpose: to help system users with problems. In fact the Help Desk also serves a second critical purpose: to tell you what your problems are so that you can adjust your training and buying accordingly.

Working in a help desk is very frustrating. It's always hectic, everything is an emergency to the people who are calling you, and it seems like the same kinds of people ask the same questions over and over. Then to make matters worse, your managers make you waste time tracking your calls. How annoying!

Why Call Tracking is Important

But call tracking is key, even if you just use a low-cost paper system. That's because the Pareto Principle—also called the 80-20 Rule—applies to Help Desk calls, and you have to track your calls to see your data. According to the Pareto Principle, 80% of the calls will be made by 20% of the people, and 80% of the calls (not the same 80%) will be about 20% of the problems. Focus on the 20% and solve the associated problems, and 80% of your volume will go away!

Call tracking is important because what you really want to do is to not just *handle* Help Desk calls but to *reduce* them, and the only way to reduce calls is to reduce or eliminate the problems that people are calling about. So if you track your calls and learn *who* is making the most calls and *what* people are calling about, then you can make improvements in your processes and systems to eliminate the need for those calls.

Let's take an example. Maybe you find that most of your calls are about a confusing aspect of a new version of an application that you just started rolling out. Since you're rolling out the new version over a period of months (please tell me you aren't using a "big bang" approach and doing it for everyone at once), you can get in touch with the people doing the training for the new version (you *are* doing training, aren't you?) and have them revise their training to cover the problem area. Or if it's your own application then you can talk to the developers and get them to incorporate a cleaner way to handle the problem in their next software release.

In another example, let's say that you find from your statistics that it's a few people who are new to your company who are making the most calls. You might want to get in touch with them (and maybe their managers) to suggest additional training to make them feel more comfortable with what they're doing. This will also help them cope with their problems without creating a dependency on the Help Desk.

Tips for Small IT Organizations

If you have a small IT organization then you may not have a separate group of people who work full-time at a Help Desk. That's even more frustrating because every time you get a call, it takes someone away from project work. If you're not careful then the Help Desk calls will overwhelm you and you'll never have any effective time for making improvements in your IT.

Here are some suggestions for small IT organizations trying to cope with Help Desk problems:

1. Try to keep your Help Desk resource separate from your project and maintenance resource. Even if the same people do both, you can maintain the separation by having an "On Call" person designated to take Help Desk calls at certain points in time. This helps the project people concentrate on projects and minimizes their disruption.

2. Rotate people through the On Call slot so that the effort and pain are shared. This also gives the project and maintenance people some perspective on the problems that their customers have, and so it provides a better incentive for them to do projects and training that meet customer needs.
3. Have the On Call person pull in project and maintenance people as needed, but try to do so only in an emergency. Otherwise the Help Desk will turn into a black hole, sucking in all of your IT resource.
4. Track *all* your calls, even if you can't afford fancy Help Desk problem-tracking software (a list of problems with tick marks beside them is better than nothing). It's the only way that you'll be able to do an 80-20 analysis and spot problem areas that you can address.

Outsourcing

It's tempting to outsource your Help Desk. But if you do, make sure that your outsourcing contract requires the outsourcing vendor to give you regular statistical reports on who's making the calls and what the calls are about. Your goal should still be to reduce or eliminate the need for calls, and you can only do that by fixing the problems that cause people to call. You need Help Desk statistics to understand those problems.

The "Hidden Help Desk"

Watch out for the "Hidden Help Desk" that intercepts a lot of the calls before they reach the Help Desk. People with system problems often ask someone else in their office before making a call to the Help Desk. That takes some of the load off the Help Desk but it conceals a lot of problems that ought to be fixed. Find out who the office "Wizards" are who are doing informal duty as a Hidden Help Desk. Get to know these people and encourage them to tell you about problems that they hear again and again. By fixing the problems you'll make these people your friends and you'll probably reduce the need for calls to the official Help Desk.

Conclusion

A Help Desk isn't just a necessary expense—it's a valuable source of information about your systems and the way that people are using them. Take advantage of the information and you'll have happier users, a better reputation for your IT organization and ultimately lower Help Desk cost. Be kind to the people working on your Help Desk; they're fighting on the front lines in the battle for better IT. And they're likely to know whether you're winning or losing the battle long before you do.

2026 Commentary

The article talks about an IT help desk, but the same logic applies to any type of help desk or call center. It could be calls to HR or to Finance or to Engineering or to Facilities Management. Any time there is a significant number of help calls anywhere in the company, there is an opportunity to track the calls and make improvements based on the information.

67. Questioning IT Realignment—Should IT be Centralized?

Originally published on APRIL 8, 2008 in #IT, #ALIGNMENT, #MANAGEMENT

One of my newsletter readers sent me an interesting question:

"I was wondering if you have insights on the attitude of managers on the business side towards the realignment of IT. I can imagine that managers at the business side tend to resist giving up (overly) customized IT, since the short-term performance of their individual units is so dependent upon (overly) customized IT. Asking a manager to cooperate with de-alignment, when no alternative for 'less IT' is available yet, means asking a manager to sacrifice their short-term performance."

This particular reader works in a large multinational consumer products company, but I believe her concern is a common problem these days. Companies are cutting back everywhere they can, and it's widely believed that centralizing IT can save money. Here are my thoughts on the subject:

1. Change can be Painful, But We Still Need to Change

There's always a transition cost to any change, as people go through a learning curve of changing from one process to another. So in a sense, all managers have to constantly cope with "sacrificing" short-term performance in the interest of longer-term gain. That's just the cost of doing business in a changing world.

2. Centralization Can Offer Advantages to a Decentralized Operation

The implication of the reader comment, however, is that a local manager who has their IT support changed as the result of an IT realignment gains nothing. The reader also refers to the change as "de-alignment," when in fact it's more of a realignment to meet higher-level business goals instead of optimizing local departmental goals at the expense of overall business objectives. If a realignment is done correctly then the local manager can actually benefit from the change.

Let me sketch out what I see as a common scenario; this may or may not match your own situation. In my scenario, IT is currently distributed across multiple organizations and probably multiple locations. It may be manufacturing plants or companies that have been acquired, or it might just be different departments in a single company and location. In any case a decision has been made to "centralize" IT. Such a decision is usually made as a move toward economy, since economy of scale is virtually the only reason ever given (or needed).

The local department manager sees control shifting from their local IT people to someone else in a central location. But what the manager doesn't see is that:

- The local IT people will no longer have to waste time handling basic infrastructure needs that are better handled by a larger central staff. This really *is* an economy of scale, since it's difficult for a small local staff to have the depth of expertise required to properly support a broad range of infrastructure products.

- Off-the-shelf systems like ERP that the local IT people couldn't afford will now become possibilities. Those off-the-shelf solutions will probably do a much better job of handling local needs than the low-budget built-from-scratch custom solutions that previously existed.

- There will now be a better career path for IT people and so it may be easier to hire good IT people into positions where they're needed.
- If systems are standardized across different departments and locations then it will be easier for local non-IT managers to move people from department to department or location to location without extensive retraining.
- With standard systems it will be much easier for people in different departments and locations to share approaches and process improvements. And system enhancements can be used by everyone—not just those in the location doing the improvement.

The problem with distributed IT is that the solutions built locally are so highly customized to suit the needs of the individual people in the local departments that they aren't ever any better than the *ideas* of those individual people in the local departments. Some of those individuals are brilliant, and their IT shows aspects of this brilliance. But most localized customization just reflects the preferences of average individuals, and there's no big advantage in that customization. Centralization offers the possibility of bigger thinking, bigger solutions, and—hopefully—better solutions.

3. Transitions should be Managed Carefully

If the transition from decentralized IT to centralized IT is done correctly, then there won't be a sharp decline in performance; there will just be a gradual shift from using overly customized solutions to using more generalized but better solutions. In an ideal transition, there will be an inventory taken of the local solutions that may be replaced, and the best ideas from each local area will be added to the overall corporate solution.

The biggest issue with IT centralization is a perceived lack of responsiveness and personalization. Local IT usually provides a lot more "hand-holding," and the locals get to know the people doing the support. Losing the hand-holding can be traumatic, but even this trauma can be minimized if local IT support is continued at some level. Centralizing IT doesn't usually mean that all of the local IT people move elsewhere, and a good IT organization will encourage local IT people to maintain the relationships that they have created, even if less time is spent on doing support.

4. Any Centralization will Likely be Followed by a Future Decentralization

The business needs to seek an equilibrium between a centralized and decentralized environment, and if the equilibrium isn't achieved, then there will be a backlash and the cycle will reverse. I talked about this in my article, "31. The Information Technology Merry-Go-Round," and human nature hasn't changed since that time.

But don't look at this centralization/decentralization cycle as just an IT problem. It happens in everything we do, from government (federal control versus local control) to religion (central control versus local control) to family (parental control). We all want the benefits of a centralized organization without giving up the freedoms we have as individuals or small groups. It's one of life's continuing struggles, and perfection is not likely to be achieved.

Conclusion

IT centralization works best when each aspect of IT is considered separately. Centralize those things that will benefit from centralization, typically things like data centers, help desks, purchasing, standards and maybe development and support of corporate-wide software and systems. But for those things that show less benefit from centralization, you should seriously consider leaving things the way they are. Or maybe you should just put in place a way to better coordinate decentralized IT people rather than replacing them with a totally centralized alternative. Centralization offers economies of scale, but not everything in IT can benefit from those economies.

2026 Commentary

It's been another 18 years since the article was written, and human nature hasn't changed a bit. The centralization versus decentralization merry-go-round continues.

68. 15 Career Mistakes

Originally published on MAY 13, 2008 in #CAREERS

I've written other newsletter articles about careers (see the references throughout this article), but until now I've never specifically written an article about career mistakes. I think the subject has been too personal—I've made a lot of these career mistakes myself, and it's hard for me to admit my failures. But in the interest of helping others avoid some of the mistakes I've made, I've decided to go ahead and create a list of the major career mistakes that I've made or that I've seen other people in IT make.

1. Focusing too much of your time on things you're not good at

I wrote an entire newsletter article, "57. You've Got to Specialize," on why you need to specialize, so I won't repeat myself.

2. Talking to non-technical people using technical language

I wrote a newsletter article on this one too: "24. Use Their Terminology—Not Yours."

3. When you're trying to sell, focusing on features rather than benefits

This is a variation on Number 2, and it's a major pitfall for IT people. As I said in my book, *Boiling the IT Frog*, the most common IT mistakes are doing the wrong things—not doing things wrong. We tend to focus on *how* things are done, and so it's very common for presentations made by IT people to the business to list a lot of features of our solution. We talk about things like the hardware, operating system, network and software we're going to use, together with the features of all that stuff: high MTBF, low latency, and other technical terms that drive business people batty.

Forget all that descriptive information! Tell the business people what's important to *them*, things like "customers will be able to access our web site 24 hours a day," "the time for salespeople to enter an order will be cut in half," or "our inventory requirements will be reduced by 30%." These are *benefits* of our IT systems, and the benefits are much more important than features.

4. Thinking that being good technically is enough

It *is* enough if you want to stay in your same job for the rest of your career, and if you're lucky enough to be in a job that won't be outsourced, offshored, or made obsolete by changing technology. For everyone else, it's important to understand that jobs require people skills, and you need to develop some of those skills even if you don't want to move into management or become a CIO. See my article, "45. 10 Rules for IT Job Success" for more information.

5. Not networking outside your company

I realize how difficult it is to take the time to get outside the day-to-day needs of your own organization. Notice that I said "*take* the time"—some people say "*find* the time" and I can guarantee that the time to do this will never be "found." This is one of those things that you have to force yourself to do. But please force yourself, because it's important. It's important to your career because you need to maintain contacts with people in other companies in case you need to make a job change. And it's important to your success in your current job because you need the perspective that comes from seeing similar problems from the point of view of people in other companies.

Some networking can be achieved by attending national conferences or training programs. But I also recommend regular networking with people in similar jobs in your city. Join local chapters of organizations like the Society for Information Management; they do a wonderful job of getting people together to discuss shared issues.

6. Not getting everything you can out of every training opportunity

It's easy to go through the motions when you're sent to a training class that doesn't interest you. But think of every training class as an opportunity—no matter what the subject. Open your mind to new ideas, and try to figure out ways to use the things you learn in your day-to-day life.

And don't ignore the networking opportunities in training classes. Many classes include a mix of people from different locations, different jobs, and different companies. Take the opportunity to compare jobs, talk to people with other backgrounds, and get a different perspective on your career and your life.

7. Not having a mentor

I've had various mentors throughout my career. Sometimes it's easy to find one; there's a friendly experienced person nearby. In other environments it's difficult, and you feel like you're all alone. But whatever your situation, make the effort to find a mentor. If you're in a technical job, then the mentor might be a more experienced technical person. As you climb the career ladder, you'll find that a better mentor can be found outside of your department—someone who can advise you on the people issues rather than the technical challenges. You can formalize a mentoring relationship, but even an informal mentor—someone you can go to for advice and counsel—is a big help.

8. Forgetting who you work for

I don't mean your immediate boss. I mean your company and its stockholders, or if you work for a nonprofit then the people who derive benefit from your nonprofit. There are a couple of reasons why it's important to remember that these "stakeholders" are who you *really* work for. First, you'll find that the lower you are in the organization chart, the more likely it is that people will ask you to do things that don't make sense from the stakeholders' viewpoint.

Sometimes these instructions will have been garbled in their translation through the organizational hierarchy, much like a message is garbled when relaying it from person to person in the "telephone game." Sometimes the instructions are the result of a misunderstanding of alternatives, maybe because someone with a non-technical background made a technical decision. In either case, it's your duty to bring the problem to the attention of your boss (do it diplomatically, please), and to try to reconcile the intent of the stakeholders with the instructions that you've been given. And although a few bosses may resent your questioning their instructions, most will appreciate your focus on what's really important.

Ethics are the second reason to remember that you really work for your stakeholders. Newspapers from the last few years are full of stories about companies where individuals acted in their own best interests instead of the best interests of the stakeholders. Many of these individuals have gone to jail, but even if they didn't, it's a sure bet that their careers hit a major stumbling block. It's important that you don't follow in their footsteps.

9. Not delegating

It's difficult to give up things that you're good at. Yet one of the rites of passage when being promoted from a lower-level job to a higher-level job is to give up detailed responsibility for doing some of the things that you've been doing in the lower-level job. A higher-level job requires you to focus on different issues, and you can't focus on these new issues *in addition to* your previous responsibilities. There's not enough time, and so you'll have to delegate responsibility for the lower-level issues to someone else. People who don't understand this are like a person with one foot on a dock and one foot on a boat that's moving away from the dock. You have to choose which foot to move, or you'll fall in the water.

10. Punishing people who make mistakes, instead of teaching them

Along with the responsibility for delegation is the responsibility for teaching. One of the basic truths underlying teaching is that the student will make mistakes. It's the teacher's job to help the student learn from the mistakes—not punish the student. As you move up in an organization, you'll be responsible for more and more teaching. Your role likely won't be called a teacher, but the goals are the same: help people learn to do better.

11. Not blowing your own horn

People who get little recognition often wonder why. It's usually because they don't make enough of an effort to bring their accomplishments to the attention of others. I'm not talking about bragging here; but it's important that others know of your successes. When you accomplish something, make sure others know. Put the accomplishment in terms of benefits (see Number 3), and make it clear that you were responsible for achieving those benefits. If others were involved, credit them as well. In fact, a great strategy to get attention for your own accomplishments is to highlight the accomplishments of others who worked with you or for you. You may be calling attention to others, but it will be obvious to everyone that you deserve recognition as well.

12. Not changing jobs when the time is right

It's tempting to stay in your rut, doing the same thing day after day because it's comfortable. Yes, there's risk involved when you change jobs, but there are incredible benefits to be gained. Look at a change with hope—not with dread.

13. Thinking of your job as something you do rather than as a process to be improved

Even if you stay in the same job for a while, there's no reason that you have to keep things the way they are. Take a good look at what you do from other perspectives. Is there something that could be eliminated? Something that could be automated? Some part of your job that should be outsourced? Try to find ways to help your company (see Number 8 above) accomplish the results of your job better, and you'll often find that you'll be rewarded with an even better job.

14. Equating your career with your life

Life is about more than your career. So no matter what happens at work, don't let it affect your *real* life: your family, your friends, your community. You have to keep your career in perspective. It's a part of your life—an important part, but it's not all there is. Balance your life and your career, and you'll find that you do better at both.

15. Not learning from your mistakes

After you finish each project, or periodically—maybe at the end of each month or so—ask yourself two questions: 1. What did I do well that I need to do again? 2. What could I have done better? Learn from your answers.

Conclusion

Life is a series of mistakes, from our first attempt at walking, through school, and throughout our career. Making mistakes—and seeing the mistakes that others make—is the way we learn, and through learning we make our lives better. Learn from the mistakes I've listed here, and have a better career and a better life.

2026 Commentary

I've made every one of these mistakes at one time or another, and I've made some of these mistakes multiple times. The key is to learn from your mistakes and from the mistakes of others.

69. 3 Things Your CEO Wants to Know

Originally published on JUNE 10, 2008 in #IT, #ALIGNMENT, #LEADERSHIP, #STRATEGY

A lot of people who move up through an IT organization are surprised to find that the higher in the IT organization they get, the less the job is about technology. At the very top of the IT organization in the CIO role, the focus isn't on IT at all—it's on doing the best thing for the business. In fact, Peter Whatnell, CIO of Sunoco and 2009 President of SIM International, recently said that most CEOs just want the CIO to provide answers to three questions:

1. Where do we need to invest?
2. How do we stand versus competition?
3. How do we mitigate risk?

If you think about it, these are three very good questions for an IT organization, yet none of the questions is technical, and none of the questions requires a technical answer. Let's look at each of the three questions in more detail:

1. Where do we need to invest

The key word here is "invest." That's because business executives consider IT an investment—an investment in greater revenue, increased profits, higher customer satisfaction, and other things that have benefit to the business. For something to be a good investment, it has to provide positive results that exceed the cost of making the investment, and that's true for IT as much as it's true for any financial investment. So don't bother to propose a project for which there's no clear business benefit. What's the point in investing if there's no return?

2. How do we stand versus competition?

It would be nice to envision a CEO as having lofty visions of the future which don't require consideration of the competition. But that's unrealistic; the world tends to compare companies that have similar products and services, and investors in a company's stock certainly look at competing alternatives.

While it's easy to compare revenue and profit numbers for competitors, it's more difficult to compare Information Technology. You have to examine the processes used for each major aspect of the business, look at the efficiency and effectiveness of the processes, and then drill down to the underlying systems and technologies being used for each process.

If a particular business process being used by a competitor is more efficient or effective than the equivalent process in your own company, then your company is at a disadvantage. If you net out the comparison of your processes to the equivalent competitor processes, and the competitor ends up ahead of the game, then there's potential cause for action, and it may be time to invest in the improvement of your own processes and systems. But don't just copy your competitors; try to do something that's better. Your competitors certainly won't sit still and leave things as they are.

And if your processes are better than all of your competitor's processes, then broaden your definition of "competitor." Look at companies that have a similar process to one of your processes, even if that company doesn't offer competing products and services. For example, L.L.Bean was at one point regarded as the leader in order fulfillment (they may still be the leader; I haven't kept up). Even if your company doesn't compete with L.L.Bean, if you have an order fulfillment process in your company then you might want to benchmark your processes against L.L.Bean's and see if you can learn something about how to improve your own order fulfillment process.

3. How do we mitigate risk?

There are two important things to notice in this question. First, there's no qualifying adjective in front of the word "risk." We're not talking about *IT* risk here; we're talking about risk in general. So this is more than just the question "How do we mitigate information technology risk?" It includes that, of course, but it also includes the question "How do we use technology to mitigate *business* risk?" It's important for CIOs to look at both of these issues: how to mitigate IT risk as well as new ways that IT can be applied in the business to mitigate overall business risk. Here's an example of the difference. You can mitigate some IT risk by having redundant servers in multiple locations. You can mitigate some business risk by figuring out a way to reduce the impact of high fuel prices on your company by implementing a system to optimize truck routes or to reduce travel by using videoconferencing.

The second important thing to notice in the question is the use of the word "mitigate" which means "to lessen the impact or intensity of." No sane executive expects to *eliminate* risk, since risk is a part of the world in which we live. We take steps to minimize risk where it's appropriate, but it gets expensive to reduce risk beyond a certain point of diminishing returns. Business executives take a two-step approach to risk: reduce risk where it's cost effective, and then mitigate the remaining risk to the extent possible.

Conclusion

Wise executives know that CIOs will do everything possible to manage their IT resources wisely. It's expected that this management will take time out of a CIO's busy day, but this day-to-day IT management role is just a basic part of the job, taken for granted like breathing, eating and sleeping. Where the CIO really contributes to the business is in the added value that IT brings to the table beyond the day-to-day. Peter Whatnell's three questions get to the heart of that contribution.

How would you answer the three questions for *your* CEO? Or are you too busy working on the day-to-day stuff to be able to answer them?

2026 Commentary

For a longer list of potential IT investment areas for the business, see the article, "4. Seven Ways Information Technology Can Improve Your Business."

70. The Right Span of Control Isn't a Number

Originally published on JULY 8, 2008 in #CAREERS, #MANAGEMENT

In an earlier article, "56. How to Organize IT," I didn't address one of the questions that keeps coming up during a bad economic climate: how do we deal with executives who want to cut the number of IT managers by increasing the span of control for each manager? Span of control, for those unfamiliar with the phrase, refers to the number of people reporting directly to a manager. So, for example, a manager with 10 people reporting to them has a span of control of 10, and a manager with 100 people reporting to them has a span of control of 100.

Cost-cutting consultants like to look at the average span of control for the managers in each organization, and any organization with a low average is immediately a target for elimination of manager jobs. IT manager jobs are particularly good targets because no one outside of IT understands the IT manager role. And what you don't understand, you tend to eliminate; at least that's the way a lot of cost-cutting consultants think.

But the reality is that the "right" number for span of control depends on a number of factors that aren't usually taken into account when these cost-cutting consultants go looking. This article will provide you with some ammunition to fight back if you're confronted with accusations that your own organization has a low average span of control.

The Basics

The idea behind span of control is fairly simple. You as a manager have a requirement to spend a certain amount of time each week doing non-management tasks, and you spend the remainder of your time managing your employees. Let's take a simple example. Let's say that you spend a total of 50 hours per week working (anything over 50 will be reserved for unplanned use). Now let's say that 20 hours of your week will be eaten up by work not related to people management. The actual number will vary greatly depending on how much "individual contributor" work you have to do as part of your job, how much time you spend with your boss, how much travel you do, and how much time you spend with your peers.

Subtracting 20 from 50 leaves 30 hours per week to manage people. If on average you need to spend 2 hours per employee per week, then you can handle 15 employees (30 / 2 = 15). If you need to spend 4 hours per employee per week, then you can only handle 8 employees. And if, in the other extreme, you only need to spend 1/2 hour per employee per week, then you can handle 60 employees. In these examples, your span of control would be 15, 8 or 60.

The Factors that Influence the Required Manager Time Per Employee

Why would one manager need to spend 4 hours per week per employee (or even more) while another manager might only require 1/2 hour or less? Here are the factors that affect the number:

1. Is the work well-defined and well-understood?

Employees who are doing simple repetitive work typically require very little time from management. The employees know what they're doing, they do it the same way each time so there's very little reason for asking questions, and they know how they're measured. A manager only gets involved in the rare cases when a new employee is needed, when an employee has an occasional problem that requires management intervention, and for periodic goal and performance discussions.

On the other hand, consider the case where employees are doing a new type of work for the first time, the work requires a high degree of skill, there are many decisions to be made that require a manager, and training must be provided to the employees before they can perform useful work. Clearly this situation requires a lot more manager time per employee to ensure productive employees.

2. Have your employees been in their jobs a long time?

Experienced employees with long tenure have very little need for management (and you'll often hear them griping about this very subject). As long as the work and environment stay the same, a large number of these employees can be managed by just one person. But make sure that no change is imminent, because when change is required to this sort of organization, it's going to take even more management time than would normally be needed, since people who are set in their ways are usually reluctant to change.

3. Are job or working conditions changing?

One of the primary duties of a manager is to implement change. When things are changing, more time is required between the manager and their employees. Even if the work is constant, change can include an alteration in physical location or environment, a change in the number of people available to do the work (a decrease due to downsizing or an increase due to business growth), or a change in the pace of the work. If there is no change, then less manager time is required, and a higher span of control will work effectively. But during times of change this high span of control will interfere with successful change implementation; more manager-to-employee time is needed.

4. Do the employees help each other?

In many working environments it's reasonable and appropriate for experienced employees to help newer employees learn the job and its subtleties. Sometimes we call these experienced people "project leaders" or "team leaders;" other times it's just an informal mentoring relationship. When help from other employees is possible, it relieves the manager of some of these responsibilities, and it allows a higher span of control. But when the workers are physically separated (for example, support people based in different remote facilities), or when pay-for-performance motivation or union regulations reduce the likelihood of this sort of help, then the manager has to spend more time bringing each new employee up to speed.

5. Is there a high variation in the type of job working for the manager?

If all of your employees are doing the same thing (for example, a help desk organization), then it's possible to use group instruction to communicate the same message to many employees at once. If, on the other hand, each of your employees is doing a different type of job, then more individualized communication is required, and this requires a lower span of control.

6. Do you have a separate training and/or coaching organization?

Some companies split the traditional management responsibilities, keeping employee accountability with the employee's manager but moving responsibility for training and coaching elsewhere. This type of organization appears to allow a higher span of control, but in fact it usually just spreads the span of control across the different parts of the organization (see article "56. How to Organize IT" for more about matrix organizations). If you use this approach in your organization, you should be especially careful when your company looks to reduce costs. Separate training and coaching organizations are easy targets for cost-cutting (even more popular than managers with a low span of control), and when those organizations are eliminated, the managers will have to pick up the slack. Sure, having a separate training or coaching organization may allow a higher span of control *now*, but make sure the span of control goes down if that training or coaching organization is eliminated.

7. Is your organization on the trailing edge of new technology or on the leading edge?

If your organization is using leading edge technology, then the number of problems associated with the new technology will require more attention, more decisions, and more management time. On the other hand, if your organization lets other companies do the pioneering, and you use technology that's already been proven, then a higher span of control might be possible.

8. What is the ratio of technology investment per employee in this organization?

In a highly leveraged organization with many millions of dollars of technology investment being installed or supported by each IT employee, it makes sense to provide more management oversight to reduce the possibility of problems. In this type of organization, the span of control should be kept lower to provide that required oversight, and to make sure that your high technology investment is protected.

Conclusion

Span of control is one of those things that's easy to attack when a company is desperately searching for ways to reduce cost. There may be certain situations in IT organizations when management jobs should be eliminated, but it's dangerous to make a generalized attack on an IT management job just because it has a low span of control.

First, consider that many IT manager jobs require a large amount of time for individual contributor tasks that have nothing to do with managing people. Many IT managers are doers; in addition to their management role they also have to roll up their sleeves and work beside their employees on the front lines. This gives them less time to spend actually managing employees, and so their span of control number is deceptively low.

Second, look at what happens when you increase the span of control above a reasonable number for a given situation:

- Employees don't get the support and encouragement they need, morale drops, and employee turnover increases.
- Employee resources are wasted due to miscommunication and rework.
- Employees don't get adequate supervision, and they sometimes end up making decisions that aren't in the best interest of the company.

When evaluating the need for a change in the span of control, consider the answers to the eight questions in this article. You'll find some cases where a greater span of control is possible, but it's important to weigh the trade-offs. Will that increase in span of control result in ineffective use of new technology? Will it make your IT organization inflexible and unresponsive to change? Will you end up costing your company more money than you'll save? It's important to know the answers. And it's important to push back at consultants who want to increase your span of control if the evidence shows that it will hurt your company.

2026 Commentary

See "72. How Does Outsourcing Affect Span of Control?" for additional discussion on this issue.

71. 9 Ways to Reduce Business IT Expense

Originally published on AUGUST 12, 2008 in #CAREERS, #IT, #ALIGNMENT, #MANAGEMENT, #STRATEGY

Last month I got a lot of positive feedback about the timeliness of my newsletter topic, "70. The Right Span of Control Isn't a Number," so this month I've written another newsletter article dealing with today's problems. In our current economy there is a high likelihood that you've been asked to "tighten your belt" and cut back on business IT expenses. If you follow some of the recommendations in this article, then you'll be more successful in that endeavor.

Before we get started, note that the title of this article is "9 Ways to Reduce *Business* IT Expense—not "9 Ways to Reduce IT Expense." I want to make that distinction because one easy way to reduce IT expenses is to stop doing things that cost IT money. You can stop supporting certain products, stop taking certain kinds of Help Desk calls, or stop helping the business with certain kinds of problems, and that will indeed reduce IT expenses. However, if those things just move the expense from IT to the business then you haven't done anything good for the business at all. That's why I include the word "business" in the title. The goal is to reduce business IT expenses—that is to say you're trying to reduce overall expenses to the business for IT-related services—not just move expense from one cost center to another.

Now I realize that this altruistic view may not be shared by the accountant or manager who's bugging you to cut your budget. But for the good of your company, it's important that you focus on reducing expenses for the overall business—not just for your own department. Make it clear that your focus is bigger than your department, and this attitude will do wonderful things for your career down the road.

TANSTAAFL

OK, now that we've got the article title out of the way, let me talk about one other key to IT expense reduction. It's important for you to get the message across that with each and every IT expense reduction comes a lower level of service. This is equivalent to the old saying, "There ain't no such thing as a free lunch," and it may seem obvious, but it's critical that you actually demonstrate it in your budget reductions. You've got to say this clearly to whoever is after you to reduce expense, "*Any* reduction in IT expense will have an impact on the business. The impact may be worth the reduction, but that has to be a business decision—not an arbitrary decision by someone in IT. No IT expense reduction can come without a corresponding business price."

Here's why you have to make that message clear:

1. If the business sees a lower level of service as the consequence of IT budget cuts, then they'll know exactly what they're losing, and they'll be on your side when it comes time to eventually increase budgets and restore those lost services or restart those postponed projects.

2. If you ever—even once—cut your budget without showing a corresponding cut in services to your customer, then you'll have set a precedent and an expectation that you'll be able to do this again and again. You'll have established the reputation that "IT budgets loosely, so we can always get them to squeeze money out of their budget without incurring any business penalty," and you will have lost the trust of your business customers.

9 Strategies for Expense Reduction

Now it's time to reduce expense. To help make the strategies clear without getting us tied up in IT issues, let me first go through an example of how we would cut expenses for a hypothetical household if our family were forced into a cost-cutting situation. Then, after you see how these nine strategies work for families, we'll go through the same strategies and apply them to IT.

Most families look at nine basic strategies for expense reduction:

1. Postpone expenses which aren't required right now.
   This isn't the best time to get a new car, and we can probably avoid buying new clothes for a while.

2. Postpone all non-essential maintenance.
   The house can probably do without painting for another year, and maybe we can get our hair cut every 6 weeks instead of every month.

3. Substitute less expensive alternatives for those things that have substitutes.
   We can buy store-brand groceries, eat more beans instead of meat, and get cheaper shoes for the kids when they outgrow the ones they have. We'll do our own yard work and stop using a lawn service.

4. Reuse old stuff.
   Eight-year-old Sarah can use ten-year-old Sally's old coat. We can make lunches out of leftovers from dinner.

5. Consolidate needs where possible, and let one thing satisfy multiple needs.
   We can reduce gasoline costs if we just shop once a week. And the new coat for Sally (she outgrew her old one) can have a zip-out liner so she won't need two coats.

6. Go to a lower standard of living.
   We can turn down the thermostat in the winter and turn it up in

the summer to reduce energy costs. If necessary, we can move to a smaller, less-expensive apartment.

7. Stop "going out."
   We'll cut back on restaurants and vacations for the time being.

8. Move to less expensive ways to enjoy life.
   Instead of a week in Florida at Disney World, we'll do some day trips to local parks.

9. Look for additional sources of income.
   Maybe we can find weekend jobs, or maybe we can take some of those baskets Jean likes to make and sell them at a craft fair.

9 Strategies for Reducing Business Expense in IT

Now let's see some examples of applying each of these nine techniques to IT.

1. Postpone expenses which aren't required right now.
   You had planned to upgrade the server, but the users can live with their current system response time, so that upgrade can be postponed. And maybe you can do without that new software development tool a little while longer.

2. Postpone all non-essential maintenance.
   You have to keep doing operating system and security patches (those are essential for company security), but some of the other application software maintenance can probably be postponed. You can also delay some of those minor software enhancements the users have been wanting; that will let you focus on what's really important right now.

3. Substitute less expensive alternatives for those things that have substitutes.
   What first comes to mind is open systems software: both for operating systems as well as applications. This can definitely be a money-saver, but it's usually not a quick decision, you and your users will need training to make it work, and there's a migration involved. Still, it's something you ought to consider

at some point. Another possibility is offshoring, but again this requires a great deal of planning and thought—it's not a quick budget fix, and you pay the price for offshoring in increased rigidity of your processes. Surprisingly, you may find that insourcing—moving some of your outsourced work in-house—may help you reduce expenses in some areas where you have people who can do the work.

You'll probably get the most bang for your buck in this area by reducing the amount of "gold plating" that you've got going on in the requirements for your systems. Try to convince your business users that they can live with a less-expensive system that doesn't have all of the features they think they need.

4. Reuse old stuff.

   Every IT shop has lots of old equipment lying around. Try to reuse some of it. Refurbish old notebook PC's and use them for new employees. Use old desktop PCs for kiosks. And definitely reuse software licenses—it's way too easy to just buy a new license for each new employee, but it's a waste of money if you have old no-longer-used licenses for the same software. Stay legal, but there's no point in giving unnecessary money to software vendors.

5. Consolidate needs where possible, and let one thing satisfy multiple needs.

   "Virtualization" is the IT buzzword in this area. Use a single server for multiple purposes. Use blade servers for remote users over the Internet instead of giving each of them their own high-powered individual PC. Share offices between two people who each work half the time from home.

6. Go to a lower standard of living.

   Re-look at those up-time goals; maybe you could save money if the business tolerated slightly lower up-time. You might also be able to reduce expense by delivering a lower level of service on your help desk: maybe a next day response to problems instead of a few minutes or a few hours. If it makes sense, share offices, consolidate computer centers, and reduce the number of square feet required for your department. Give the space to someone who really needs it.

7. Stop "going out."
   Postpone unnecessary training. Try to find local training when you absolutely need it, and avoid conferences for a while to minimize travel costs. Many of those conferences can be accessed via the Internet now anyway. And many vendors can give you training over the Internet too.
8. Move to less expensive ways to enjoy life.
   You might have to cut back on bonuses, but maybe you can provide other perks instead. If you go to a staggered four-day workweek (each person works four days, but someone is always in the office) or let more people work from home, then your employees will save money on gas and waste less time commuting. That's worth a lot these days.
9. Look for additional sources of income.
   Take a hard look at the skills in your department. Is there any way that those skills could be the basis of a paid service done for your customers (that's the customers of your business—not just the IT customers)? This is probably the least likely of the 9 strategies, but give it at least a little bit of thought.

Conclusion

You can be creative about cutting costs, or you can turn the cost-cutting process into pure drudgery. It's your choice, and your attitude toward the problem will have a big impact on your career. People get rewarded when they have a good attitude and help the business. People who complain in this economy often get targeted as layoff candidates.

Whatever happens, try to be positive about the situation. Turn the process of cost-cutting—and of then living with the subsequent reduced budget—into a challenge for your employees. Help them be creative in coming up with ways to do more with less, and in reusing things that already exist. Cost-cutting is itself a marketable skill, and if you do it right then you and your employees will have fun while you all enhance your careers.

Think about each of the nine cost-cutting strategies, and figure out ways to use them in your own situation. It's much better to be the leader in cost cutting than to fight it and end up being the cost that someone else cuts.

2026 Commentary

This was written during a downturn in the economy, and cutbacks were being made in many businesses. Downturns go through cycles, so if this isn't relevant for you today, then I'm sure it will be important at some point.

72. How Does Outsourcing Affect Span of Control?

Originally published on SEPTEMBER 9, 2008 in #MANAGEMENT, #PROJECTS

Two months ago I did an article called "70. The Right Span of Control Isn't a Number," pointing out some of the variables that make an IT manager's span of control so unlikely to conform to any consultant-recommended norm. I received a number of positive emails about the article, but one email in particular asked about a different kind of span of control: the span of control for managing outsourced projects. Here's an excerpt from the email:

I enjoyed your article on "The Right Span of Control isn't a Number", and agreed with the factors you set forth when considering the right SOC [span of control] for one's organization. Large organizations, such as the one of which I'm a part, are increasingly outsourcing work to vendor partners. To illustrate this point, the organization I support has more onshore & offshore resources through vendor partners (more than 600) than internal employees. While our managers don't personally manage these resources, they do manage the overall work product, and much of their time now gets split between managing their internal employees (and that body of work), and the status of deliverables being accomplished by our vendors. In many cases, it now takes managers more time and coordination to effectively execute the work than in our former non-sourced model.

Do you have any thoughts on how to factor vendor partnerships into a span of control analysis? As more and more companies are leveraging these types of arrangements to reduce expenses and gain efficiency, I believe the IT industry will increasingly need to understand the impact of these arrangements on our managers.

Thank you in advance,

[name and company withheld by request],

an HR person who supports IT organizations for a large insurance company

This is an interesting question, and probably one that has relevance to a number of you in companies that are doing more and more outsourcing. The question is about extending the span of control concept to take into account what some people call "virtual staff"—the people who you manage but who don't specifically report to your organization. They're not employees, but they have similar requirements on your time. You "hire" them (select a vendor), "fire" them (switch vendors if performance isn't up to your expectations), do performance reviews (periodic evaluations of the vendor's performance), manage their performance, and generally have to spend time keeping them up to date on changing business needs. They take all the time of an employee—sometimes more—and yet they aren't usually counted in any span of control calculation.

If we're going to accurately measure the workload of a manager, the calculation of a *true* span of control should count each outsourced responsibility in addition to the normal span of control count of employees. And all of the factors discussed in my July article (how well-defined the work is, tenure in the job, degree of change, etc.) apply to these virtual employees as well. But it doesn't stop there, because there are additional factors that affect the amount of time you have to spend managing an outsourced project or service.

The Factors that Influence the Required Manager Time Per Outsourced Project

The remainder of this article lists some of the additional factors that you need to consider when looking at this span of control for outsourcing.

1. Are the outsourced projects complex?

More complex projects require more management time because there are more decisions, more things that can go wrong, and a higher probability for misinterpretation of requirements. On the other hand, less complex projects and well-defined services like off-site data storage require less management time.

2. How clear are the project deliverables of the outsourced projects?

Can the deliverables be measured objectively, or are they more vague and undefined? The more objective the measurement, the easier the project will be to manage, and the less manager time it will require. But vague project deliverables will require constant clarification and discussion, and will often lead to rework and late projects.

3. How independent are the outsourced projects from each other?

To what degree is the project result from one project a factor in the success of another project that's outsourced to a different vendor? If you're managing a combination of projects that interrelate, then your time is going to be eaten up resolving conflicts between the two vendors. And if the second vendor is being managed by someone other than you, then add even more time for trying to resolve disputes across organizations in your company.

4. How critical are the project due dates of the outsourced projects?

If some part of your business—or some key financial objective—lives and dies by having the project done by a certain date, then count on spending lots of extra time making sure the vendor stays on track.

5. How risky are the outsourced projects due to things like new technology, new processes, lots of unknowns, etc.?

If you're lucky, your contract will require the vendor to absorb all of the risk. But the more risk that's passed back to you, the more time you'll need to spend dealing with the consequences of a risky project. And if the risk impacts the project due date, then see #4.

6. What kind of track record does the project resource have?

Have they done successful projects for you before? Or do they need lots of help to be successful? Vendors with good track records require less manager time. And new vendors always require more manager time when you're first developing a relationship, setting the ground rules, and getting to know each other's style of working.

7. Does the vendor have a single contact person for the project?

Managers who deal through a single "lead" or vendor "project manager" require less time than managers who have to communicate with multiple vendor resources for the project. But there's a downside to the single lead. You had better be sure that your single lead person truly understands your requirements. With multiple people to talk to, you're more likely to discover any miscommunication because the people are likely to talk to each other and discover any differences in interpretation. With one contact point at the vendor, everything is dependent on that lead project person communicating correctly with everyone else in the vendor organization.

If you go to the other extreme, and your vendor just supplies contract resource for you to manage (sometimes called staff augmentation or contract labor), then you can pretty much count on every individual contract person requiring the same amount of manager resource as would an employee. You have to give them individual direction, you have to give them individual performance feedback, and you may even have to deal with their personal problems. Your company gains because you don't have a long-term commitment to keep the person around, but in almost every other way your manager requirements are the same as for a full-time employee.

8. How stable are the project teams assigned by the vendor to the outsourced project?

Are people constantly changing roles at the vendor location? Is there high turnover? In theory, it's not your problem (unless your vendor contact leaves), but in reality it's difficult for the vendor to keep things moving in the right direction with constantly changing staff. And ultimately you're the one who will have to deal with the resulting problems.

9. Are there any communication issues with the outsourced project team?

Are there any language barriers between you and the vendor contact? Between the vendor contact and the rest of the project team? Communication of requirements between business people and IT people is difficult enough without having to deal with translating those requirements from one language to another. And remember that American English isn't the same as English taught anywhere else (nor is any native language the same as that language taught to students elsewhere). You'll have to translate idioms and reconcile alternate definitions and interpretations of the same word. This will require more manager time.

10. How familiar is the project team with the nature of the work done by the project recipient?

Project requirements tend to omit things that project recipients take for granted, and sometimes it's those hidden assumptions that cripple a project result. I've always had my IT employees spend some time doing the user's job or at least watching the user perform the job. It makes a huge difference in IT's understanding and appreciation of why the user might want a particular kind of system. Getting an outsourced team exposed to actual user work is sometimes tough to do, and even tougher if the team lives on a different continent. But if the team doesn't get that exposure, then count on the manager being required to solve problems that are caused by that lack of understanding.

11. Is the outsourced project being done under a fixed price bid or is it "time and materials"?

With a fixed price bid, vendors try much harder to totally clarify requirements before the bid is submitted. They obviously don't want to bid on a project with mushy requirements and then get held to the original fixed price when the requirements change. So *if* you agree with the requirements as defined, and *if* the requirements don't change, then a outsourced project with a fixed price should be easier to manage, and should require less manager time. But—and this is a huge issue—if the requirements *do* in fact change after the fixed price bid is accepted, then this type of vendor arrangement is likely to require *more* manager time than a time-and-materials arrangement. That's because the change process for a fixed price vendor contract is usually very cumbersome and bureaucratic, and every change requires customer and vendor sign-off with a corresponding renegotiation of price.

And finally,

12. Is the outsourced project succeeding or failing?

Failing projects absorb management resource like a sponge. It's the manager's responsibility when a project fails, so the manager has to do everything possible to keep a project from failing, or at least to minimize the amount of failure (e.g., only one week late instead of a year late).

Conclusion

Some companies have the naive view that if you outsource your work then you can get rid of managers. While it's possible that a few lower-level manager jobs may be moved to the vendor, in most cases the demand on a higher-level manager's time will go *up* when work is outsourced—not down. Outsourcing something doesn't just move the work from your organization to another—it creates a huge gap in the middle of your business process that has to be bridged between two companies by extraordinary efforts from the manager responsible for the outsourcing. This effort is often neglected when considering the manager's ideal span of control, and there are a lot of overloaded managers as a result. We need to remember that managing an outsourced project or service takes *more* time than managing an employee—not less.

2026 Commentary

In addition to the issues mentioned in the article, if any part of the outsourced labor is working in a politically unstable country, then the complexity of the outsourcing goes up. Provision has to be made for guaranteeing the safety of the people and their work. Even if the vendor claims that they will deal with it, there's significant risk of losing work that has already been performed. The responsible manager has to put processes in place to frequently save work and its associated documentation, and this adds to the workload of that manager.

73. No Programmer Left Behind?

Originally published on OCTOBER 14, 2008 in #CAREERS, #IT, #MANAGEMENT

1. Tests

I've always been pretty good at taking tests; I guess you could say I have a gift for "quizmanship." But on a 1-10 scale, I would probably rate myself a 6 or 7 on the geek-o-meter. I'm not up there with some of the kids I once went to school with, who, when the teacher lapsed into a personal recollection, would immediately stick up their hand and ask whether this would be on the mid-term.

The nice thing about taking tests is that you're dealing with a fixed environment in which there are hard and fast rules. You know the subject area to be covered by the test, and you have a general feel for the types of questions that will be asked. Reality is different: there are no rules, and you're just as likely to need some skill that you've long ago forgotten as you are to be asked to take advantage of some recently acquired knowledge

Tests don't measure reality—they measure a miniature subset of reality projected into a multiple-choice, short-answer, true-or-false environment. In our quest to be objective about the measurement of learning, we've zeroed in on the closest approximation to objectivity that our teachers can find. But it's a poor reflection of the reality that we all have to live with after we graduate.

2. Creativity

A couple of decades ago I taught a class to help people be more creative and to help teams work better together on creative solutions. The class included some exercises, and two of those exercises are applicable to my topic today. In one exercise we asked each person in the class to write down their best recollection of which letters are associated with each number on a phone keypad. This was in the days when phones were for making calls—not sending text or email messages—and so the exercise was difficult for most people.

After each person had recorded their own remembrance, we broke the group down into small teams, and each team was given the goal of developing a single "best answer." Not surprisingly, when we compared the individual answers to the team answers, the teams generally were more likely to be correct than the average individual score. The team discussion helped people fill in the gaps in their knowledge. One person might remember one part of the keypad but be foggy on the rest, but the team was able to pull individual memories together into a perfect list of letters and numbers.

The second exercise was a little more complex. The exercise described a scenario in which you are part of small team of survivors from a plane that crashes in the desert 100 miles from the nearest civilization. The plane burns shortly after the crash, and you were all lucky to escape with just a few potentially useful items, including a compass, overcoats for each survivor, a cosmetic mirror, a gun, and a map of the area. Your task in the exercise is to rank the items in order of their usefulness to your situation. Again, we asked the participants to first individually rank the items, and then to get together into teams and agree on a team rank for the items.

The second exercise had a different result than the first exercise. In the first exercise, the best team score equaled the best individual score—this was to be expected since the problem was easier and a perfect score was more likely. But in the second exercise the team scores were generally better than the worst individual score but not as good as the best individual score. In other words, the teams gave an adequate performance but never as good as the most outstanding individual on the team.

Why, you might ask, would this happen? In the second exercise the best ranking of items depended on survival skills that are somewhat unconventional. For example, of the items I mentioned, the correct highest ranking item in terms of usefulness is the cosmetic mirror (for signaling a plane), and the second highest ranked item is the overcoat (spread over a hole in the sand to collect moisture from the morning dew)! But these results seem so outrageous that it's difficult for an expert to convince the other members of the team, and so invariably the team achieves a compromise result that isn't as good as the expert ranking.

3. Putting these two observations together

In a recent issue of ComputerWorld, Judy Estrin, author of the book, *Closing the Innovation Gap*, made a comment with which I strongly agree. She said,

"It's much easier to stifle innovation than it is to allow it to grow, and a natural inclination in the country and in the business world is to put in rigid processes, policies and metrics—things like Six Sigma or No Child Left Behind. Some of those things actually discourage the behaviors you need for innovation."

I've written in the past about how bureaucracy cripples our ability to react to change. Here's an excerpt from my article, "23. Julia Roberts, Training Wheels, and Bureaucracy":

When the word "bureaucracy" was first invented, it was supposed to be a good thing. Corporations were trying to cope with growth in situations where it was no longer possible to do enough hands-on management to ensure that the company implemented processes consistently across the enterprise. Bureaucracy was a standard set of processes and procedures which created that consistency even when implemented with different people, in different locations, and in different cultures. But the strength of bureaucracy—consistency—is also its biggest weakness; a bureaucratic environment can't cope with changing conditions or with situations for which the rules weren't designed.

Take another look at Estrin's comment. Do *you* think of Six Sigma as bureaucracy? Many people don't, and yet it fits the definition perfectly: Six Sigma accomplishes its goal of consistent high quality by using standard processes that minimize the deviation in the way things are built. The word bureaucracy has a negative connotation and the phrase Six Sigma generally has a positive connotation, but both have the same result: a rigid process.

What about No Child Left Behind? No Child Left Behind is a US government mandate to "improve" education by subjecting each child to a series of standardized tests. But the result of the law has been a change in the US education system to motivate teachers to "teach to the test"—to emphasize quizmanship rather than actual learning.

In my opinion, Estrin is absolutely correct in her criticism of both Six Sigma and No Child Left Behind. Both programs make overly simplistic assumptions about the real world. Six Sigma assumes that the right thing to do is to fine-tune your existing process or product to the point where there's no deviation at all (sometimes valid, but less so in a constantly changing world). And No Child Left Behind assumes that a child who can do well on tests is best prepared for the real world.

Conclusion: The Cost of Oversimplification

How has the oversimplication philosophy of Six Sigma and No Child Left Behind affected *your own* IT processes? Are you measuring your software developers using crazy metrics like lines of code per week or—worse—by using "standardized tests" and certifications? Are you forcing your developers to use standardized processes that are long on documentation but short on creativity?

Think about that desert plane crash survival exercise. When you make heavy use of standardized processes and unrealistic measurement, you'll get a result that is better than the worst individual but not nearly as good as the best that any one individual could offer. But what's even more upsetting is that when your truly creative people are faced with overly rigid processes, they will get impatient and leave. You'll end up with a bunch of mediocre and sub-par developers who somehow manage to collectively turn in an average performance.

You'll end up with the IT equivalent of No Child Left Behind: everyone will be able to pass the test, but you'll kill creativity, and you'll have—at best—an *average* performing IT group. No programmer will be "left behind," but neither will you have the outstanding developers who differentiate your business from the competition. Is that what you want? Is that what's important to the business? I don't think so.

2026 Commentary

There are three references in this article that may be unfamiliar to current readers. First, there's the group exercise of figuring out the letters associated with each number key on a phone. Those letters would be familiar to users of a Touch-Tone keypad in 2008, but maybe not so familiar now, even though the virtual keypad of an iPhone still shows the letters.

The second and third references are to Six Sigma and No Child Left Behind. Both of these programs are still around, but I haven't seen them in the recent news.

Whether you understand the references or not, the point of the article still holds: We try to use "objective" measures for things, but often end up biasing our results toward the mediocre, while discouraging and possibly eliminating the creative superstars we most need.

74. Turn Left at the Last Traffic Light

Originally published on NOVEMBER 11, 2008 in #MANAGEMENT, #STRATEGY

It happened a while ago—before GPS devices, and before Google Maps or Mapquest. I was trying to drive to an address in a small town and I did what reasonable people did back then: I asked someone at a gas station for directions. One detail from the directions stands out in my memory: the gas station attendant told me to "turn left at the last traffic light." I was so astonished that it took me a few seconds just to react. Then I asked him, "How do I know which traffic light is the *last* one?"

Maybe you've never gotten directions like that, but I'll bet you've seen or heard similar statements. Here are a few of my favorites:

Crazy Statements

- Buy at the bottom.

Of course we'll only know we were at the bottom after the market has climbed for a while, but the advice sounds good.

- Cook it until it's done.

This is actually meaningful to some people, but not to someone who doesn't know how to recognize the appearance of food that is fully cooked. I fall into the latter category.

- Keep testing until you find the last bug.

I guess in theory there is a "last bug," but I've never found it. There always seems to be one more hiding out there somewhere.

- Just make the new system do the same things the old one did.

It's not obvious when you first hear it, but this kind of requirement includes three parts:

1. The documented things that the old system does
2. The things the old system does that aren't in the documentation (some of them probably just work that way by accident—they weren't designed that way at all)
3. The things the old system *doesn't* do that are in the documentation

Good luck finding them all. And you'll need even more luck to have the new system recreate them.

- Give me a disaster recovery plan that anticipates every possible disaster.

I guess that should include the possibility that I go crazy and destroy all of the backup hardware.

- The IT Strategy should align with the business strategy (but I won't tell you the business strategy).

A passive-aggressive classic. The "Catch-22" of IT.

What do all of these statements have in common?

They all sound meaningful at first glance, but they provide advice that is useless to the listener without additional information—information which isn't being provided, or which maybe doesn't even exist.

What should you do when you're given one of these crazy statements?

Ah, that's the important question! I'll answer that for each of the statements I've included so far.

How to Answer Crazy Statements

- Turn left at the last traffic light.

Ask how you'll recognize the "last" traffic light. Is there a street name? A landmark? Does it have a sign that says, "last traffic light"?

- Buy at the bottom.

This one's a matter of risk. Statistically, you can assume that there's been a short-term bottom when the index or stock you're following gains a certain amount. The more the stock gains, and the longer it stays above the bottom, the more likely it is that a worthwhile bottom was reached, and that it's time to buy. It all comes down to how much of a premium you want to pay for risk minimization. Wait a day or a week and the stock may go below its last bottom. Wait a year and the bottom is more assured, but you'll have to pay a higher price for the stock.

- Cook it until it's done.

Ask for hints: About how long will that take? What should I look for? Bubbles around the edge? Browning? Smoke coming out of the oven?

- Keep testing until you find the last bug.

I've heard people suggest that you base this on time: that you test until you haven't found a bug within a certain amount of time. But time isn't a good predictor if you're just running the same tests over and over. I once heard a better solution (I think it came from Jerry Weinberg) that suggests that you take a piece of software that is being tested and you artificially create a certain number of bugs in the software—in effect you "seed" the software with bugs. Then you see how many of the seeded bugs are discovered during testing, and how long it takes. This should give you some idea of how many real bugs still lurk out there waiting to be discovered. In most cases—assuming the people who plant your seed bugs try to make them subtle—you'll probably find it difficult to discover all of the seeded bugs. (Remember to take the seeded bugs out of the software—and then retest—before you release the software.)

- Just make the new system do the same things the old one did.

The best way to deal with this requirement is to refuse to accept it. It is virtually impossible to exactly duplicate the way a piece of software works without developing it using all of the same tools. Since in most cases you're replacing the software because it uses obsolete tools or technologies, you need to go back to square one when redefining the requirements. Work with the system users to extract the essence of what the old software does. Focus on the process the users go through in using the old software, and try to find ways to improve the process. Then develop new software that optimizes the performance of the users with the new improved process. If the user process is improved, then no one will care whether the new software works exactly the same way as the old software.

- Give me a disaster recovery plan that anticipates every possible disaster.

Disaster recovery is a risk/reward game. It's OK to brainstorm a set of crazy risks; in fact, it's a good idea to try to list anything that might possibly go wrong (I say "try" because you'll never anticipate everything). Then try to categorize all of the risks on your list into scenarios where the disaster recovery can be done the same way. You'll eventually arrive at a short list of scenarios that you need to plan for. Evaluate the cost of providing for each scenario and compare it to the probability and cost of not having any disaster recovery plan available. Then work with the business to make decisions about how much to spend on preparing for each scenario.

- The IT Strategy should align with the business strategy (but I won't tell you the business strategy).

This happens more often than you'd think. I tell IT executives that when they can't get business executives to tell them the business strategy (or when the IT executives can't be involved in the creation of the business strategy in the first place), then the IT executives should write down what they *think* the business strategy is, and what the corresponding IT strategy should be. Then they should go back to the business executives and give them this speech:

"I haven't been able to get a clear idea of what our company's business strategy is, so I've put together my own assumptions. Here they are (show them the key points of your version of the business strategy), and here's what I think IT should do to support this business strategy."

What you'll find is that most business executives who are hesitant to reveal what they consider "privileged" information are much more likely to correct your own assumptions. So through their corrections you'll end up getting the business strategy information you need.

Conclusion

If you've been watching any of the political debates lately, you know that although most debates start with a question, a good politician knows how to use the question as a springboard to answer another question that better suits their agenda. When you're confronted with one of these crazy statements, treat it the same way. Use it as a starting point, and then ask clarifying questions to get at what the speaker really means. This is an obvious approach for many people, and yet it's not obvious to most IT people (see my July, 2006 newsletter article "44. 5 Reasons Why IT People Love Lists" for a little bit about why IT people think that every question needs to be answered). When confronted with a crazy statement, you need to "uncrazy" it before it can become useful.

2026 Commentary

Of all of the items covered in this article, probably the most important is the item on IT understanding business strategy. I've used the approach I talk about—writing down my assumptions and then having the business executives revise them—and it definitely works. You see the same thing now in social media. If you want the answer to a question on social media, the best way to get the answer is to post something that has the wrong answer. People will immediately jump in and correct you.

75. IT Alignment is Simple, Part 1

Originally published on DECEMBER 9, 2008 in #IT, #ALIGNMENT, #LEADERSHIP, #STRATEGY

Results of the latest SIM CIO Survey were announced at SIMposium 2008. Once again, "IT and Business Alignment" was number 1 on the list of top IT management concerns (it's been number 1 for a lot of years). It's amazing to me that CIOs haven't yet learned how to solve the IT alignment problem, so in this article I'll explain what IT alignment is all about, and in the next article I'll explain in more detail how to achieve business/IT alignment.

Unaligned IT is like driving blindfolded

First, I'll use a metaphor to describe the problem. Suppose you're a CIO driving a car, with your CEO in the front passenger seat. But in this example you're blindfolded behind the wheel and so the CEO has to give you extremely detailed instructions on how to drive: "Turn just a little bit left. Prepare for a right turn. OK, turn NOW. Slow down just a little bit. A little bit more. Now speed up."

Do you get the picture? How frustrating and nerve-wracking would this be for you as the driver? And how would it be for the CEO? If you were CEO wouldn't you be incredibly frustrated that you have to spend so much time telling the CIO what to do? Why can't the CIO just figure it out for him/herself?

Now take the blindfold off, and do some advance planning. Suppose you—the CIO—get together with the CEO prior to your trip and make plans for your travel. Together you pick a destination, and you jointly decide on criteria for picking a route: fastest, most scenic, most economical, whatever is important to the two of you. Then you—the CIO driving expert—recommend a route that will get the two of you to the destination in the best way possible, optimizing those selection criteria on which you just agreed. You discuss the recommended route with the CEO, make changes as necessary, and then start on your trip. You're in the driver's seat, but you're no longer blindfolded.

Along the way things will happen. Your destination might change. Your route selection criteria might change. You may have to revise your route to adjust to unexpected weather, bad roads or traffic. But you make all of these adjustment decisions jointly with the CEO, and so you both agree that you're doing the best that can be done.

This second approach—advance planning with no blindfold—is an example of alignment, but the first approach—driving blindfolded—is an example of what most unaligned IT organizations do instead. IT is driving blindfolded because most CIOs don't have full visibility on where the CEO and other senior executives are trying to take the business. Instead, the CIOs are given short-term tactical statements of direction (e.g., implement this system, expand this data field, improve Help Desk service, improve transaction performance) rather than letting the CIOs drive IT for themselves with full knowledge of the destination, an agreed upon route, and the ability to respond to local driving conditions along the way.

Just like the CEO passenger in our blindfolded CIO driver example, the CEO and senior business executives are highly dissatisfied with an unaligned IT organization. They're frustrated with having to constantly give detailed directions, they're dissatisfied with the slow rate of change in IT, and they dislike the high "accident" rate (failed projects) that comes from blind driving. Everyone—in both IT and the business—hates operating under the blind driver model of IT. But in most organizations no one has ever taken the steps required to change the approach.

What is Business/IT Alignment?

In my experience, you achieve alignment between business and IT when all seven of the following criteria are met:

1. The Business strategy is understood by the business and by IT.
2. The IT strategy is understood by IT and by the business.
3. The two strategies agree, and don't contradict each other.
4. The IT strategy supports the Business strategy.
5. The Business strategy supports the IT strategy.
6. Business people and IT people trust each other to follow the strategies they've agreed on.
7. Business people and IT people agree that the business and IT are aligned.

Let's go through them one by one:

1. The Business strategy is understood by the business and by IT.

This is one of the problem areas most cited by CIOs when they're asked about alignment. Many CIOs—particularly CIOs who don't report to the CEO—claim that they aren't part of the executive "inner circle" and that they frankly don't know where the business is going. Although I understand their concerns, I don't see membership in the inner circle as a necessary prerequisite for IT alignment. Often it's enough to have a "pretty good feel" for where the business is going—a *total* understanding of the business strategy isn't required. In fact, many CIOs would be surprised to find out that even the CEO doesn't have a total understanding of the business strategy, since some of business strategy is "played by ear" and adapted to business conditions as they change and to business opportunities as they appear.

The most important aspect of this criterion is the communication of business priorities. The CIO must understand what's important to the business, and what's not as important. Alignment can be achieved only if IT is able to focus on the things that are most important to the business.

2. The IT strategy is understood by IT and by the business.

This criterion, on the other hand, is probably the *biggest* obstacle to IT alignment. Many IT organizations don't have an IT strategy at all, and when they do, it's often a document that's used for communication *downward* to the rest of the IT organization. Seldom is the IT strategy put together in such a way that it can be clearly communicated *upward* to the business using business language. Yet upward communication of the IT strategy is even more important than downward communication. If business executives don't have a good feel for IT's overall direction, then they won't trust IT (see #6), and they'll try to micromanage the IT organization. This will bring you right back to the driving blind metaphor, and your IT organization will go into "order taker mode" where IT doesn't have a strategy—it just does whatever it's told.

3. The two strategies agree, and don't contradict each other.

Once #1 and #2 are true (a pretty big step), then it's usually fairly easy for business and IT leaders to discuss differences in strategic intent. Without those business and IT strategies, it's impossible to carry on a rational dialog, since you're trying to compare two things that are both nebulous and unwritten. But once you've got a business strategy that IT can understand and an IT strategy that the business can understand, it's pretty easy to see if there are contradictions between the two documents.

When contradictions are detected, you can figure out why. Maybe a business assumption changed and IT didn't know, or maybe an IT assumption about a certain system environment is causing problems in supporting a new direction for the business. Whatever the reasons for contradictions, the important thing is to revise the business strategy and/or IT strategy to eliminate the contradictions.

4. The IT strategy supports the Business strategy.

Having resolved #3, you have now eliminated contradictions between the business and IT strategies. But that doesn't mean that the two strategies support each other. First look at the way the IT strategy supports the business strategy. Are there elements of the business strategy that cannot be performed given the current strategy for IT? (e.g., a move toward Internet selling that can't be supported by your legacy systems) Are there elements of the business strategy that could be performed *better* with a shift in the IT strategy? Are there elements of the business strategy that have no support at all in the IT strategy? (e.g., the business strategy says acquire internationally, but the IT strategy doesn't address the issues of multinational, multi-language, multi-cultural, multi-currency systems) Modify the IT strategy as required to support the Business strategy.

5. The Business strategy supports the IT strategy.

Now that #4 has been accomplished, look at the two strategies from the other direction. Are there elements of the IT strategy that cannot be performed given the current business strategy? (e.g., business strategy calls for IT spending cuts but the business strategy also requires increased IT capabilities) Are there elements of the IT strategy that could be performed *better* with a shift in the business strategy? (e.g., business strategy calls for separate IT organizations for each business division, but IT project efforts for certain key business objectives will be duplicated in each division) Are there elements of the IT strategy that have no support at all in the business strategy? (e.g., the IT strategy calls for replacement of inadequate infrastructure but this investment hasn't been factored into the business strategy) Negotiate modifications to the business strategy as required to support the IT strategy, or revise the IT strategy accordingly.

6. Business people and IT people trust each other to follow the strategies they've agreed on.

I've listed this as #6 but in fact trust is critical to every stage of the alignment process. As I've said in my book, *Boiling the IT Frog*, if your company doesn't have a mutually trusting relationship with its IT organization, then IT won't be successful, no matter how good you are at everything else. Read more about how to build trust in that book. There's a whole chapter on the subject.

7. Business people and IT people agree that the business and IT are aligned.

This is the last step in achieving alignment. After you've done everything else and met all six of the other criteria, it's important for business and IT people to specifically recognize the fact that they're aligned. There are two reasons: (1) It will drive a stake into the ground to recognize that on a given date you actually had alignment, and (2) It will help motivate your organizations to put processes in place to *keep* business and IT in alignment.

You need to recognize that alignment is a transitory thing; it's not enough to achieve alignment once—you have to keep working at maintaining that alignment. You can do this by periodically going through an analysis of the seven criteria, and by putting change management processes in place to deal with changes in business and IT strategy as they occur.

Conclusion

IT alignment is simple, but it's not easy. Good strategies are short and to the point—we're not talking about 100 page documents here. Unfortunately, many strategies are long documents that are just brainstormed lists of things we'd like to do someday. It's hard to take something like strategy and make it simple, but if your strategy—for *either* IT or business—is too long or too complex, then it won't be understandable, and alignment won't be achievable. The real effort in alignment comes from focus: focusing on the things that are most important for the business and on the things that are most important for IT. By definition, focus means spending *less* time and money on other areas that aren't in your focus. And that's the key to alignment: spending most of your business time and most of your IT time on the same things.

In my next article, I'll get more specific about some steps you can take to make an IT strategy understandable to the business, and I'll talk about how to deal with common problems in achieving alignment.

2026 Commentary

Of all of the articles I've written, this subject is probably the most important in making business IT successful. And as the conclusion says, it's simple, but it's not easy. It takes effort from both business and IT executives.

And while I've obviously written this about IT, the process applies equally well to any other business organization. It's important for each business organization to have a strategy that's aligned with the overall business direction.

76. IT Alignment is Simple, Part 2

Originally published on JANUARY 13, 2009 in #IT, #ALIGNMENT, #STRATEGY

In the previous article I talked about business/IT alignment, explained the problem, and gave you the 7 criteria for achieving IT alignment. In this article I'll give you some tips for better and easier IT alignment.

I'll start with a five-step process to change your IT budget to better support alignment. First, let's clean up maintenance and infrastructure so they don't distract us from our project budget.

Step 1. Follow the advice in my book, "Boiling the IT Frog," and organize your IT budget into separate categories for infrastructure, projects and maintenance.

Over time the percentage of an IT budget devoted to infrastructure and maintenance tends to rise because of the increased ongoing cost of all of the projects you've completed. So breaking out your budget into these three categories makes your budget more transparent and understandable to business executives. When they see that the percentage of money available for new projects is declining, they'll better understand why you have trouble delivering more projects as time goes on. And they'll be more receptive to increasing your total IT budget to fund additional projects.

Step 2. Now focus your maintenance budget on software maintenance.

Put hardware maintenance under infrastructure where it belongs. Also move to infrastructure the cost of certain software contracts that are required to make the infrastructure work. This includes server operating systems, database and network software licenses that are required by existing systems, and the cost of user software licenses like Microsoft Office that are installed on user PCs.

Step 3. Make your maintenance budget even clearer by taking any major software maintenance projects out of the maintenance budget category and putting them in the project category instead.

Maintenance projects are things like increasing the size of an order number, beefing up system security, or interfacing a new system to a bunch of old systems. These could all legitimately be considered maintenance, but they're also projects that the business can appreciate for their business value. Accordingly, put these maintenance projects into your project budget where they belong, and reserve your maintenance budget for smaller things that are tougher to link to specific business benefits. Keep your maintenance budget category as small as you can.

Step 4. Tie your infrastructure budget to user-related variables.

Business users understand that increased volumes require increased cost, so tying various parts of your infrastructure budget to user volumes (e.g., transaction counts, network traffic, Help Desk calls, number of supported PCs) will transfer the responsibility for volume increases back to the user. For example, if the business doesn't like spending more each year for increased Internet usage, then the business needs to work together with IT to find a way to control the Internet volumes. Or if the business doesn't like your increased Help Desk cost, then the business can trade off lower support levels, or can discontinue support for certain products.

Make sure, however, that you use legitimate variables that are directly related to infrastructure cost. And anticipate the possibility of a *decline* in a volume variable. Will it be possible for you to cut costs in that infrastructure area if volumes decline?

Now for our last step in the five-step process, we'll focus on aligning projects with the business.

Step 5. Connect each IT project to a specific strategic business objective that has a business deliverable and a business benefit.

That way they're not IT projects—they're projects for the business with some of the work done in IT. This is absolutely key to alignment; you cannot have alignment if business people believe that the IT organization does projects for its own sake.

The Advantages of this Five-Step Process

It's a good idea to go through this five-step process even if you don't go further and attempt IT alignment. First off, revamping your budget this way takes away most of the pain of budget cuts. When someone in the business wants to cut your budget, just go back to the business and ask what infrastructure service they're willing to sacrifice, what systems they would like to discontinue (to reduce maintenance and infrastructure costs), and/or what projects they're willing to cut or postpone. Since every part of your newly revamped IT budget has a specific business benefit, it becomes a business decision to choose which of the business benefits to defer or live without. IT budget cutting is no longer just an IT problem—it's a joint business/IT problem.

On the other hand, if someone wants to add a new project, then it's up to the business to find the funding to pay for the IT cost of the project and for the increased IT infrastructure and maintenance cost that the project requires. And if the business is actually going to enjoy a benefit from the project, then that's only fair.

Business/IT alignment means that each and every part of IT activity is coupled to a specific part of the business (or coupled to multiple parts of the business in certain cases. But watch out for finger-pointing when you do this). Where the business goes, IT goes, and vice versa. Linkage leads to alignment.

Common Questions & Answers about Business/IT Alignment

1. My IT organization is a small part of a much larger corporate IT organization. How do I get business/IT alignment when I don't have control over the larger corporate IT organization?

Answer: Just as a large system rollout is more successful when it's done in a series of phases, you'll find that getting business/IT alignment can also be done in phases, and often works better that way. Focus on your own IT organization, and do your best to get alignment for the areas of IT that are under your control. The nice thing about alignment is that you can get the benefits even if you just get alignment for a subset of IT. And when your local business partners start bragging about how good their IT is, now that it's aligned (I know this from experience—it's happened to my own business partners), you'll find that the corporate IT group will become very interested in learning what you're doing right, and they will want to do the same thing at a corporate level.

2. The business people won't tell me their business strategy. How can I align with the business if I don't know what their strategy is?

Answer: First of all, you probably know more about the business strategy than you think. Most senior IT people have a better view of the overall business than someone in any other business organization. That's because IT interacts with so many different parts of the business, and it gets involved in the key processes of each organization.

So if the business people won't tell you their strategy then put together a summary of what you think their strategy is and what the corresponding IT strategy should be. Then go back to the business executives and give them this speech:

I haven't been able to get a clear idea of what our company's business strategy is, so I've put together my own assumptions. Here they are (show them the key points of your version of the business strategy), and here's what I think IT should do to support this business strategy.

What you'll find is that most business executives who are hesitant to reveal what they consider "privileged" information are much more likely to correct your assumptions. So through their corrections you'll end up getting the business strategy information you need.

3. I understand what you're saying about linking projects to business objectives, but my business partners insist that I cut my IT budget without sacrificing any of my planned project deliverables. What can I do?

Answer: Your problem is that your business partners either (a) don't trust you, (b) don't think your IT organization is productive enough, and/or (c) have too much of a sense of magic about IT (see Chapter 1 in my book, *Boiling the IT Frog*) and therefore expect the impossible.

First you need to decide whether you're worthy of their trust. Have you been honest with them about anticipating project successes and failures, or have you been giving them surprises? Have you been accurate in your project estimates, or when estimates were off, have you explained how changes to requirements caused the error (assuming that's the real cause)? Do your business partners think you're sandbagging on estimates (deliberately estimating extremely high to give yourself more flexibility)? Until you develop a trusting relationship with your business partners, they're going to keep squeezing you to do more and more with less and less.

Second, take a hard look at the productivity (deliverables/dollar) of your IT organization. Are there areas in which you could make order of magnitude improvements (10x or more) in your budget by outsourcing, by adopting new techniques, or by eliminating redundancy? Maybe there *are* ways to cut your budget without sacrificing deliverables.

Last, if you have the trust of your business partners and you've done everything you can to improve productivity, then you're dealing with a magic perception: your business partners are expecting the impossible. See my *Boiling the IT Frog* book for suggestions on how to reduce magic in your IT. There are also articles in this book on that topic.

4. I'm in a small company with a very small IT organization. Is IT alignment relevant to me?

Answer: Yes, IT alignment is relevant and important for *all* IT organizations, no matter how big or small they are. In fact, I can make the case that IT alignment is even *more* important for small IT organizations. Small IT departments have no resources to spare, so it's critical that your limited resources are focused on the things that are the most important to the business. That's what IT alignment is all about: aligning your IT resources with what the business needs.

5. We have an IT strategy, and we have a business strategy, but we still can't achieve alignment. What are we doing wrong?

Answer: There are a lot of possibilities. First, take a hard look at the business and IT strategies. Are they short (usually less than 10 pages) and do they convey a real sense of focus (e.g., "we're going to do this and this, but not this and this")? Many strategy documents are worded in "consultant speak" because they're written by consultants (or with the help of people who think like consultants) who are more focused on creating something that sounds good than on creating a strategy that can be useful. Consultant speak is a strange language that looks like eloquent English but in fact strings together a bunch of business buzzwords without having any real meaning.

One way to do a quick evaluation of a strategy is to go through the strategy and try to create a hypothetical alternate strategy that does the exact opposite of what your strategy says. For example, if your strategy says we're going to "take advantage of competitive opportunities as they arise" (typical consultant speak) then the opposite would be what? To avoid competition? To ignore opportunities? You'll find that if the opposite of a strategy statement isn't meaningful then there's a good chance that the original strategy statement isn't meaningful either.

Good strategy statements are specific, things like "we are going to expand into Canada by looking for companies in our business that we can buy" or "we're going to focus on increasing same-store sales and slow down our growth of new stores." Good strategy statements actually help you focus on doing the things that are important for the business. Bad strategy statements just give you platitudes and generalizations that aren't helpful at all. Good strategies are short and to the point. Bad strategies are long—just a collection of wish list items that don't help you focus.

If your business and IT strategies are good, then you'll have something to build on for IT alignment. There's a lot of communication required, but this is something that is very achievable if you have mutual trust and a willingness to cooperate. Go back to part 1 of this article and look at the seven criteria for business/IT alignment. When you can achieve all seven criteria, then you'll be aligned.

2026 Commentary

These budget changes are critical to IT success. If you can't transparently tie the IT budget to business deliverables, then you'll be at the mercy of arbitrary budget cuts.

I mentioned in my 2026 Commentary on the first part of this article that the same alignment process can be used for non-IT organizations. Since Part 2 is so IT-specific, that's less true for this article. But for certain non-IT organizations that provide services to other parts of the business (e.g., HR, process engineering, accounting, etc.), there are enough similarities that a modified version of this approach will probably help.

77. How to Buy an IT Product

Originally published on FEBRUARY 10, 2009 in #IT, #MANAGEMENT, #PROJECTS

Suppose your IT organization is in the market for a new IT product (or service, but I'll use the word "product" here to simplify the discussion). It could be a computer, network device or other hardware item, or it could be a software package or SaaS (software as a service). Regardless of what you're looking for, you'll probably go through a process that you've used over and over again for product selection. And there's a good chance that your process is flawed. You won't select the best product, and you'll be burdened with having to support the wrong product for years to come.

It doesn't have to be that way. In this article I'll go through a list of common criteria for IT product selection, and I'll give you my recommendations for change. Use this article to fine-tune your product selection process, and the result will be the selection of a product that's a better fit for your true business needs.

The Most Common Product Selection Criteria

Here's my list of the most common criteria for product selection, in descending sequence by the popularity I've seen. There's no science here, and I haven't taken a survey. These ratings are based on my own experience:

1. A checklist of "requirements"

This sounds like it ought to be a good way to select a product, but unfortunately the checklist is usually based on a wish list generated by the business users, with no distinction among what's essential now, what's important in the future, and what's a nice-to-have. In simple pass/fail product selection, you narrow your list of product candidates by eliminating all candidates which have omissions on this checklist. The vendors know how this game is played, so most vendors have included capabilities in their feature lists which match the most common checklist items. Unfortunately, the vendor need to provide these checklist items is so great that most vendors stretch the truth. If their product really doesn't provide a certain capability, they'll figure out a narrow case in which the product sort-of fits the checklist description. Or they'll provide a clumsy workaround to address the need, and then claim to have the capability.

My advice: Using a checklist of requirements is fine if, and only if, you:

- Limit the requirements to what you really need, plus what you think you'll need in the future. Include nice-to-have features in a separate list if you want to, but don't use the nice-to-have items in your product selection process to eliminate potential products.

- Push the vendors to show you exactly how each checklist item is provided by their product. If the explanation seems the least bit contrived, then talk to current customers of the vendor who are using the capability the vendor claims to have. Make absolutely sure that the vendor isn't trying to satisfy the requirement in name only, with no commitment to provide the complete capability in an easily usable way.

2. The one your boss wants

This is the second most widely used selection criterion. It might even be a good way to pick a product, but you won't know unless you know how your boss made the choice.

My advice: Don't just take your boss's word for it. Follow through with the other selection criteria and decide independently whether you agree with the boss's recommendation/edict. If you end up agreeing, then fine; now you'll have some reasons to support the decision. And if you disagree, then you'll need all the ammunition you can get to make the case that you ought to go with an alternate vendor.

3. You trust the vendor

This is a popular selection criterion, but it isn't usually stated as the explicit reason for the decision. Realistically, your job is on the line based on the product decision you make, and you'll feel more comfortable going with a vendor you trust. That trust can come from the size of the vendor, the vendor's financial stability, or a proven track record of vendor sales to your company in the past. It will often come from personal relationships that you've built with vendor representatives.

My advice: Make an effort to build a relationship with competing vendors. Find out whether you can trust them as well. If you only go with vendors you already have a trusting relationship with, then you'll always go with the *same* vendors. That's not ncccssarily a bad thing if your IT group is wildly successful with its current technology, if none of your competitors are doing any better than you, and if there are no technology breakthroughs in the future which might make your current technology obsolete. But given that all thrce of these things probably arcn't truc, you owc it to yourself to look at other vendors and products.

4. The vendor or product is a "name brand."

This is a variation on the "trust the vendor" criterion. If you hear a lot about a vendor or product in the press—even if the vendor is using its marketing skills to generate most of that publicity—then you're more likely to trust the vendor or product. For my advice on this, see #3.

There's a second aspect of name brands that ought to be considered. Vendors and products which are recognized as leaders often have a larger availability of skilled talent in the use of the product, and there's likely to be more third-party training available.

My advice: think about how you will install, customize, support and train for the product if you buy it. If external resources are in short supply, what alternatives do you have? Factor their price (including any travel required to bring in support or training from outside your city) into your total cost decision.

5. Recent publicity for the product

This makes the name brand even more apparent. In fact, recent publicity for a product is often the impetus for looking at a product in the first place. Very often your CEO, COO, or other executive will hear about the tremendous advantage that a certain product is providing to another company, and will request the product for your own business.

My advice: The recent publicity tends to make the publicized product the front runner in an evaluation, but try to be objective. Look at the other criteria and decide whether the publicity is justified. Sometimes a less publicized product will be a better fit for your business.

6. You like the salesperson, or—perhaps worse from an objective evaluation viewpoint—your *boss* likes the salesperson.

This is once again a variation of the trust/name-brand/publicity criteria. It's not really a good reason to buy a product, but it can end up being a reason why you don't buy a competing product. See #3 and #2.

7. The price of the product.

There are three variations of this criterion. Some people like to go with the least expensive product, feeling that they're getting a bargain. Others like to go with the most expensive product, feeling that they're paying extra for the best. And still others like to go with a product that is neither the most expensive nor the least expensive, like the idea of throwing out the highest and lowest score in judging certain Olympic events.

My advice: Don't look at price; look at total cost and return instead. The total cost of a product is the price of the product plus all of the installation, customization, migration, training, adaptation, maintenance and support you'll have to do or pay for. That is often quite different from the simple price. For example, a product that is less expensive might not have a key capability that your business needs, and you'll have to pay for obtaining or developing a custom version of that capability. Make sure the work on that capability is factored into your total cost.

Similarly, look at the return to your business from buying and installing the product as a ratio to the total cost. Business people understand this as Return on Investment (ROI), but be careful to calculate ROI based on *real* expectations of costs and benefits. See other articles in this book, "20. Why ROI Isn't Working" and "21. How to Improve ROI and your Project Selection Process," to get a good fccl for how to calculate a valid ROI.

8. The product fits your current or future architecture

Isn't it sad that this criterion—possibly one of the most important criteria for determining the long-term impact of the product on your IT organization and budget—is relegated to number 8? In my book, *Boiling the IT Frog*, I say that, "The fewer [IT] products you have, the better off you'll be, as long as you've chosen good products." One of the primary contributors to excessive IT cost is the maintenance and support of a hodgepodge of divergent architectures and platforms due to a bunch of product decisions that have been made without taking architecture into account. And picking a product that doesn't fit your current or future architecture will lead to inordinate amounts of time and money spent on integration, interfaces, training, wasted effort, and eventually migration to another product.

My advice: Know your current architecture, and decide how you want to change that architecture for the future (that's part of your IT Strategy, which is an absolute necessity for a good IT organization). Don't buy a product that conflicts with your architectural direction unless there's an overwhelmingly compelling reason. To help justify an architecture-based product selection to your boss, estimate the long-term cost of a divergent platform. Don't cave in on this issue—it's important.

9. The reliability, resilience, security and auditability of the product

These are all factors which should be in your requirements (see #1) but are often omitted. The business tends to take them for granted and so everyone assumes that all products are equal in this area, but that's not the case. New products (including even operating systems like Microsoft Vista) are often released before they're mature enough for critical business use. They may work 99.9% of the time, but that .1% failure rate will wreak havoc on a business that requires a reliability of virtually 100%. And even if the product works most of the time, how much trouble is it to deal with the occasional failure? Can the product be properly supported in a resilient environment that automatically deals with hardware or network failure? Will the product increase the likelihood of security breaches or data leaks? And how easy is it to investigate transactions which appear to be invalid, inappropriate, or even illegal?

My advice: Include minimum standards for these factors in your requirements, and insist on hard evidence from the vendors that your standards will be met. And don't just do this for new products—some products never mature.

10. Ease of Use

Here's another factor that is often ignored in product evaluations. Our focus on requirements checklists (see #1) often neglects the way in which a particular requirement is met, even though the usability of the product will be critical in the long run. The problem with considering ease of use in product selection is that there are just too many different product capabilities in large products, and it's impossible to investigate ease of use for every capability in every prospective product.

My advice: The answer to the overwhelming ease-of-use problem lies in the 80/20 rule, which states that 80% of the time using the system will be spent in 20% of the system's capabilities. So look at what the users of the new product will do with the product, possibly by observing how time is spent in the current business process. Identify the processes and capabilities that will get the most use and attention, and focus on determining the ease of use for *those* capabilities. In extreme cases, you might even do benchmarks on two competing products (for example, measuring the time it takes to enter a batch of new orders using each product). At a minimum, you'll assure your product's future users that you're paying attention to their needs. And you may prevent a potential disaster by choosing to stay away from a product that has a bad or inefficient user interface.

11. Flexibility

There are two kinds of flexibility that you ought to look for in any prospective products. First, you want the flexibility to be able to fit the product into your current environment. Vendors typically provide two ways to do this:

- Parameterization (this is my term—vendors may call it something else). This allows the adaptation of the software or hardware by changing settings, values or parameters. No actual change to the product is required, and no software code has to be written.
- Customization (again, this is my term). This will require actual software to be written, possibly as subroutines, extensions, or interfaces. Or for hardware, it will require add-on devices, fittings or fixtures.

A product which can be adapted to your needs using parameterization is preferable to a product which requires customization. The fundamental problem with customization is that you're adding on to the product beyond the product itself. Then when the vendor modifies the product in future releases, your customization may no longer work correctly. In the long run, customization will be a big pain, and maintenance of customization will be a significant expense.

The second kind of flexibility you ought to consider is the flexibility associated with unknown future changes in your company, your business and your requirements. Businesses have a habit of evolving to meet the needs of their customers and their environment. This is a good thing, but it plays havoc with processes and systems that have become rigid in a quest for temporary optimization. The second kind of flexibility—which you may want to call adaptability—is extremely important for any industry. It protects your product investment against future needs.

My advice: make flexibility and adaptability important criteria, particularly as a tie-breaker between two more-than-adequate alternatives. Go with the product that supports your current needs through parameterization rather than customization, and that seems to be more adaptable to handle any business changes you might see on the horizon.

12. Do you really need the product at all?

Last, we have to wonder whether we're asking the right question. Should the question be, "Which product do we want?" or should the real question be, "Do we need a product at all?" There's a natural tendency among human beings to want to simplify their lives by employing the latest and greatest tools. This tendency leads business people to want an IT product because—they believe—it will simplify their lives. But we have to remember Larry Tesler's Law of the Conservation of Complexity: "Every application must have an inherent amount of irreducible complexity. The only question is who will have to deal with it." To apply the law in a slightly different context, every attempt to simplify the life of business people by adding new IT products will increase the complexity of life for the IT organization. It will also make the business more vulnerable to systemic errors that can lead to major business outages and vulnerabilities.

My advice: Try not to buy a product for emotional reasons. Look at the *real* total cost of the product (see #7) and the *real* total benefit from getting the product. Make sure that the difference (benefit minus cost) exceeds the risk of all of the issues possibly introduced by using such a product. If the difference doesn't exceed the risk, then stay with what you have—don't buy a new product.

Conclusion

Product selection is one of the most important aspects of a CIO's job. This is particularly true for cornerstone products that form the basis of your IT architecture. The biggest mistake you can make is to turn the product selection process entirely over to the business users of a product. They know what bothers them about their current process, but they don't understand the architectural environment in which the system must operate, they frequently don't understand how their business processes fit into the overall corporate picture, and they often get distracted by product bells and whistles that won't matter in the successful day-to-day business use of the product.

Focus on core requirements, work jointly with business users, get a product that fits your architecture, and buy from a stable vendor, and you'll be rewarded with happy users and fewer headaches. Sure, some of the business complexity will be shifted to the IT organization, but that's what they're paying you for.

2026 Commentary

The reference to Microsoft Vista is obviously dated, but even today it's still true that products are often released before they're truly ready for large-scale use. Think twice before you choose a recently released product for use in a mission-critical business process.

Another vendor ploy not described in this article is the offer of an extreme price break if you buy a large number of licenses for a product now. When buying licensed software for corporate use, it's always wise to do a pilot project using the product in a small part of the organization. That's the only way to determine whether the product will really satisfy your needs. The vendor tries to get you to make a bigger commitment before the pilot so you'll be trapped into paying for a much larger number of licenses no matter the outcome of the pilot project.

I've seen a number of companies waste huge amounts of money taking "advantage" of this volume pricing, and then finding out that the product didn't really perform as expected.

78. Why CIOs Have to Work Harder Than Other Executives

Originally published on MARCH 10, 2009 in #CAREERS, #IT, #LEADERSHIP, #ALIGNMENT

I was having dinner with a CIO who had retired after a long and distinguished career. The conversation had been mostly small talk, but suddenly a thought occurred to him, and he sat up straight in his chair.

"Why is it," he asked vehemently, "that CIOs have to work so much harder than every other executive at making their role in the business understood?"

It was a rhetorical question, but I made an attempt to answer it anyway. "It's partly because of the 'magic' around IT that I've described in my first book," I said. "Business executives don't have a fundamental understanding of what IT is all about, and they compartmentalize IT and use the CIO as an intermediary. They talk to the CIO, and they rely on the CIO to translate what they need into the technical double-talk they think is used in the IT organization. It's like the wizard role I talk about in the book; the CIO is perceived as the chief wizard, but everything inside the wizard's organization is too mystical for business people to understand."

"But that's crazy," he answered. "You don't see the same sort of attitude toward other complicated parts of the business. Business executives go out of their way to learn all of the important aspects of finance and accounting. They don't view the CFO as an intermediary—they know what the CFO has to do and they hold the CFO accountable for doing it. Finance is just as complicated as IT, but business people take classes in it until they understand it."

I paused to reflect on his statement. He was right—you never hear about CFO's looking for "alignment" with the business, nor do you hear about alignment issues in other parts of the business. I searched my brain for an answer. What's fundamentally different between finance and IT?

I thought about the classes he mentioned. "I don't think the classes in IT really teach the right things to business people," I conjectured. "Introductory finance classes cover fundamental accounting principles, income statements, balance sheets, double-entry accounting, budgeting and capitalization. The more advanced finance classes cover sources of business funding, and the use of various approaches to underwrite the day-to-day financial needs of the business, and to reduce business risk through diversification and the use of offsets."

"IT classes on the other hand," I continued, "cover things like computers and data storage, networking and software development. If finance were taught the same way then they would teach you how to use a spreadsheet, how to close out a financial period, and how to write an annual report. IT classes emphasize *tools*, while financial classes emphasize *principles* and how to apply the principles to the business. IT principles that are important to the business—like the ones in my book—seldom get taught."

I went on, "Finance also has a uniform way of presenting its current position—a standard scorecard using well-understood definitions of balance sheets and income statements. IT has nothing like that; there is no uniform way to present the status of IT. There is no standard for an enterprise architecture view of the systems in a business. There is no standard way to represent the status of IT projects. There is no standard for showing the state of IT infrastructure, revealing areas of risk and areas of excessive investment."

"OK," he answered, "Maybe you're right about finance, but what about other complicated parts of a business? What about operations in a manufacturing business? Or research and development in a pharmaceutical company? Or even marketing in a consumer products company? There are no standard ways to show the state of *those* parts of a business. They've each got complicated things going on, but *they* aren't viewed as mystical by the CEO."

"That's an interesting point," I said. "Let's go through your examples one by one. In a manufacturing business you can portray the process with a diagram showing inputs and outputs: you put raw materials in, you 'turn the crank,' and you get manufactured products out the other end. Sure the turning-the-crank part is complicated, but it can usually be represented in some sort of high-level process diagram, and then you can 'drill down' to show more and more detail underneath each part of the process diagram."

"When a CEO looks at the diagram," I went on, "they can see the fundamental manufacturing process. Bottlenecks—or limits to production—can be represented in the diagram, and anyone can see where you could invest to make the process smoother. So even though manufacturing is often complicated, in almost every case someone goes to the trouble of using a pictorial representation of the manufacturing process to convey simple information about the status. The CEO may not understand all of the technology used in manufacturing, but each piece of manufacturing equipment can be viewed by the CEO as a 'black box' in the diagram with inputs, outputs and operating characteristics."

"Your second example was R&D in a pharmaceutical company," I continued. "R&D is generally viewed by business executives using the classic 'investment portfolio' model: each R&D investment can be viewed as having an investment amount, a risk, and a probable return. The probabilities of the success of each R&D investment will vary greatly, but taken together, the R&D investments will generally provide a payback to the company. It's like drilling oil wells: You know you're not going to hit oil with every hole you drill (although you use science to help you pick the very best spots), but you know that you'll hit oil on some reasonable percentage of the holes you drill, and that makes it all worthwhile."

"What was your last example again?" I asked.

"Marketing in a consumer products company," said the CIO. "Surely you'll agree that marketing isn't viewed as mystical."

"Well, it's complicated," I said. "But again, it's like R&D investment or drilling oil wells: You invest in various marketing campaigns, and you hope to find a few that will generate a significant increase in your sales revenue. Then you fine-tune them to maximize the revenue."

"The trick with marketing," I said laughingly, "is to figure out whether a revenue increase is in fact related to a marketing campaign at all. At least with oil wells, you know which wells are producing oil. With marketing, you're never quite sure which marketing campaigns cause your sales increase. It's mystical, all right, but it's more like gambling than investment."

"I see your point," said the CIO, "but I still don't understand why CEOs can't relate to IT. If they can deal with manufacturing and all its complexities, and if they can cope with R&D and marketing with all of their uncertainties, then why can't they just treat IT like any other part of the business?"

"Maybe it's because they just don't want to," I ventured. "It's like having a car mechanic to take care of your car. If you trust the mechanic then you don't have to worry about how the car works, and so you don't bother learning."

"But that's terrible," exclaimed the CIO. "IT isn't just about mechanics—it's about all the data and information processes in the business. Isolating IT as if the IT organization is just a bunch of infrastructure mechanics is like closing your eyes to all of the wonderful contributions that can be made by IT throughout the business. It's as if the CEO is assuming that everything is good enough 'as is,' and IT's job is just to take requests from various business units. That's not the way to get the most out of IT."

"You're right, of course," I replied. "but in a lot of companies, the view of IT as a trusted mechanic—dismal as it is—is just an aspiration. In many companies IT is treated as a mechanic, but without the trust. That makes things even worse: You take requests, you're not involved in business strategy discussions, and your organization isn't trusted to deliver what's been requested. In those companies there's an adversarial relationship between IT and other business organizations. I've worked with a few companies like that, and it's no fun."

"I've seen a few of those myself," said the CIO. "In a situation like that you have to first build a trusting relationship between IT and the business. Only then can you try to take IT into more of a strategic role."

"So let's get back to the original question," continued the CIO. "Why is it that CIOs have to work so much harder than every other executive at making their role in the business understood?"

"Your guess is as good as mine," I concluded. "But it's a shame, isn't it?"

[Note: I really was at a recent dinner with a retired CIO, and he really did ask the question mentioned in this article. The conversation started out the way I describe, but the rest of this hypothetical conversation took place later in my head. It's a lot harder to come up with good answers on the spur of the moment.]

2026 Commentary

I've lost contact with most of the IT people I know, but I suspect that nothing has changed in this area.

79. IT in a Bad Economy is like Playing the Game Jenga

Originally published on APRIL 11, 2009 in #IT, #MANAGEMENT

Have you ever played Jenga, a game marketed by Hasbro? In the game you have 54 wood blocks and you start by stacking them in rows of three at alternating right angles to build an 18-level tower. Then you take turns removing a block from a lower level and adding it to the top to make the tower taller. Eventually the unstable and weakened lower levels won't support the height of the tower, and the whole structure falls over. The loser of the game is the person who made the tower fall. The winner is the person who played just before the loser.

I see a lot of parallels between this game and our current situation in IT. In a bad economy we have to survive with a lower budget, yet the demand for IT projects hardly slows down at all. The only way we can extend our IT capabilities is by cutting back somewhere else in our budgets. So we cut back on maintenance or on infrastructure or maybe on support, and we hope that we aren't weakening our IT organization too much. Just as in Jenga, we take something from a supporting area and then try to extend our IT capabilities by building on that increasingly weak area. The tower grows taller, but since the number of blocks hasn't changed, the tower structure gets weaker and weaker.

So what's the equivalent in IT of our tower falling over? Sometimes it's catastrophic: our systems fail at a critical time, we have a major security breach, or a critical person leaves the company and we don't have a backup. That's when we see how thin and frail our capabilities really are. That's when we see the error of playing Jenga with critical parts of the business.

Just as in Jenga, the loser is the person who was in charge when the tower fell over. But the winner isn't their predecessor.

No, in the business version of Jenga there are no winners. Everyone loses—the business, the IT organization, and your customers.

You wouldn't play Jenga with a real skyscraper. You wouldn't deliberately sacrifice strength in the lower levels of the structure in order to make the building taller. So why would you sacrifice strength in an IT organization in an attempt to "satisfy" the business needs of your company?

You've got to recognize that there is a limit to how much you can safely cut back on mission-critical areas of your organization. Your CEO won't know what that limit is—you'll have to determine the limit yourself. But a CEO can definitely understand that there *is* a limit, just as there's a limit in other parts of your company. A key aspect of your job as a CIO is to assess those limits, and to make sure that you don't weaken your organization or your infrastructure too much.

4 Things IT Should Do in Bad Economic Times

Here's my advice for bad economic times:

1. Deal with the situation as best you can.

If you're asked to cut your budget, then do so (see tips in my August, 2008 newsletter article, “71. 9 Ways to Reduce Business IT Expense"). But don't go too far (see the Jenga discussion above), and watch out for things that will cut IT costs but raise costs in other parts of the business. A smart corporate cost-reduction plan will look at the big picture and not just impose a standard percentage cut on every department. If you're forced to make an arbitrary budget cut, make it clear what services and projects you're cutting back as a result. Be transparent—don't take a cut without passing some of the pain back to your business users.

2. If resources allow, then build for the future.

Business is down for a lot of companies, so it's a buyer's market. If you've got the funding to do it, this is an ideal time to invest in your company's future. Vendors are eager to negotiate lower prices, and consultants are widely available to help. If you're sure you are going to need new systems and capabilities for the future, then spend the time and money to build or buy them now.

3. Emphasize flexibility.

Flexibility is critical right now. Ideally you want systems that will scale up to handle much larger volumes, but also scale down to support lower volumes with a reduction in infrastructure cost. An automation project often replaces a variable labor cost with a fixed infrastructure and maintenance cost. That might make sense if the fixed cost is much lower than your labor cost under most conditions. But if the project justification requires a high volume to break even, then the project won't be a good investment if your volume drops. Avoid projects like this, and focus on projects which *increase* the flexibility of your work force instead.

4. Be optimistic.

A recession is a downward spiral in the economy caused by negative expectations. Things are bad, so people cut back on their spending. Then these spending cutbacks make things bad for other companies. The resulting downward spiral is only reversed when people begin to see improvements in the economy and in their business. In effect, a recession is a state of mind. Make sure you make your own contribution to halting the recession. Be optimistic, encourage others to look forward to better things that will happen in the future. Be realistic, but don't panic, and don't make cuts that will cripple your business when the spiral reverses.

Conclusion

There are a lot of people playing Jenga in their companies right now. It gives the illusion of growth at no cost. But the hidden cost of playing Jenga is the loss of structural integrity of your systems, your infrastructure and your support team. There's a huge risk in playing this game, and there are no winners.

To deal with a bad economy, focus on the four things I've listed above. Then when the economy turns around, you'll be positioned for growth without having to rely on a weak foundation.

2026 Commentary

It’s probably obvious from context, but there was an economic downturn in 2009. This article, and a few others from this same time period, offered advice on how IT organizations should best deal with the downturn.

It’s possible that you’re reading this during a booming economy, and that’s great. But don’t skip the warnings about how to deal with downturns. Economies are cyclical, and sooner or later you’ll have to face your own downturn. It’s best to be prepared.

80. The Quest for a Quest

Originally published on MAY 4, 2009 in #LEADERSHIP, #STRATEGY

I'm fond of fiction; I read a lot of books and watch a lot of movies. I think that a large part of the appeal of fiction comes from the single-minded focus of the principal characters in the plot. When the hero of the book or movie is trying to track down a secret or elude their pursuers or find the murderer, they are solely focused on the task at hand. You seldom see the hero stop to pay their bills or clean up the house or even go to the bathroom. No, our hero has only one goal for the duration of the book or movie, and they are totally absorbed by the effort of achieving that one goal.

This single-minded focus sometimes occurs in real life. I've been involved in a number of projects over the years that made me feel like I was part of something bigger—the kind of thing that they make movies about. You get a wonderful feeling when you're part of something like that. You start each day at work knowing your purpose and knowing what you're supposed to do that day. You see the goal in the distance and you see your team getting closer and closer to the goal as time goes by. You feel the excitement as the goal approaches. You don't stop the everyday stuff—caring for the kids, mowing the lawn—but you go through those things on autopilot because your thoughts and attention are focused on the goal.

A Quest gives you Productivity

Whether it's putting a man on the moon or something much more ordinary like delivering a new software product, the focus of a quest seems to boost your concentration and fuel your productivity. Petty issues get resolved quickly and smoothly when everyone understands the real purpose of your mission. When your entire team is heading in the same direction there is far less disagreement over mundane trivia. With a quest comes a new perspective, a different frame of reference. Everything is defined in terms of whether it takes us closer to our goal or further away from it. Life is simple—at least for the duration of the project—and decisions are easy when the objective is clear.

A Quest gives you "Flow"

Mihaly Csikszentmihalyi talks about "flow" in his books and articles. He describes an altered sense of time where we get so caught up in what we're doing that we don't notice time passing. We start working on something and then "wake up" hours later, wondering why it's so dark outside. Flow comes from focus, total involvement, and deep concentration on something that completely absorbs us. Flow is part of the reason that a quest is so appealing. We go into a flow mode when we're working on something associated with a quest. Flow is enjoyable; it feels good to be so focused.

This focus also lets us use both our conscious and subconscious mind to deal with challenges we encounter during the quest. We work on a problem during the day, and finding the solution to the problem becomes so important to us that we devote our subconscious to the task as well. We have ideas while we're sleeping, while taking a shower, or when we're doing something that has nothing to do with the quest itself. Thus the quest helps us devote some of our subconscious energy to the achievement of our objective.

How to Lead a Quest

It's really pretty simple to lead quests of your own. Here's how:

1. Define an extremely clear objective, one that can't possibly be misunderstood or misinterpreted. The objective has to be significant but it also has to be simple: something that can be communicated in a single phrase. Examples: "before this decade is out … landing a man on the moon and returning him safely to the Earth," "reduce the amount of labor required for the xyz process by 50%," "sell our products over the Internet by the end of the year," "be the first company to ship a product that does xyz."

2. Frame the objective in a way that makes the achievement of the objective an emotional experience for the participant. This can be tricky. You have to tie the objective to a basic human emotion—preferably a positive emotion.

Pride is often a good choice. Maybe the objective is something that no one else has done, or that no one else has done using your approach. Pride can be a powerful motivator if you can fuel that emotion in quest participants.

Competition is also a good way to make an objective emotional. "We're going to beat our leading competitor in introducing this type of product."

Survival is also possible as an emotional motivator, although with survival you're using a negative emotion: avoidance of a catastrophe.

Often there's a mix of emotions. "Getting this product on the street before the competition is absolutely required for our company's survival. There's no second place—only winners and losers."

3. Include a time component in the objective, even if it's arbitrary. Make sure the deadline is aggressive but achievable. If the deadline is too easy to achieve or too far out in the future then it will be difficult to motivate day-to-day effort. If the deadline is impossible to achieve then everyone will give up before they start.

4. Define the project steps for the achievement of the objective in a way that clearly delineates the work that each sub-

team or person needs to accomplish. You don't want a free-for-all; you want a concerted effort by each member of a team. You'll only get that concerted effort by clearly defining the way in which each team member will contribute to the achievement of the objective.

All Four of these Requirements are Critical

If any one of them is omitted then you won't have a quest. Leave out the clear objective and people will constantly argue about what you're trying to accomplish, and eventually they'll lose interest. Leave out the emotion and no one will care about the project—it will be just another stupid objective that management gives them. Leave out the time component and there will be no urgency, no need to prioritize the project ahead of others. Leave out the delineation of effort and everyone will argue about who's doing what. All four requirements have to be included for your quest to be successful.

People Can't Participate in Multiple Quests Simultaneously

You can have multiple quests going on in your company at the same time, but each individual employee can only be a part of a single quest. If you give an employee a role in two quests then they won't be able to emotionally commit to either of them. You might be able to get away with two simultaneous quests if you clearly prioritize one quest over the other. But the second quest distracts the employee, and it doesn't allow the employee to devote subconscious time to solving the challenges that are encountered in the first quest.

The Connection between Quests and IT Strategy

I've talked about IT strategy in other articles and in my book, *Boiling the IT Frog*, but I haven't previously explored the connection between IT strategy and quests. I believe that the key to an effective IT strategy is focus: being able to communicate a few key objectives that your IT organization is trying to accomplish in order to increase its contribution to business success. Using a quest approach is an ideal way to achieve those focused IT objectives. Each major IT objective becomes a quest, and through a series of quests your IT organization will be able to make significant progress by focusing your employees on one thing at a time. You'll be successful, and your employees will get the motivation they need and deserve.

Conclusion

The quest approach is used by many great companies. Look at the history of Apple, for example, and you'll see that all of its major products were built using a quest approach. You'll see the same thing at Pixar, at Sun, at GE and at IBM. The best products were built using a quest approach, and the best employee experiences came from participating in those quests.

If you have a really big objective—something that's important to a significant number of people—then I highly recommend the quest approach. But while it's extremely valuable, the quest approach can also be misused if you violate the rules I've defined in this article. Make sure you include a clear objective, an emotional tie, a time component and a clear delineation of work. And make absolutely sure that employees only participate in one quest at a time. Otherwise you'll just discourage your employees, and you'll hurt your reputation as a leader and as a motivator.

Life can be a series of quests, just like in books and in the movies. It can be fun to focus if we're truly allowed to focus on one thing. We've still got a life to lead, and we'll still have to do all of those mundane day-to-day things in our non-work time. But participating in a quest gives us something to be proud of, and it contributes to the meaning in our lives.

2026 Commentary

When I think back to the projects I did during my career, all of the ones I remember most fondly were quests. Participating in quests is probably the thing I miss the most about leaving the corporate world.

And to be absolutely clear, quests are not just for IT. Any organization can have them. In many cases the quest is at a higher corporate level, and multiple organizations have specific objectives within the quest.

81. Don't Let the Notes Get in the Way of the Music

Originally published on MAY 15, 2009 in #CAREERS, #MANAGEMENT

I went to my grandson's piano recital last weekend. He's just seven (almost eight) and he's really good for his age. His part of the recital was only a few minutes, but the entire recital lasted two hours so we listened to a lot of other kids before and after my grandson.

What struck me in listening to all of these kids is how few of them actually produce something that feels like music. They can put the notes together in a way that *sounds* like music, but there's no feeling involved. Most of them are in the early stages of learning and they're still focused on getting the notes right. If they trip over a note, they just back up and play it again, sometimes two or three times. There is music to enjoy somewhere underneath the notes they're trying to play, but only a few of the kids actually convey the spirit of the piece they're playing.

I see the same thing happening in many business organizations—both in IT and outside of IT. We have a tendency to get so caught up in our day-to-day activities that we (a) stop enjoying what we're doing and (b) lose track of our overall objectives. And I think the two things are related: when we lose our focus we lose our enjoyment.

Stephen Covey has written about this problem in his books. He says that we tend to confuse what's urgent with what's important. The ringing phone is urgent, the boss asking you a question is urgent, but creating a new business process for your company is important. Turning in a revised budget on time is urgent, filling out a performance appraisal is urgent, but hiring and developing the right employees is important. Getting a paycheck is urgent, getting a promotion is urgent, but spending time with your family is important.

Things You Can Do to Help Your Focus

It takes conscious effort every minute of every day to focus ourselves on what's important. And since we often get distracted, we need to take steps to keep our focus. Here are a few things I've learned over the years:

- Schedule a small block of time every day or every week. Explicitly book the time in your calendar so nothing will be scheduled on top of it. Use the block of time to think about how you can better focus what you're doing on what's important.

- Set a periodic alarm on your PC or on your iPhone or PDA or even your watch. When the alarm goes off, think to yourself, "Is what I was just doing the most important use of my time?" Do this for a while and you can gradually decrease the frequency of your reminders.

- Keep a log of how you're spending your time. Then after a week or so take a look at the log and see if you think you're spending your time on the most important things. If you have trouble keeping a log, then combine this with the periodic alarm idea. Whenever the alarm goes off, write down a few words to say what you were doing since the last alarm.

  Or if you want to simplify this process, make up a log page in advance for each day (Use spreadsheet software, then print it out). Have each row correspond to the time interval for your log (start with an interval of 15 or 30 minutes), and have each column correspond to a list of common things you do (e.g., talking to customers, generating new ideas, teaching employees, answering email, budgeting, social media, etc.). Then when the alarm goes off, put a check mark in the correct column for the row corresponding to the current time.

The last two items will take some extra effort on your part, but they're primarily learning exercises. Once you've done the exercises for a few weeks and you have a pretty good idea what you're doing wrong, you can discontinue those periodic alarms and just schedule daily, weekly or monthly "self-focus sessions" on your calendar.

Conclusion

Life is full of urgent tasks, but to get the most out of our lives we have to focus on what's important. In music it's the feel of the song that's important—the way it affects your emotions. Don't let the notes get in the way of the music, and don't let the urgent get in the way of the important.

2026 Commentary

Time goes by whether or not you're spending it on important things. The only one who can judge what's important is you.

82. Organization Transplant Claims Another Victim

Originally published on MAY 22, 2009 in #CAREERS, #ALIGNMENT, #STRATEGY

How a New CEO Affects the CIO

A few weeks ago I ran into a CIO I hadn't seen in a few months. The last time I had seen him we talked about his job situation. The CEO he worked for had just been replaced, and the CIO and the new CEO were butting heads.

This situation is pretty common and takes place all over the world. A new CEO is brought in by the Board of Directors. There's a reason—usually it's an attempt to take the company in a new direction or to pull the business out of a slump. The new CEO knows that they only have a few months to make a difference. That's because after the first six months the board will begin to have second thoughts and will give the new CEO a tighter leash.

Two Approaches for the New CEO

Put yourself in the position of the new CEO. What do you do? There are a couple of basic approaches:

1. **Problem-Solving: Figure out what's wrong, then fix it.** This is a logical approach to improving the business but it demands patience and steady nerves. You're going to need to talk to key customers, key employees, and anyone you can find who can give you an impartial opinion. The evidence will conflict. There will be a lot of finger-pointing. You will hear a lot of "noise"—negative comments about things that are working well and positive comments about things that aren't working. A high level of judgment will be required, but if you can gain focus on the key issues, then you'll be able to take

clear, decisive action to solve problems quickly. This approach is like diamond cutting; it requires a lot of study and then nerves of steel when making the change.

2. **Organization Transplant: Replace the old organization with a new one that's worked for you before.**
   This is a more brute force approach. It doesn't require great intellect or very much study. The concept is simple. You've run a similar organization before, and you have a list of people who have worked for you in the past. When you're hired into a new CEO position you just bring in your own people for all of the major roles. You superimpose your previous business culture on the new company. You discard all of the top-level executives reporting to the old CEO and replace them with people from your list who have worked for you in the past. They run the new company exactly like they ran the old company you all came from.

Which Approach is Better?

I guess it depends on your point of view. The Board of Directors want a change, and either approach will lead to a change. The Organization Transplant approach will probably produce faster results if—and it's a big if—the new transplanted organization is actually appropriate for your business. But there's a catch: the kind of organization that can be transplanted is usually bureaucratic and mechanical. You can transplant a heart but not a soul, and an organization transplant tends to snuff out the cultural spark that may have existed in the company prior to the transplant.

On the other hand the Problem-Solving approach, when it's successful, is able to save the existing cultural spark in a company and focus that spark in a new direction. Problem-Solving removes the things that were preventing the spark from growing, and fans the existing sparks into bright flames of growth and development.

So the choice of approach comes down to the question, "Is there something in the company worth saving, or should we just overlay the company's culture with a new one?" And if we choose the latter approach, what donor do we use for the transplant?

What if this happens to you?

Put yourself in the position of the existing CIO. You get a new boss and you obviously want to make a good impression. You also want to keep your job. What do you do? Regardless of the approach your new CEO will take, you need to clearly communicate three messages to the new CEO:

1. **You're on top of what's going on in your company.** You know the business inside and out, you understand the business processes that are being used throughout the company, and you know where the best opportunities are for improvement. Make the new CEO believe that they would be a fool to bring in someone else for your job—you're just too valuable.
2. **You're willing and eager to alter your IT strategy to fit the new CEO's interests.** You're not a problem for the new CEO—you're an asset.
3. **You understand and acknowledge the weaknesses of your current processes, systems, and infrastructure.** You know why there are weaknesses (e.g., lack of budget emphasis), and you're eager to fix them. You accept criticism as a way of identifying opportunities for improvement. You're not someone who will fight progress—you'll be an ally in the battle.

What's the result?

All you can do is your best, but you're not in control in a situation like this—the CEO is. In the case of my friend the CIO, his Board of Directors chose the Organization Transplant approach. He made a strong effort to communicate with the new CEO and help solve the problems facing the company, but it wasn't enough. He's being replaced with someone from outside, as are many of his counterparts in other business organizations. Will his company survive? We'll see. But one thing's for certain: the company's soul will never be the same.

2026 Commentary

Another more extreme version of the Organization Transplant approach is used by private equity firms when they take over a company, strip it of its assets, and add enough debt to the company to pay off the private equity firm shareholders. I say “more extreme” because most Organization Transplant CEOs at least want to keep the company running. Private equity firms don’t really care if the company continues in the long run.

If you’re the CIO of a company bought by private equity, my advice would be to start looking for another job immediately. Your days are numbered.

83. See People for Who They Are—Not for the Roles They Play

Originally published on MAY 29, 2009 in #CAREERS, #MANAGEMENT

We all have a tendency to define ourselves by the roles we play. The first part of almost every new conversation between strangers is asking the question, "What do you do?" We then use the answer to that question to apply a stereotype to the person. If the person answers, "I'm a doctor," then we apply the stereotype of a respected but conservative individual who works long hours caring for people. We also immediately associate the person with a higher income level and the advantages that come with it.

If the person answers, "I'm a CIO," then we apply the stereotype of a corporate executive with an inner nerd—someone who spends their time overseeing major projects and spending long hours in corporate boardrooms.

Why Stereotypes?

Why do we use these stereotypes? Because it simplifies our lives. It's complicated to truly understand an individual, and it's a time-consuming process. We may not have time to spend getting a good understanding of the person, and we certainly don't want to spend that much time on everyone we meet. So we take shortcuts, make gross assumptions, and then figure that we'll fill in the blanks later if we really want to know someone.

I'm not saying that's wrong, but it's deceiving. Our oversimplification causes us to miss out on learning the subtleties of an individual that truly make a person unique. A "doctor" may be an accomplished artist with a scalpel who creates new faces for burn victims. A "CIO" may be a former machinist who got into IT because it offered a better paycheck, and then built on their experience to rise through the ranks. Most people have complicated back stories, and their overall life experience has contributed more to their current sense of self than any current title.

We probably do the most disservice to those with more mundane titles. A "waitress" may be a single mom putting herself through school. A "UPS delivery man" may be a passionate coach of amateur athletics or a committed family man who enjoys running a tight schedule and meeting new people.

Students are Different

When people introduce themselves as students, we cut them a little more slack. We inquire about aspirations, career plans, life direction. So why does a non-student answer give us any less interest in career growth and future direction? Maybe the doctor longs to be a politician. Maybe the CIO wants to start their own company. What people are is not the same as who they will become.

People Serving in Other Roles

Our typecasting even extends to people who in fact have good "titles" but serve in other capacities. We go to a meeting of a business organization and we assume that the person checking us in for the meeting serves in some menial capacity. They may in fact may be president of a company—you just don't know.

I've run into this consistently in one of my roles for the Atlanta Chapter of SIM (Society for Information Management). For the past six years or so I've done the web site for the group. It was kind of a fun project and it gave me an opportunity to contribute. I extended the role by taking pictures at the meetings. The photos made good additions to the web site and they were helpful in soliciting new members.

What amazed me, however, is how people's perception of me changed when I picked up a camera at SIM meetings. No longer was I an experienced IT executive attending the meeting—I was a "photographer" who faded into the background like the waiters who run around in their white coats removing dirty dishes. On several occasions I was standing with a group of my peers talking about IT-related subjects when someone new came up to us to introduce himself. Invariably the person would shake hands with each of the people standing with me but bypass me because of the camera hanging around my neck. Apparently it's an amazing camera—it makes me invisible.

Conclusion

I know it's easier to simplify your view of new people by applying stereotypes. But try very hard for the next week or so to go beyond the stereotype and get a better feel for the real people underneath. When you meet someone new, ask them about their passion, ask them about the best day of their life, ask them about what they want to do with their lives (You might even consider doing this with people you already know.) You'll begin to see how different people are from their stereotypes, and it will open up your eyes to some of the real beauty inside people. And if you do this for a week, maybe you'll like doing it well enough to keep doing it for the rest of your life.

Oh, and if you see someone at a meeting with a camera around their neck, say hello. They're probably a lot more than a photographer.

2026 Commentary

It was definitely the camera that made me invisible. When I stopped taking photos for the group, people started talking to me again. But I've tried to apply my own advice and avoid stereotypes in my day-to-day interaction with people I meet. I've gotten a better understanding of some amazing people.

84. Words and Phrases I'm Sick Of

Originally published on JUNE 5, 2009 in #ACQUISITIONS, #MANAGEMENT

I see and hear these words and phrases all the time, but it has gotten to the point where they've lost their meaning for me—maybe for you as well. Let's start with some oldies but goodies then work up to something more current.

Synergy

This used to be the consultant's ultimate goal. Synergy is a type of magic: it's creating something from nothing. Or to be more accurate, it's creating something new out of two old things. Let's say you have a company and it's worth $x, and I have a company and it's worth $y. Using synergy, our two companies together are worth more than $x + $y. That's the theory anyway.

But wait a minute. Wasn't there a benefit from two companies working together before the word synergy was invented? What about old phrases like "horizontal integration" and "vertical integration"? Weren't they about synergy? Sort of. Most people tend to use horizontal and vertical integration as words related to an acquisition—one company actually buys another. What's different today is that companies are more likely to try to partner with one another without M&A being discussed.

So synergy is like the polite way to do vertical or horizontal integration without all of the nasty complications associated with M&A. It's like "friends with benefits" instead of marriage. You can play together without the long-term commitment.

Paradigm Shift

The phrase was first used by Thomas Kuhn in 1962 in his book *The Structure of Scientific Revolutions*. Kuhn was describing a fundamental change in basic assumptions: something so significant that it created a whole new paradigm. Since then the phrase has grown so common that it has been used in marketing almost any kind of new product.

"Paradigm shift" is now used like any other set of magic words. We wave our magic wand and—presto chango paradigm shift—we see this remarkable result. My rule of thumb: be suspicious of anything using the phrase. It's probably a variation on the same old paradigm you're used to.

Perfect Storm

The phrase came from Sebastian Junger's 1997 book, *The Perfect Storm*, and was popularized by the the 2000 movie of the same name. It refers to a combination of weather coincidences which come together (using *synergy*, I guess) to create a storm of much higher magnitude than is typically seen. It's like the *paradigm shift* of storms—everything is on a much higher level: winds, water, waves, general devastation.

I first heard "perfect storm" used in a non-weather context when I attended the *Gartner Symposium* in 2000. Gartner used it to describe a remarkable confluence of economic and social factors which had created an ideal environment for Information Technology progress, jobs and business influence. Not quite a year later on September 11, 2001, that ideal environment ended.

These days I've heard the phrase used to describe the economy. But let's face it—it only took bad mortgages and widespread sale of mortgage derivatives to trigger our latest economic reversal, and mass panic took care of the rest. Our current economic climate isn't anywhere near as bad as any perfect storm situation I could envision. So let's stop exaggerating.

Downsizing

This is a euphemism, pure and simple. We say downsizing when we want to depersonalize the act of firing people. Yes, we may have an economic motivation for takings jobs away from these people, but in a lot of cases the economy is just used as an excuse for getting rid of people we wanted to get rid of anyway.

I understand the need for using euphemisms—I really do. I know that it's easier to explain things to employees using this word, and this sort of explanation makes it easier for employees to accept their fate. But that doesn't make it right.

There are worse words. "Rightsizing" is much worse—it implies an ethical basis for firing. I also dislike "force shaping" which seems to compare the whole process to a haircut—“we'll just take a little off the top." "Work force reduction" seems like a better phrase—at least it states a fact: the work force is being reduced.

Recession

There are economic definitions of this word, but I'm more concerned about the emotional content of the word. While not as bad as the word "depression," recession is a word that makes people afraid. I've said in a previous article, “79. IT in a Bad Economy is like Playing the Game Jenga” that the biggest key to economic recovery is optimism. With an optimistic attitude, people will buy things they need and want, and the economy will flourish. So if you're using the words recession and depression a lot, please stop. You're screwing things up for the rest of us. Be optimistic instead, and let's get the economy back on track.

2026 Commentary

This article hasn’t aged well. Buzzwords from 2009 aren’t popular today, so you probably don’t see these as a big deal. It’s only when I heard them over and over that I became annoyed and wrote the article.

Anyway, I’m including the article for completeness, but there’s nothing particularly significant about it.

85. Why Do You Want to Be a Manager?

Originally published on JUNE 12, 2009 in #CAREERS, #MANAGEMENT

There are a lot of bad stereotypes associated with management—the TV show "The Office" illustrates many of the stereotypes on a weekly basis. But there are advantages to being in management, so I thought I would write a bit about management for those of you who are still in individual contributor roles. And for those of you in management roles, you can use this article to help you with career discussions you have with some of your employees.

Management Advantage 1: Scope

From my own experience the principal advantage of management is that you get to do bigger things. There are limits to how much you can do by yourself, and so by necessity you won't be able to accomplish as much in an individual contributor position. Of course you can still be part of a team and contribute to a group goal (see my article "80. The Quest for a Quest" for more about this topic), but it feels different when you are in charge of a team and your team is able to accomplish something significant. Frankly, it feels wonderful.

Management Advantage 2: Control

The second advantage of management is control. This is a loaded word, and it can be used both positively and negatively. Let me first use the word in the positive sense: as a manager you have more control over some of the project decisions that need to be made. I won't give you false hopes, though. As long as you report to someone else (e.g., a higher level manager, a director, the CEO, the Board of Directors, the stockholders), you never have complete control. But still, it's generally true that you have more control the higher up you rise in the management structure.

Control means that you can envision a project result and actually make it come true. It means that you have fewer situations where you have to argue over small details—you just make a decision on the small details and then you don't have to worry about them anymore.

Notice that I singled out the *small* details. That's because you still have to deal with higher-level management to work out the broader issues. The higher you go in the management hierarchy, the fewer details you're focused on, and the more you're focused on broader issues.

Negative Aspects of Control

Now as long as I'm talking about control, let me talk about the negative aspects of control. First off, you're kidding yourself if you think that being a manager means being some sort of dictatorial snob who tells everyone what to do. Barking orders at people might work in the military (and I'm not sure it even works well there), but it doesn't work in most companies. When people are given orders without any explanation or motivation they tend to get annoyed. Annoyed employees fight back—both actively and passively—and the end result is that you end up failing as a manager.

So if you want to be a manager because you see it as some sort of power trip, then forget it. Wait a few years until you have a more mature attitude about management, or maybe you should leave your company and join the military (Good luck with that!).

A second negative aspect of control is accountability. Yes, you can determine the details, but you have to take responsibility for anything you decide. If you make a decision and it doesn't work out, expect to incur the wrath of upper management. Make enough bad decisions, incur enough wrath, and you'll be fired.

What this means is that there's a cost to control, and that cost is the pressure you'll feel to make the right decisions. Some people can't deal with the pressure. They get upset, they neglect their families, and they have a lot of trouble coping with the ambiguity of decisions (most decisions aren't black and white—they're shades of gray). If pressure bothers you, don't become a manager.

Key Management Skill: Working with People

It sounds obvious, but I'll say it here anyway: Being a manager is about working with people. In a nutshell that's what management is all about: getting results through other people.

Some people mistakenly believe that a management position in a technical specialty like IT is just another rung in the technical skills ladder. They see a job progression like "programmer, senior programmer, project leader, manager" or "analyst, senior analyst, project leader, manager" or [substitute your own list of titles with the last item being manager] and they think it means that as you get better at one job it's natural to jump up to the next job.

What you need to understand is that while the scope of your job does in fact get bigger as you move through this progression, there is a change in the skills you need and the things you spend your time on. Lower level technical jobs require you to be good at the technology. Higher level jobs like project leader and manager require you to be good at dealing with people: bringing out people's strengths, communicating with them and influencing them.

There's a fundamental contradiction here that a lot of people can't get their heads around. Being a manager means spending *less* time on technical stuff. If you *really* enjoy doing the technical stuff, don't become a manager.

Conclusion

I'll write more about management in upcoming articles, but for now let me just recap what I've told you. Most technical people who become managers do so because they want more scope and control. They in fact get that scope and control, but they get it only by leaving behind some of the detailed technical skills that they developed as individual contributors. But perhaps most important, you don't become a good manager by being good technically—you become a good manager by being able to get things accomplished through other people. Management is a people skill—not a technical skill.

2026 Commentary

This was one of the more popular articles when these articles were on a web site. What I describe in the article seems obvious to me, but I guess it's not so obvious if you haven't been a manager before.

86. How to Justify Security Investment

Originally published on JUNE 18, 2009 in #LEADERSHIP, #MANAGEMENT, #PROJECTS, #STRATEGY

In March of 2009 I was the speaker at a SIM Meeting in Dallas, Texas. In the Q&A at the end, a young security manager asked me a question. He said that he has succeeded in making his company's infrastructure good enough that they have no problems with security breaches, data leaks, viruses, or any of the other security issues that plague many companies. Now the security manager is having increasing difficulty in justifying additional infrastructure investment in the area of security. He knows that continued investment is necessary—the bad guys get better all the time—but his business customers feel that their security problems are solved. He asked me what he should do about this.

I've been pondering this question off and on for the last several months. Then I was reading a book, *Brimstone* by Robert B. Parker, and I ran across the following words: "safe is more how you feel than how things are … Safe and not safe is mostly in your head."

Three Different Parts to the Question

When I read those words it struck me: There are three different things going on with the security manager's question. First, there is the issue of making your business customers *feel* secure. Clearly, the Dallas security manager has been very successful on this front. Second, there is the issue of actually implementing an infrastructure that provides a reasonable level of security for the amount of money the company is willing to invest. The Dallas security manager seems to be successful in this area as well. Third, there is the challenge of recommending the right level of infrastructure security investment and getting agreement from the business. And that's what seems to be at the crux of the problem.

You Can't Eliminate Risk—Only Mitigate It

In June of 2008 I wrote an article entitled "69. 3 Things Your CEO Wants to Know." Two of the three things were these questions:

- How do we stand versus competition?
- How do we mitigate risk?

I won't repeat the text of that article here, but one of the key messages in the article was the idea that businesses can't hope to *eliminate* risk—only *mitigate* it. It's similar to the idea of Secret 6 in my book, *Boiling the IT Frog*: "There is no 'right' amount of money to spend on IT infrastructure." No matter how much money you spend on infrastructure, you'll never be totally safe and secure. So the "right" amount of money for a company to spend on IT infrastructure—whether it's for security or for something else like database reliability or resilient servers—depends on the amount of risk that the company is willing to tolerate.

Four Guidelines for Determining the Right Level of Security Investment

So how do you recommend the "right" level of security investment? I think you should use these guidelines:

1. What are other companies doing who have a similar risk tolerance to your company? That might be your direct competitors, or companies in the same industry, or companies in the same broad category (e.g., financial services or consumer products or industrial suppliers). These are the companies who you will be compared to if you have a security breach, so your security level needs to be at least as good as these other companies.

2. Does your company deal with confidential information from your customers? Medical history? Credit card numbers? Payroll or investment information? Personal details of their lives? Extra security should be implemented to protect this valuable customer information. Government regulations may require it, but it makes sense even without government regulations.
3. Does your company differentiate itself from its competition based on an *enhanced* level of trust or risk avoidance? If you want to be viewed by customers as "more trusted" than your competition, then you need to take increased security measures to justify that view.
4. Does your company hold a proprietary advantage over its competition which could be lost if confidential company information was revealed? This information could be secret designs, secret formulas, or even the contents of a proprietary marketing database. If your company's competitive advantage depends on the secrecy of this information, then your security investment level should reflect the need for increased protection for that confidential company information.

Good Security is Invisible

It's difficult to justify security when it's working. The biggest investments in security usually come right after a security breach —one in the news or a breach in your own company's security. Many aspects of IT are like that—they're unnoticed, unrewarded and invisible until something goes wrong. It's human nature to ignore the things that are working and to focus instead on the tasks that need to be done.

This is where trust comes in. If things are going well then your boss has to trust that you're doing a good job, and has to trust your recommendations for doing a better job.

But trust has to be earned, and even trust has its limits during tough economic times. You may be recommending the right things, but you have to be able to convince others that *your* right things are a more important investment for the company than *someone else's* right things.

Making People Dissatisfied is the Only Way to Justify Investment

In my first newsletter article, "1. How to Get People to Change —the Human Side of IT Projects" in April, 2003, I wrote about the three things required to get a person to change. The first one is dissatisfaction with the status quo, and that's the one that's most important when you're trying to sell security investment. To justify additional security investment you have to convince the business that your current security infrastructure is inadequate.

You can do this by:

1. Providing a factual comparison of your security infrastructure to the infrastructure used by other comparable companies, pointing out the areas where these other companies are stronger and better
2. Making the case for increased security for your business due to the company's unique security needs, including any security needs required by government regulation
3. Helping business executives to visualize what the cost of a security breach might be

Conclusion

A security manager has five strategic roles:

- Recommend the right level of security investment, and get your business customers to invest at that required level
- Implement an infrastructure that provides a reasonable level of security for the amount of money the company is willing to invest

- Figure out how to refine the processes and products used in your infrastructure to optimize their performance and reduce their cost
- Determine how you'll respond when there *is* a security breach, and prepare for that possibility
- Make your business customers *feel* secure

To be successful as a security manager you have to be successful at all five strategic roles. There is no "right" amount of money to spend on security infrastructure—or any other type of infrastructure. You have to fight your way out of invisibility to show the business the risk of having an inadequate security infrastructure. And you have to make the case to justify the level of spending that you think your company needs.

So let me revise the wording on that last strategic role. It should say:

- Make your business customers feel secure—but *only* if you think that feeling is justified

If you don't think the business is secure *enough*, then it's up to you to make sure they know it.

2026 Commentary

I had forgotten all about that meeting in Dallas. I hope I did an adequate job of answering the question in the meeting, but I think this is a better answer.

87. How to Become a Manager—13 Skills You'll Need

Originally published on JUNE 24, 2009 in #CAREERS, #LEADERSHIP, #MANAGEMENT, #PROJECTS

In a recent article (#85), I wrote about *why* you might want to be a manager. If that's what you want, here's my list of the 13 skills you'll need:

1. Communication

There's a lot of communication when you're a manager. You have to communicate with each of your employees. You have to communicate "sideways" with your co-workers and customers. And you have to communicate upwards with your own manager or executive. You need some substance in the communication, of course—you need to have something worthy of being communicated. But substance isn't enough—if you know what you're doing and can't properly communicate it to anyone else, then you'll never be a good manager.

2. Listening Skills

This is a part of communication, but I want to single it out because it's so important. Some managers get so impressed with themselves that they spend much more of their time *telling* people things than they spend listening. But no matter how high you go in the management hierarchy, you need to be able to listen. It's the only way you're really going to find out what's going on in your organization, and it's the only way that you'll ever learn to be a better manager.

3. A Commitment to the Truth

You'll find that the higher you are in the management hierarchy, the less likely you are to be in touch with reality. Managers get a lot of brown-nosing, and people tend to sugar-coat the news and tell managers what they want to hear. The only way you'll get the truth is if you insist on it. Listen to what people tell you, and ask questions to probe for the truth. Develop information sources outside of the chain of command and regularly listen to those sources as well. Make sure you know the truth—even if it's not good news.

4. Empathy

This is the softer side of listening and truth. You should be able to understand *how* people feel, *why* they feel that way, and *what* you can do to make them feel differently. Empathy is especially important when you're dealing with your customers. And whether you think so or not, you'll always have customers. Customers are the people who derive benefit from the work you do. If no one derives benefit from your work, then what's the point of keeping your organization around?

5. Persuasion

Put all four of the preceding skills together, because you'll need them when you try to persuade someone to do something you want done. You could describe this as "selling" but it's more general. Whether you're trying to convince your employees to give you a better effort, your boss to give you a bigger budget, or your customers to agree to something you want to do for them, your persuasion skills will be strained to their limits.

6. Leadership

Leadership is a specialized form of persuasion focused on getting other people to follow you in the direction you want to go. It's assumed that the leader will march into battle at the head of the army, so be prepared to make the same sacrifices you're asking your employees to make.

7. Focus

The key to successful leadership is focus. You can't lead in a hundred different directions at once, so setting an effective leadership direction depends on your decision *not* to lead in the other directions. Focusing light rays means concentrating the light energy on one spot. Focusing effort means picking the most important thing to do and then concentrating your team's effort on doing it.

8. Division of Work

This is the ability to break down large tasks into sub-tasks that can be assigned to individual employees. It's a tricky skill—maybe more an art than a science, almost like cutting a diamond. Ideally you want to figure out how to accomplish a large objective by dividing the work up into manageable chunks. The people working on each chunk should be as autonomous as possible so that the tasks don't get bogged down in endless discussion and debate. You have to pay careful attention to the interdependencies among the chunks. And you have to carefully assess each employee's strengths, weaknesses and interests so that you can assign the best set of sub-tasks to each employee.

9. Obstacle Removal

Inevitably, problems will occur. Your ability to solve them is critical to the ongoing success of your organization. Part of your job is to remove the obstacles that are preventing your employees from doing their best.

10. Heat Absorption

Not all problems can be solved. When upper management complains about certain things that can't be avoided (e.g., an unavoidable delay in a project deliverable), it's your job to take the heat. But what's more important, it's your job to *absorb* the heat to keep it from reaching your employees. It's the manager's responsibility to meet objectives. If the objectives aren't being met, then it's the manager's responsibility to:

- Make sure that upper management knows about the problem as early as possible.
- Take all possible steps to solve the problem with the resources you've been given.
- Suggest alternatives to management that will either solve the problem or minimize it. These other alternatives may propose the use of additional resources beyond the current budget, or they may propose a change in the objective that's more achievable.
- Keep the problem from affecting the performance or morale of your employees.

11. Uncertainty Removal

When higher management can't give you consistent direction in a certain area, it's up to you to shield your employees from the confusion, remove the apparent uncertainty, and lead your employees in a consistent direction until there's a good reason to change that direction.

12. Project Management

This is a more advanced skill that formalizes some of attributes 7-11. Although both "Management" and "Project Management" contain the word "management," they aren't the same thing. Management implies a focus on people, while Project Management implies a focus on the project objective. You can be a Manager *and* a Project Manager, or you can be a Manager *without* being a Project Manager. You can also be a Project Manager without being a Manager (in which case you don't have people reporting to you—you just deal with overseeing the project-specific tasks).

13. Administrative and Financial Skills

Most managers have a budget, and you'll have to be able to set the budget and then manage to it. You'll also have to deal with hiring, firing, rewarding good employee performance, dealing with unacceptable performance from some employees, and generally making sure that your employees have the environment and tools they need to do their work. It's ironic that this is skill number 13 (an unlucky number in some cultures), because a lot of managers hate this part of the job the most. But if you're good at budgeting, you'll find it much easier to do the things you want to do. And hiring and dealing with employees on a day-to-day basis is one of the key skills to give you the best, happiest and most productive employees.

Conclusion

This article explains some of the things you'll need to learn before you become a successful manager. You can probably become a manager without having all of these skills, but you'll need all of them to be really successful and to get promoted to higher levels of management.

For every one of these skills, there are various levels of performance. No one expects a new manager to be superior at every one of these skills, but you should be aware of all of them, and you should do everything you can to learn more about each skill. Some of that learning will come through education (like reading the articles in this book). But much of the learning will come through experience—trial and error.

Just learn as much as you can about each skill, take nothing for granted, and focus on doing the very best that you can do. Learn from your mistakes and try not to repeat them. And ask for feedback—in many cases you won't know what you could do better unless someone tells you.

2026 Commentary

This was the most popular article on my web site, getting more than a million views. Apparently a lot of people thought it was a good description of what's needed for a new manager.

88. What's the Opposite of ROI?—How to Compare Cost-Cutting Opportunities

Originally published on JULY 1, 2009 in #MANAGEMENT, #PROJECTS

A lot of companies routinely use ROI (Return on Investment) to compare multiple projects competing for limited investment resource. But what do these same companies use during a bad economy when they're trying to figure out where to make cuts? How do you compare multiple opportunities for cost cutting?

Recently I moderated a group discussion at an Atlanta SIM meeting on the topic "What do we do differently in IT as a result of a bad economy?" Early in the discussion, I asked the question, "What process do you use for determining the areas to cut?" but all of the answers I received were very general. It got me thinking that there ought to be a specific approach like ROI that can be used during a bad economy, so I set out to define such an alternative.

I decided that I'm looking for something that's the *opposite* of ROI. ROI is essentially Return divided by Investment. The opposite of Return is Loss. In this situation we're not trying to make money from a cost-cutting project—we're trying to minimize our loss of existing capability. For example, we might want to cut back on our help desk, but we want to minimize the impact of the cuts on the people being supported. That impact is the Loss associated with this opportunity.

Looking at the other part of the ROI formula, the opposite of Investment is Savings. We're not going to contribute additional funds to this effort—we're looking to save money instead: we'll reduce the amount of money we spend. In the help desk example, we want to cut the cost of the help desk, and that cost reduction is the Savings.

OK, so now we've got a ratio that compares Loss to Savings. But with ROI we know that a higher ROI is better—we get more return for our investment. With our bad economy formula, however, we want to maximize Savings and minimize Loss, so if you use a Loss/Savings ratio then a *smaller* number is better. To change this, we can swap the numerator and denominator in the formula, and instead of dividing Loss by Savings we can divide Savings by Loss. That gives us a number that—like ROI—gives a better result for a higher value.

Let's plug some numbers into our formula to try it out. Suppose we can cut $100,000 out of our help desk budget and have a loss of service of only $50,000 (estimated by looking at the value of the time lost by help desk users who have to wait longer for a solution or who have to solve minor problems on their own). That would give us a Savings/Loss ratio of 2, which we can then compare to the Savings/Loss ratio of another cost-savings opportunity. Voilà! We now have a ratio that works like ROI—we can use it to compare multiple cost-saving opportunities. Opportunities with a higher Savings/Loss ratio offer better returns, and ratios less than 1 don't offer enough Savings to make up for their Loss.

My Measure for Cost-Cutting Opportunities = Savings / Loss

Let me end with a few disclaimers. First, this new ratio has all of the limitations that ROI has. It gives you a deceptively precise result that's based on the inaccurate estimates that are usually put together by people who have a vested interest in the outcome. It neglects any consideration of whether the losses and cost savings are strategic or non-strategic. It doesn't consider interrelationships among multiple opportunities. And it ignores the impact of cost-cutting decisions on people—whether jobs are kept or lost.

What do we call our new ratio? If Return on Investment is called ROI then it seems logical that this new ratio should be called

SOL, or Savings on Loss. And maybe that's appropriate. Because when you decide what areas of your company to cut, you really are deciding which employees are SOL.

Conclusion

Despite the unfortunate name (feel free to use a different one), SOL is an interesting way to compare cost-cutting alternatives. And if you use the measure you'll get a very solid benefit. Most people who look at ways to cut cost don't consider the impact of cost-cutting on the customers of the organization being cut. Using SOL lets you take the needs of those customers into account, and so it considers both sides of each alternative.

Cut costs if you must, but make sure you're cutting costs in a way that doesn't just shift the cost onto your customer or into a different part of your business. Using SOL will allow you to compare the true impact of different cost-cutting alternatives, and select the most appropriate ones.

2026 Commentary

This is another article with a subject that reflects the poor economy in 2009. But cost-cutting doesn't just occur during a poor economy, so the measure can be used during any attempt to save money. The biggest problem with a measure like this—as it is for ROI—is the neglect of intangibles that are extremely difficult to measure. But maybe it's better than the alternative of arbitrarily cutting things without evaluating their impact of the rest of the business.

For articles about the issues with ROI and how you can deal with them, see "20. Why ROI Isn't Working" and "21. How to Improve ROI and your Project Selection Process."

89. The Meaning of "Scalable" Needs to Change

Originally published on JULY 8, 2009 in #IT, #MANAGEMENT, #PROJECTS, #STRATEGY

The word "scalable" has been used in IT for over forty years. IBM used the word when they first talked about their 360 series of mainframe computers in the 1960s. Since that time scalability has been a consideration in every aspect of computing: mainframes, minis, personal computers, servers, networks, proprietary systems, open systems, even smartphones.

But up until recently, "scalable" usually meant that you could make a system scale *up*—you could add more computing power and run the same application with much higher transaction volumes. That's what most people were interested in: being able to reuse their application architecture for higher volumes as their companies grew.

We began to see the need for downward scalability about twenty years ago during the beginning of the outsourcing era. As IT organizations outsourced some of their applications, they needed less processing power on the computers they had been using for those applications. But they found that it wasn't very economical to scale down. Vendors usually offer you great deals to trade-in your smaller hardware for bigger boxes, but you won't find such a good deal when you want to replace a large box with a smaller one.

In the last ten years, virtualization has had a large impact on scalability. With virtualization you can carve up a large computer into a series of smaller virtual computers, then alter the size of those virtual computers at will. Virtual computers can scale both up and down, but you're still stuck with the large computer that's underneath all of those virtual systems. So while virtualization can help you balance computing load across your different systems, it doesn't help much if your overall enterprise computing requirement is shrinking.

More recently, "cloud computing" has given us an outsourced version of virtualization. Now our virtual computers can exist at a vendor location or even at a combination of onsite and remote facilities. Scalability has gotten many times more flexible, with its limits determined more by what we can negotiate than by the limits of physical computer hardware.

Yet despite all of the advances that have been made in virtualization and cloud computing, I still hear the word "scalability" primarily referring to the ability to scale *up*. It's like our IT people have been trapped in the scale-up definition for so long that they forget that downward scalability is important too.

Take a look at IT departments that are doing well in the current bad economy. You'll find that in many cases their success is due to their ability to provide the same systems capabilities at much lower volumes with corresponding reduction in cost. Downward scalability is helping them be successful in a time when revenue and system volumes are declining.

So when you look at a new application system as a candidate for use in your company, make sure that it's scalable both upward *and* downward. Your success—and the success of your company—may depend on it.

2026 Commentary

This is yet another article about dealing with an economic downturn, but downward scalability doesn't just happen during a bad economy. You may see the need for downward scalability when you're switching from one systems architecture to another, or from one platform to another, or even from one data center location to another. So there's definitely a need to consider downward scalability, even in the best of times.

90. First-Time Manager Stories of Failure and Success

Originally published on JULY 15, 2009 in #CAREERS, #LEADERSHIP, #MANAGEMENT

I've promoted scores of people into first-time manager positions. Some did well and some didn't. Here are a few of their stories, with names changed and a few relevant facts altered to protect the individuals involved.

Fred

This was early in my career, and I didn't have any experience in promoting people into management. But I was working for a rapidly growing company, and my organization had gotten too big for me to manage all by myself. I needed to add a manager below me to supervise some of the programmers.

Fred had already demonstrated project leadership skills, and he seemed to relate to people pretty well. I didn't have a lot of other candidates in my organization, and I wanted to promote from within, so I decided to give Fred a shot as a new manager.

Two things sabotaged Fred's chances from the start. First, he seemed to think that making him a manager made him always right. Where previously he would have open and honest discussions with project team members about technical details, now he didn't want to get their feedback. The way he saw it, he was manager so he told his people what to do. If he didn't know what to do, he just made something up. Big mistake!

Second, he couldn't handle the pressure. This was probably related to the first reason, but it made matters worse. He would freeze up and become almost unresponsive in certain confrontational situations. It's as if he was trying to hold himself back from exploding. Or maybe he just froze out of fear.

I tried to help Fred in his new role, providing advice and coaching. But it was clear that things weren't going to change very quickly, and so I took Fred out of the manager role and looked elsewhere for a solution. Fred calmed down once he was off the hot seat, and he went back to being a good project leader. I lost track of him over the next few years, so I don't know whether he ever got over his initial problem and went into management again. His first attempt certainly didn't work out well, so I could understand him being reluctant after that.

Lessons Learned (by me, by Fred)

- You can't always tell in advance who will make a good manager and who won't. Obviously I was wrong about Fred.
- Some guidance from my own manager about how to pick a new manager would have been useful, but I was pretty much left to fend for myself.
- Being in management doesn't give you any special knowledge or insight that you didn't have before. And it's important that you don't act like it does.
- Being a manager gives you some authority and the opportunity for some different behavior. It's up to you how you use it.
- It helps if there's a way to reverse a bad promotion. In some companies you couldn't move Fred back to an individual contributor role—you'd have to fire him. And in other companies Fred's pay grade would have changed with the promotion, and it would be an administrative nightmare trying to change him back to his old position and salary. Few companies allow you to "try out" a manager.

Pam

Years later, with a lot more management experience under my belt, I got a promotion but kept responsibility for my old department too. I needed to "replace myself" in my old role. Again I wanted to promote from within if at all possible (it's always my first choice unless I want to take the organization in a radical new direction). My existing employees all had relationships with our customer organization, and industry experience was important. I made it known to the employees that I was asking for internal candidates, and I talked to three or four employees who wanted to apply.

I decided on Pam, who had years of experience with the customers, and who had an in-depth understanding of the industry. She wasn't as technically knowledgeable as some of the other candidates, but she had good people skills and she was respected by all of the employees.

In preparation for the management turnover, I spent a lot of time with Pam talking about each of the employees in the department, pointing out strengths and weaknesses, and giving advice on how to best motivate each employee. Pam absorbed the information eagerly, and she was very prepared for the turnover when it happened.

Once Pam had officially taken over the department, I got together with Pam every day to see how things were going and to offer coaching advice where needed. After a week or so I stopped the daily meetings, but I still kept in touch and I made myself available whenever needed for advice and coaching. A few times Pam got herself in over her head on particular issues, but I worked with her to help her get herself out of trouble—I tried very hard not to "take the wheel" and make her look incapable. It was important that her employees viewed her as the manager in her own right—not as a figurehead representing me.

Lessons Learned (by me, by Pam)

- The support I gave Pam really helped. I used it as a model for future management turnovers.
- Good management skills can be taught as long as the new manager wants to learn.
- I had to clearly communicate to employees that Pam was the new manager and that they had to work through Pam to solve any problems. Otherwise the employees would have done an end-run and come to me. This would have put me in the middle between Pam and the employees—not a desirable place to be.

2026 Commentary

I think the key to performing well in any position is to keep listening and keep learning. I made a mistake with Fred, but learned from my mistake.

91. New Tagline—IT's all about the Business

Originally published on JULY 21, 2009 in #IT, #LEADERSHIP, #MANAGEMENT, #ALIGNMENT

I've changed the tagline in my blog header. It used to be "Insight for Current and Future IT Leaders." Now it's "Insight for Current and Future Business Leaders." I took out the word "IT" and replaced it with the word "business." Here's why:

For over six years I've been writing about lessons I've learned from my business IT experience. In looking back at what I've written, I've noticed that while many of the posts come from an IT viewpoint, most of the lessons are related to the business. In some of the posts there isn't a single mention of IT or technology. Only a few of the posts are technical in nature, and even then there are business lessons in the article.

If I look at the most common ways that IT organizations fail, most of them are due to a misunderstanding of human nature. IT organizations fail because of lack of communication or because they emphasize features of their projects instead of benefits or because the people in IT don't set expectations appropriately. These are not IT issues—they're *people* issues.

So I figured, "Why not make my tagline reflect my beliefs?" IT doesn't succeed because of technology—it succeeds because of its contribution to the *business*. So the new tagline is "Insight for Current and Future Business Leaders." This puts the emphasis where it's needed: You can't be an effective IT leader without being an effective business leader as well.

And the new tagline also calls attention to the fact that many of the things I write are appropriate for non-IT people as well. IT doesn't have its own set of rules independent of everything else in the world. As I've said in my book, *Boiling the IT Frog*, there is no magic in IT. So let's not act like it.

2026 Commentary

I debated whether to include this “article” in the book, but decided to go ahead because it marks a change in emphasis in the things I wrote. It also restates my belief that most IT failures are due to a misunderstanding of human nature, rather than technical issues.

92. 18 Things I Believe about Business—a Manifesto

Originally published on JULY 22, 2009 in #CAREERS, #LEADERSHIP, #MANAGEMENT, #STRATEGY

I Believe ...

1. The best businesses are honest with their customers and their employees.
2. The best companies have a win-win relationship with their customers and with their employees.
3. Management is about focusing the work of the employees by assigning tasks which best align the company's interests with the employees' interests.
4. Executive management is about focusing the work of the company by setting objectives which best align the customers' interests with the stockholders' interests. These objectives define the company's interests.
5. Management and leadership are different. Management makes sure that things are done right. Leadership ensures that the right things are done by focusing on the right things and diverting effort from the wrong things.
6. Both management *and* leadership are performed at all levels in the organization. The scope of your decisions is greater when you're higher up in the organization, but there is both management and leadership going on at all levels.
7. We can and should motivate employees without making them afraid.
8. It's possible for employees to balance work and family life, and still achieve company objectives. This balance should be an

employee goal, and companies should encourage and support their employees in the achievement of this goal.

9. Bad companies try to take inappropriate shortcuts. An appropriate shortcut is an ethically responsible improvement to a business process that is planet-friendly, customer-friendly, employee-friendly and generally people-friendly. On the other hand, inappropriate shortcuts use resources without replenishing them—ecological resources as well as people resources. They pollute, they destroy, they burn out people and throw them away. This results in short-term profitability but long-term loss—for our planet, for our people, and for the companies themselves.

10. Every business, ecological or financial failure can be traced back to attempts to take inappropriate shortcuts. I have no evidence for this, but I believe that it is true.

11. A company that refuses to take inappropriate shortcuts is *not* at a competitive disadvantage. Indeed, I believe such a company has a competitive *advantage*. Customers want to deal with a company they can trust. Employees want to work for a company that treats them fairly.

12. Most new companies start off on the right path and don't take inappropriate shortcuts. But most new companies reflect the good values and ethics of the individuals who start those companies.

13. As companies get bigger, they tend to lose their connection with individual people. The reputation of a company no longer reflects the values of its founders, and the executives and boards of directors lose their way. Success leads to overconfidence and feelings of invulnerability. The executives and board members begin to believe their own hype, and they think they can get away with violating ethical rules. Groups of people (e.g., boards of directors, gangs, mobs)—especially groups of

people who consider themselves superior—will do things that individuals would never do. Ethics are diluted in groups.

14. Greed and impatience are the motivation for most inappropriate shortcuts. People want a result without having to pay the fair and appropriate price. And they begin to think that they're entitled to take these shortcuts because of their reputation, their status or their wealth.

15. A company that takes inappropriate shortcuts is essentially saying, "My company is more important than the planet, more important than sustainable life, more important than employees, more important than *anything*." That's wrong.

16. An inappropriate shortcut is first proposed by someone in a company—particularly someone in a leadership position—who has lost their way, at least temporarily. It's up to other people in the company to say "no." I can forgive a temporary lapse in judgment for the person proposing the inappropriate shortcut. But how can we tolerate the other people who agree to go along?

17. In a sense, lack of integrity in business is like cancer in the human body. Cancerous cells are constantly being created in everyone's body—in each and every one of us. But healthy bodies quickly shut down the cancerous cells to prevent them from growing bigger. The difference between a healthy body and a cancerous body isn't in the creation of the cancer cells—it's in the body's ability to shut down the cancer cells before they grow bigger. Lack of integrity in business works the same way. It's human nature to occasionally want to take an inappropriate shortcut. But in the good businesses those thoughts of inappropriate shortcuts are shut down before they're implemented.

18. All employees have the ability to say "no" to inappropriate shortcuts. In many cases it's difficult to do so, but we *do* have the ability.

I've painted this as a black-and-white issue, but it's really more like shades of gray. Each executive, board member and employee in a company influences the direction and overall integrity of the company. Higher-ups in the organization have more influence, but everyone has *some* influence.

I'm not suggesting you immediately start saying "no" to the least little thing that seems wrong to you. This is more of a statement of direction. So pick your battles, and through a gradual change in approach we should be able to make things better.

Conclusion

I believe in the fundamental goodness of people. And I believe that we can have fundamentally good businesses as well. We just need to make it a priority.

There are a number of good businesses in existence right now, and there are also bad businesses. Which kind do you work for? What are you going to do about it?

2026 Commentary

The world has changed a lot since 2009 when I wrote this, but I still agree with everything I wrote back then.

93. What is Cloud Computing? And Why Should You Care?

Originally published on JULY 27, 2009 in #ALIGNMENT, #MANAGEMENT, #PROJECTS, #STRATEGY

Cloud Computing sounds like something vague and amorphous. After all, it's a cloud, right? So that means it's something that's insubstantial, floating above us in the sky.

If you think of the clouds as the Internet, and think of Cloud Computing as using computers over the Internet, then you can see where the phrase came from. But that doesn't tell you very much about what Cloud Computing is or why you should care.

Another Step in a Progression

The concept of Cloud Computing is simple. It's just another step in a progression that we've been taking with our technology. We started with the idea that all of our systems run on computers which belong to our company—everything runs "in-house." Then we gradually caught on to the fact that maybe it's not the most efficient thing to use our own employees and computers to do this sort of stuff, especially the mundane things like accounting and payroll and basic business processes. After all, it makes more sense for our company to spend its money creating value for our customers. So if there's a way we can outsource some of the day-to-day work then we ought to seriously consider it.

Step 1: Outsourcing

Outsourcing is great for processes that are easy to do at arm's length. Payroll is the classic example. But there are other day-to-day processes that require more interaction with our managers and employees. And classic outsourcing—send the whole process out the door for someone else to do it—isn't very conducive to lots of employee interaction.

So we're caught in the middle: we want to outsource a process, but we also want our employees to be able to interact with the process. What do we do?

Step 2: Outsourced Hosting

Up until recently, the only solution was to outsource the "hosting" of the process. Essentially, we hire an outside vendor to set up one or more computers for us, install the software we need, and then take care of all of the "back room" technical activities. Our employees use the software by connecting remotely to the hosted computers. In the early days our employees would connect using a phone and a modem. But ever since the Internet this has been largely unnecessary, and now it's quite common for us to connect to outside systems using the Internet. It's secure too, especially if you use an encrypted connection via Secure HTTP (HTTPS).

OK, so this approach seems to do the job, but it's not quite as nice as we would like. For one thing, we still have to buy a fixed amount of computer capacity. There are still dedicated computers that are used for running our software—they just sit in another company's location. If we need more capacity then we have to pay to expand. If we need less capacity due to a decrease in demand (the economy has been doing that to a lot of us lately), then we're stuck—the computers have already been purchased by our hosting vendor, and it's not very economical to cut back.

Step 3: Software as a Service (SaaS)

Enter "Software as a Service" (SaaS). SaaS lets you share your hosted computing resource with other companies, but without revealing your secure secrets. It's as if you're each logged in to the same (virtual) computer but you're each using a secure ID that keeps your data separate. Think of Amazon.com. When you're logged in to Amazon you can see the orders that you've recently placed, but you can't see other people's orders. That's like SaaS.

A company that offers an SaaS product doesn't put you on dedicated hosted computers. You might share a computer with other SaaS customers, but you'll never know it. You get most of the security of having your own computers, but you don't have to worry about buying a fixed amount of computer capacity and then upgrading it. It's kind of like the way you get mobile phone service. You don't have to buy your own cell tower—you just share the towers that exist, taking advantage of the buying power of all of the other customers who share your resource. If you make a lot of calls, you pay more. If you make fewer calls, you pay less. You can sign up for different types of contracts based on your planned usage, and the price varies based on your contract and your usage.

Notice that I said "you get *most* of the security of having your own computers." I won't kid you: There's slightly more risk in sharing a computer than in having your own. But good companies that offer SaaS products have taken some remarkable steps to build a software wall between customers, and in practical terms you're very, very safe.

Step 4: Cloud Computing

So how is Cloud Computing different from SaaS? Quite frankly, to 99% of you there isn't a visible difference. The difference is in the back room where the computers reside. Cloud Computing lets you distribute the computer power across multiple computers that may be in multiple locations. It's like setting up your checking account with an overdraft option. If you accidentally write a check for more money than you have in your checking account, the overdraft option will automatically move money from your savings account to make up the difference.

Similarly, Cloud Computing allows your computer center to run your work on one computer but then automatically farm out some of the work to other computers when there's an unusually high need for computer resource. It helps alleviate the response-time problems that come with peaks in demand. In car terms it's like a turbocharger—kicking in when you need it, but sitting quietly in the background when you don't.

Conclusion

Software as a Service (SaaS) lets you outsource a process but still do lots of interaction with the outsourced systems. And it gives you a great deal of flexibility because you do not have to pay for an entire computer. Cloud Computing is like turbocharged SaaS—it gives you the power of SaaS with the flexibility to support periodic peaks in demand. They both let you access your data over the Internet, and they both let you keep your data secure. Most importantly, using both approaches you can pay for the results you need—not for the cost of the technology it requires.

2026 Commentary

This probably seems like pretty basic stuff to current readers, but Cloud Computing was very new in 2009. While the SaaS term doesn't seem to have survived in today's common usage, the term Cloud Computing is still in use, and is taken for granted by most people these days.

For a related article, see "123. Why Cloud Computing is Good for Your Financials."

94. 4 Reasons We Disagree, and What to Do About It

Originally published on JULY 29, 2009 in #CAREERS, #ALIGNMENT, #LEADERSHIP, #MANAGEMENT, #PROJECTS, #STRATEGY

You're trying to get a new project approved, and you're having trouble. Or you're trying to get an employee to do things *your* way, and the employee keeps fighting you. Both these situations are disagreements, and the process to deal with them is similar.

Why Do We Disagree?

Let's start with individual disagreements. When two people disagree, it's usually for one or more of these four reasons:

1. **Objective**
   What one person wants to accomplish is different than what the other person wants.
2. **Method**
   They may agree on the objective, but they have differing opinions on the best way to achieve the objective.
3. **Communication**
   They may agree on both the objective and the method, but they don't realize it. They keep arguing, and because of miscommunication they each get the impression that the other person disagrees.
4. **Emotion**
   They may agree on the objective and the method, and they may even understand that they agree on these things. But they choose to disagree anyway (or at least one of them does). There are a lot of possible emotional reasons: everything from

distrust, dislike and hate, to embarrassment, fear, peer pressure, face-saving or just a steadfast commitment to a prior position.

Group Disagreement

These same four reasons apply to individuals when they're in groups. Each individual in the group has an opinion, but group dynamics influence those individual opinions and make the situation more complicated. A group opinion is more than a simple sum of individual opinions. Some members of the group are more vocal and tend to contribute more to the group opinion. Some members of the group have more clout (due to position, rank, respect, trust or just habit) and so their individual opinions carry more weight in a group discussion. Some members of the group are apathetic or afraid to express their opinions, and so their personal opinions go unheard and unrecognized.

How to Resolve an Individual Disagreement

1. Start with reason #4—emotion. You need to understand whether or not there's an emotional response to you or to your issue, and figure out how to cut through the emotion and get to an objective conversation. Sometimes this can be done by putting the emotional concern out on the table, often in a self-deprecating way that acknowledges and values the other person's position. Here's an example: "I know that I can sometimes come across as dictatorial and abrupt, but we both want this project to be a success. I'll do my best to try and understand *your* point of view, and I would really appreciate it if you would try to understand *my* point of view. And maybe by working together we can do what's best for the company."

The approach you take in dealing with the emotion is going to vary from situation to situation. In some cases you won't be able to deal with the emotional issue, and you'll have to skip this step—at least temporarily—and move on.

2. Next, work on reasons #1 and #3 simultaneously—objective and communication. Go back to basics and find some basis of agreement, however small. Maybe it's a part of the objective, or even the context of the objective. The idea is to start with something you both know you can agree on, and then build from there.

This process helps communication, but here's something else you can do to improve communication even more: Try to restate the other person's position in your own words. Then have the other person restate *your* position in their own words. Go back and forth on this until you can agree that you're both accurately representing each other's opinion. Many times you'll discover that you're closer to agreement than you thought. And if not, at least you have a better understanding of why you disagree.

3. Build a logical case on top of the things you agree on. You may find that you don't agree on objectives, and that's as far as you can go. Or you may find that you disagree on objectives but that your different objectives still support the same solution or method. For example, you may disagree on *why* the health care system in the United States needs to be revamped, but you can still agree on a course of action that will lead to changes.

4. During all of this, keep reevaluating the emotional side of the disagreement. Watch out for emotional reactions to what you perceive as logical statements, since not everyone sees certain words and phrases the same way. Address emotional issues as you can, or tiptoe around them if you need to.

How to Resolve a Group Disagreement

The best way to resolve a disagreement with a group—or to prevent a potential disagreement—is to talk to each member of the group one-on-one. If you can convince every member of a group to agree with you, then you only have to deal with face-saving when the group reconvenes to discuss the issue. And you can facilitate the face-saving process by suggesting ways that a group member may be able to explain or justify a change in their opinion. Help the group members get out of the situation the group has put them in.

If a group is too large for you to talk to each member one-on-one, then talk with certain key members who hold the most influence over the group opinion. Make sure you single out the influential members who you think might be opposed to your idea—getting them on your side will be key. When you talk to a group member who agrees with you, ask for help in convincing the others. And ask for advice—they may have ideas on things you can do to sell your point of view and win over other members of the group.

Conclusion

Dealing with disagreements is a part of being human, and a key part of being a manager or leader. Use the process suggested above, but remember that you're not always right. Sometimes the best way to resolve a disagreement is to recognize that you ought to change your mind. In that case you may end up agreeing with what the other person has been saying all along.

2026 Commentary

In today's world we seem more divided than ever. I don't know if this approach can be used to bridge political divides, but it certainly works for day-to-day business disagreements.

95. The World's Economic Problems are Like 2 Cross-Wired Thermostats

Originally published on AUGUST 5, 2009 in #LEADERSHIP, #STRATEGY

Many years ago I did some work at a newly built manufacturing plant in Phoenix. The new plant was having trouble with its air conditioning system—the administrative offices were too cold and the manufacturing shop floor was too hot. While I was there it was discovered that a mistake had been made during installation: the air conditioning system for the shop floor had been wired to the thermostat in the administrative offices, and the air conditioning system for the administrative offices had been wired to a thermostat on the shop floor. It's funny in hindsight: the people in the administrative offices would be cold so they would turn the thermostat up. The people on the shop floor would be hot so they would turn the thermostat down. Thus the temperature problem was created by cross-wiring the two thermostats.

I was thinking the other day that our world economic problems are similar.

Consumers

On the one side we have the consumers who are being battered from every direction. They're losing their jobs, their 401(k) retirement accounts have lost much of their value, gas prices are high, home values have dropped, and adjustable rate mortgages are increasing their payments.

Business

On the other side we have businesses who are being battered as well. Consumers are spending less, so revenues are down. Investors are living in fear, so it's difficult to fund new development or—in some cases—even generate enough cash to pay the bills. Other countries aren't doing much better—it's not just a U.S. problem. Businesses are in a slump, and the future doesn't look great.

Like 2 Cross-Wired Thermostats

We have the same cross-wired situation between consumers and business. Consumers control the "temperature" of businesses, since their purchases make businesses successful. On the other hand, businesses control the temperature of the consumers, since they provide jobs and since business investment forms the basis for the stock market and those 401(k) investments.

See the problem? It's just like in that manufacturing plant: The consumers are feeling "hot" so they're turning down the heat on business by refusing to buy. And the businesses are feeling "cold" so they're turning up the heat on consumers by laying off workers, tightening consumer credit and in some cases raising prices.

A Solution

Now in the manufacturing plant we had the option of rewiring, but that's not an option for the economy. No, we have to live with the idea that consumers and business are interdependent—that's what capitalism is all about. So what can we do instead to solve the problem?

In the manufacturing plant we needed to provide a short-term solution for our cross-wiring problem until the electrician arrived. So we set both thermostats to a "normal" position (one that wasn't too hot and wasn't too cold). Then we fine-tuned the thermostat settings by making small changes, waiting to see how it affected the other part of the plant, and then revising as needed.

How the Government Fits In

If you think about it, this is the same thing that the U.S. government has been trying to do (and probably other governments as well). The government has tried to take steps to make both business and consumers act like things are "normal" even though it's somewhat artificial to do so. "Stimulus payments" and "recovery rebates" are ways of providing short-term money to help consumers begin to feel normal again. Government "bail-out" of banks and a few key industries is a way of helping business get back to normalcy.

So the solution is to have the government help consumers and businesses feel like things are normal on a temporary basis. And if things go right, then both consumers and businesses will begin to count on things being normal in the longer term, and the economy will recover. The stock market will recover, consumers will spend, businesses will once again become profitable, and out-of-work consumers will get jobs.

You've Got to Believe

Will this work? It worked for our thermostat problem in Phoenix, and I believe that it will work for the economy as well. I encourage each of you to believe it because *believing in this solution is what makes it work.* Like Dorothy in the *Wizard of Oz*, we've had the power to go where we want to go all along—we just have to use it.

So close your eyes, click your heels together, and repeat after me, "There's no place like a good economy. There's no place like a good economy. …"

Sidebar: A Business IT Application of this Principle

And while we're on the subject of cross-wired thermostats, this metaphor works for Business and IT as well. Business controls the perception of IT, and IT has a large impact on the effectiveness and efficiency of Business. When Business views IT as failing, then Business tends to micromanage IT. This usually results in further IT failures, since most people in Business don't really understand IT. Thus we get a downward spiral, just like a downward economic spiral.

Sometimes we need to "push the reset button" on Business and IT views of each other, just like we're trying to reset the economy. This gives both parties a chance to develop a new working relationship, and things can improve from there.

So if your company's perception of IT is stuck in a downward spiral, call a truce, get everyone to agree to put the past behind you, and try to move forward with a new view of IT. If you've got the right people leading IT, then this will stop the downward spiral and start an upward spiral in its place.

2026 Commentary

It's difficult to remember exactly what was going on in the economy in 2009, and how this approach actually seemed to work. And the current U.S. government seems to be taking the opposite approach: pointing out problems that many people don't think we have. It will be interesting to see what happens next.

96. Irreconcilable Differences and Runaway Projects

Originally published on AUGUST 12, 2009 in #ALIGNMENT, #MANAGEMENT, #PROJECTS

A runaway project is like a married couple on the brink of divorce. There are two opposing points of view, both sides are usually angry, each side blames the other, legal action is imminent, and a lot of time and money is being wasted.

So why do projects go into a runaway mode? It's usually because requirements are changing faster than the project team can keep up with them. It's like swimming upstream in a tidal wave—no matter how hard you swim you still get swept back by the overwhelming water.

But that just raises a different question, doesn't it? Why is it that requirements are changing so much? Shouldn't requirements change the most at the *beginning* of the project while they're still being defined, and then settle down once the requirements are written down and agreed to? Yes, of course, but that "normal" way to define requirements is usually based on some erroneous assumptions.

The Usual Erroneous Project Assumptions

Most runaway projects start by making one or more of these assumptions:

1. It makes sense to freeze requirements at a moment in time and then stick to them, no matter what happens.

2. The requirements process is perfect and what's written down is exactly what the customer wants.

3. The requirements as written are totally unambiguous, consistent, and doable.
4. The customer fully understands what's been written down and agrees that it's correct.
5. The project team members fully understand what's been written down and plan to do exactly what the requirements say.
6. The project estimates are in fact based on the requirements—not on a similar project done elsewhere for a different set of requirements.
7. The project will be completed quickly enough that no requirements will change during the project work.

None of these assumptions are valid. In most cases we base a project on incomplete requirements that are likely to change during the project—requirements that aren't totally understood by either the customers or the project team. In fact, when you think about it, it's a miracle that *any* IT projects are ever completed on time and within budget. The latest statistics from the Standish Group show that only 32% of IT projects are completed on time and within budget, so in fact it does seem to take a minor miracle to make a project succeed.

The Path to a Runaway

Let's look at a typical runaway project. Some of them start on time and even look pretty good for a while. But pretty soon minor things start to get on people's nerves. A requirement changes, and the impact of the change on the project is underestimated. A new technology turns out to be more difficult to use than anyone thought. A higher-up agrees to a change without understanding the huge effect it will have. A late delivery of a key subproject causes everyone to reevaluate what's going on. Changes begin to accelerate, making the project later. The later the project gets, the more changes there are. The change rate exceeds the capacity of the project team, and someone naively suggests that more project resource ought to be added. Of course the additional resource just makes everything even later due to coordination issues and getting everyone up to speed, and the late project builds up momentum. The "burn rate" of the project increases to the point where it gets more executive attention, and this leads to extra executive reviews, which tie up critical project resource and make the project even later.

Runaway projects aren't called "runaway" for nothing. They burn money like crazy, they get later with every passing day, and they often can't be stopped except by killing the project outright.

Go back to the married couple on the brink of divorce. What could be done to try to save the marriage? Many people try counseling—bringing in a third party to help with communication [more on that subject in the next article]. I'm told that most marital problems have their root in communication, and if two people can really talk to each other and *if* they really want the marriage to be saved, then they can usually work things out.

Why don't we attack runaway projects the same way? It's the same issue—communication. The customers want something, and the project team is having trouble delivering it. There are really only two choices: revise the requirements to deliver what's possible, or figure out a way to deliver what the customers want. Usually the solution is a little of each.

What's different about the runaway project is its scale. Big projects involve big egos, politics, and all of the baggage of previous projects gone wrong. Working things out between customers and the project team isn't just working out a disagreement between two people. It's working out many disagreements among huge groups of people. It's owning up to mistakes that were made. It's accepting the fact that you may not get everything you want. It's making compromises.

Conclusion

Marriages tend to collapse when the husband and wife lose respect for each other's goals. Many marriage ceremonies recognize three interests in a marriage: the husband's, the wife's, and the interest of the married couple together. All three interests have to be recognized, discussed, and prioritized.

It's the same thing with projects. There are three interests in a project: the customer's, the project team's, and the interest of overall project success. Runaway projects happen when the focus shifts off of overall project success and instead onto the individual interests of the customers or the project team. Irreconcilable differences and runaway projects go hand in hand. And after a divorce or a cancelled project, it's only the lawyers who end up ahead of the game.

Next article: 97. IT Marriage Counseling—how to do it, and what to look for in a counselor

2026 Commentary

This is the second perspective on runaway projects that I have written. The first one was two years earlier in the article, "63. How to Stop a Runaway Project." The earlier article identified complexity as the chief cause of runaway projects, while this article focuses on communication.

Both views are correct—just different in their point of view. For further discussion on this, see the 2026 Commentary for the next article, "97. IT Marriage Counseling."

97. IT Marriage Counseling

Originally published on AUGUST 19, 2009 in #IT, #ALIGNMENT, #LEADERSHIP, #PROJECTS

I've been comparing the IT/Business relationship to a marriage for a while now. In Chapter 12 of my book, *Boiling the IT Frog*, I said:

Secret 28: The Information Technology organization is your partner in creating and managing systems and data, with shared responsibilities.

That partnership can be like a marriage, with both marriage partners working together to make a better life. But like all marriages, there are times when problems cause disharmony between the marriage partners, and if you're not careful, that disharmony can lead to name-calling and blame.

Primary Causes of Business/IT Marriage Problems

I believe that problems in the Business/IT relationship stem from the same sorts of things that cause many of the problems in a marriage:

- Disagreement on objectives and/or methods to accomplish those objectives (see my article, "94. 4 Reasons We Disagree, and What to Do About It")
- Miscommunication (see my article, "24. Use Their Terminology—Not Yours")
- Loss of trust (see my article, "65. Information Technology is like the Stock Market")
- Emotional issues like egos, territory disputes, and who gets credit (see my article, "48. The Politics of Information Technology")

- Not living up to expectations, or expecting too much (see my article, "53. How to Fail as a CIO")
- Trying to cope with stressful limitations like finances and time
- Fighting over what's best for the "children," where Business/IT children refers to the systems and processes that have been created through collaboration between the Business and IT

Signs of an Impending Need for Marriage Counseling

Here are some of the things you'll see when a Business/IT relationship is beginning to fail:

- Communication breaks down. One side or the other (or both) won't listen. Or what they seem to hear isn't what is being said.
- There is a lot of blame and finger-pointing
- The phrase "IT is failing" keeps coming up in conversations
- You'll hear a lot of "You don't tell me what you need" or "You keep changing your mind"
- IT projects are out of control. Or at least they're perceived that way (see the previous article, "96. Irreconcilable Differences and Runaway Projects")
- The business won't sign off on project deliverables

What to Do

Here are some of the steps I take when I'm asked to do IT Marriage Counseling:

- Interview both sides, capture each side's perspective on what's going on
- If communication is an issue (and it usually is), then facilitate communication between the two sides. Help each side begin to understand the other side's point of view

- If one or more individuals are at fault (whatever side they're on), then work with those individuals to help them solve their problems. Or—if all else fails—help them understand that this isn't the right job for them
- Be an impartial third-party to arbitrate disputes
- Try to find some basis to rebuild trust between the opposing parties, based on the things that are working well
- Present a summary of the situation to both parties, pointing out the things that are working well and the things that need improvement
- Make specific recommendations for changes in processes, communication styles and methods, organizational structures, and—in some cases—people

Conclusion

No matter what side of the marriage you're on—or even if you're just an interested bystander—if you see signs of "marital difficulties" between IT and the Business then you need to diplomatically bring it to someone's attention. Sometimes the two parties can work out the disagreements between themselves, but occasionally a third-party is needed to facilitate communication and help the disputing organizations resolve the problem. There's no shame in needing the help of a third-party, and using a third-party will often help you reach a resolution quicker.

If you know of someone who may need the help of an IT Marriage Counselor, share this book with them. There's a lot of material in the book to help you with various aspects of a Business/IT dispute, and my other book, *Boiling the IT Frog* is a good tool to help business and IT people understand each other.

Sidebar: 10 Things to Look for in an IT Marriage Counselor

Whether you use a third party or someone inside your company, an IT Marriage Counselor should have these characteristics:

1. Ability to listen, and to hear both explicit and implicit messages in what's being said
2. Impartiality—no axe to grind
3. Capable of communicating with both business and technical people (this is often the most difficult requirement)
4. Experienced in dealing with common IT and business problems
5. Able to put things in the proper perspective
6. Even-tempered
7. Non-judgmental
8. Experienced in arbitration and negotiation
9. Technically knowledgeable enough to see through people issues disguised as technology necessities (e.g., "we can't outsource because our technology won't allow it")
10. Business knowledgeable enough to see through people issues disguised as business necessities (e.g., "we can't roll out this system in phases because our users can't handle the disruption")

2026 Commentary

I've written two articles on runaway projects, each with a different perspective. The first one, "63. How to Stop a Runaway Project," took the point of view that runaway projects are caused by overwhelming complexity. The second article, "96. Irreconcilable Differences and Runaway Projects," took the point of view that runaway projects happen when the customer and the project team aren't communicating well.

These two points of view don't conflict. That's because complexity is often the by-product of lack of communication. In an earlier article, "40. The Blind Men and Information Technology," I used the story of the "Blind Men and the Elephant" to point out that when you don't have a high-level overview of something, you tend to describe it using lower-level information. That's what often happens with project requirements when communication breaks down. Instead of defining high-level needs, and breaking the project up into subprojects that fulfill those needs, you end up with lower-level requirements that you expect to add up to the unstated high-level requirements.

Let's take a quick example. Suppose we're building an accounts payable system. High-level project requirements would be things like "Optimize cash flow while taking optimum advantage of vendor discounts." Low-level requirements would be things like, "I need a screen to enter vendor information, and another screen to enter invoices."

If your project definition doesn't get any high-level requirements but just gets a huge list of low-level requirements, then your project probably won't succeed. That's because it's highly unlikely that your low-level requirements list will meet all of the unstated high-level requirements. Furthermore, the project will be more complex than it should be, because you'll find that the requirements will keep changing as people realize that certain low-level requirements have been omitted.

On the other hand, if you get the primary high-level requirements, then you can use those requirements to create the low-level requirements. And since the high-level requirements will guide the project, the complexity will be minimized.

98. 13 Ways to Spot Lies and Omissions in Due Diligence

Originally published on AUGUST 26, 2009 in #ACQUISITIONS, #ALIGNMENT, #STRATEGY

We all do due diligence. Some of us do it in mergers and acquisitions (M&A). Some of us do it when we're getting ready to make a major purchase like a house or a car, or when we're getting ready to sign a contract for major home repair. Due diligence is the research you do to make sure that what you buy is what you think you're buying. It's verifying that the "facts" are correct as represented by the seller.

When witnesses are sworn in before their testimony in a court of law, they swear to tell "the truth, the whole truth and nothing but the truth." It's the "whole truth" part that's usually the problem. The whole truth means that they have revealed all aspects of the truth—they haven't just answered a question in a truthful but deceitful way. This is also the primary area of concern in due diligence: someone may answer a question truthfully, yet they've held back the information that you really need to hear. President Bill Clinton's famous statement, "I did not have sex with that woman," is an example of the truth but not the whole truth. He interpreted the words the way he wanted to, and told the truth as he wanted you to see it. And in due diligence that's exactly the same thing that sellers are likely to do.

From my own experience in M&A due diligence for over twenty acquisitions, here is my list of ways to spot lies and omissions:

13 Ways to Spot Lies and Omissions

1. **Less detail in one area**

   When information is provided to you, there's a consistent

amount of detail provided in most areas, but less detail is provided in a few areas. There's usually a reason they don't want to go into detail in those few areas, and it's in your best interest to find out why.

2. **Too much detail in one area**
   This is the opposite of #1, but the intent is the same. When you're getting too much information about one area, you tend to pay less attention to the areas where the real concern is. They may be using a red herring approach—distracting you from the real problem areas. Or they may be concealing the problem area by overwhelming you with other information in the same area so that finding the truth will be like looking for a needle in a haystack (worse, actually, since you don't know that it's a needle you're looking for).

3. **They keep changing the subject**
   Whenever the discussion gets around to a certain area, somehow the subject gets changed and the area never gets covered.

4. **People in one area don't ever seem to be available**
   Isn't it interesting that the experts you really need in an area never seem to be available? There's probably a reason.

5. **Discussion on one area keeps getting pushed to the end of the agenda**
   When the seller knows you have a time constraint (like a scheduled flight), they keep pushing the problem area to the end of the agenda, hoping that you won't take the time to do a thorough job.

6. **Accomplishments and plans blur together**
   This is particularly common when you're discussing the capabilities of a system or process. The seller talks about planned capabilities as if they're available today. This overstates the capabilities and gives you the impression that you're getting more than you really are.

7. **The verb tense keeps changing**
   This is usually a tipoff that the seller is doing #6. Sometimes the seller says "the system *will* do something" and sometimes the seller says "the system *does* do something." Or perhaps different people talk about the same capability: some using current or past tense, and some using future tense. Make sure you know exactly what capabilities exist *today*.

8. **Conflicting answers from different people**
   If different people give different answers to the same question, then the answers are suspect. Some sellers avoid this problem by making sure that all of your contacts stay in the room for all of the questions. That's why it's better to break up your due diligence into smaller groups for some of the detailed questioning.

9. **Physical cues: speech, tone, body language**
   Listen to the seller's answers, but also notice *how* the seller is answering. In particular, look for *differences* in physical cues when certain subjects are being discussed. Sellers often exhibit stress when they're talking about a subject where they know there are some concerns. The stress is visible when they act differently during discussions on those subjects.

10. **Web site and advertising don't match what they're telling you**
    If the seller is telling customers one thing and potential acquirers something else, then there's probably a problem.

11. **Denial without explanation**
    The seller denies certain things, but makes no attempt to explain, probably because there isn't a good explanation. Don't take "no" at face value—ask why, and ask for examples to clarify the issue.

12. **Everything is perfect**
    I don't mean to be cynical here, but the reality is that all acquisitions and purchases involve trade-offs. If your due diligence shows that everything is exactly as you would like it to be, then look closer, because something is wrong. I've never

been involved in due diligence where everything was perfect; in all cases we had to decide whether the weaknesses we saw were outweighed by the strengths.

13. **Lost in translation**
When doing due diligence in a non-English-speaking country, the discussions usually have to be done through a translator. It's best (although more expensive) to use your own translator in addition to the one provided by the seller. If your translator is part of the due diligence team and committed to your success, then your translator should be able to listen for differences between what the seller says and what their translator tells you. Your translator should also be able to pick up on foreign-language discussion among the selling team participants. Often this discussion can reveal seller weaknesses or issues that the seller is hiding.

In the next article: "99. 12 More Ways to Spot IT Lies and Omissions in Due Diligence"

Sidebar: 7 Types of M&A Seller Behavior that Don't Point to Lies

If you're new to M&A due diligence, then some of the things on this next list may look suspicious. They're normal, and perfectly acceptable in most cases.

1. **The seller wants to meet offsite, away from the office**
This is normal. They probably don't want employees to know, and they don't want to start rumors. This is particularly true for M&A of public companies, where insider trading rules are extremely strict.

2. **Seller wants a confidentiality agreement signed before any discussion**
This makes sense, and it goes both ways. If the deal falls through, then neither the seller nor the buyer want a lot of press, and they don't want their secrets used against them.

3. **Seller wants you to destroy your notes if the deal falls through**
   This is reasonable, especially if the buyer and seller are competitors, since trade secrets may be revealed. It's extremely important if the buyer or the seller is a public company.

4. **The seller hesitates in including certain people in the discussion**
   Most sellers want to keep the possible M&A deal on a need-to-know basis. The more people they include, the more likely it is that news of the deal will leak out. Note, however, that this conflicts with #4 in the first list. If you *really* need to know about a certain area, then you should insist on getting the answers you need.

5. **The seller doesn't want its people to be seen in public with you**
   Again, this is to keep the potential deal a secret.

6. **Code names for the buyer and seller are used**
   This is common. On the M&A deals I was involved in, we made it a habit to *never* refer to the seller except by code name —even in private conversations or emails among ourselves. You never know who might be listening in. And make sure that the code names can't be used to figure out the real name of the company (e.g., Redmond for Microsoft). Pick a name that has nothing to do with the company, their location or the business they're in. Hint: Some companies who do a lot of M&A make up a long list of code names in advance, then for each transaction they just use the next code name on the list.

7. **If a site visit is needed, then the seller wants to use a cover story**
   Again, this is common. You might be introduced to employees as potential customers doing some background work before making a major purchase. Technically, you're lying to the employees, but it's closer to acting—you're playing the role of a customer to get the answers you need to be a different sort of customer.

2026 Commentary

An article on acquisitions may not be relevant to many of you, but the 13 ways can be useful when making a major purchase. Most of the 13 items apply to conversations with salespeople.

99. 12 More Ways to Spot IT Lies and Omissions in Due Diligence

Originally published on SEPTEMBER 2, 2009 in #ACQUISITIONS, #IT, #ALIGNMENT, #STRATEGY

I talked about Due Diligence in a previous article, and gave you 13 ways to spot lies and deception. Here's an additional list that's specific to Information Technology, although you can probably see parallels in other types of due diligence:

12 More Ways to Spot IT Lies and Omissions

1. **The current solution doesn't scale up**
   Maybe there's a small-scale version of a certain capability working, but it's not ready for the large-scale production you're going to need. For example, a certain process is done by hand, and isn't yet ready for mass production.

2. **The current solution doesn't scale down**
   This is less common, but still a concern. Sometimes a system only works for large volumes, and isn't cost effective when scaled down to lower volumes. That can be a concern in a fluctuating economy.

3. **The seller is evasive on version numbers and specific types of hardware and software**
   This is often a tip-off that the seller is using outdated equipment or unsupported software versions. It can also mean that the seller is using unlicensed software.

4. **The number of people in an area doesn't make sense**
   You're given an organization chart that shows a certain number of people in an organization, but you know from personal experience that this number is too high or too low for the work that needs to be done.

5. **Omitted business functions**
   Think about the processes and business functions that need to be performed in a company like this. Have you gotten a description of the people and systems for every one of them? If not, then how are they done? Are they outsourced? Has certain information been held back?

6. **Customer list doesn't match revenue**
   Either the seller is claiming more revenue than can be accounted for, or the revenue seems low given the number of customers the seller is claiming. Either way, further investigation is needed.

7. **Lack of clarity around ownership of rights**
   For any proprietary software or hardware that is being used by the seller, it's important to get a legal document from the seller stating that they have complete intellectual property rights for that software or hardware. If any part of the development of that software or hardware was outsourced or subcontracted, then the seller should be able to produce legal documents that show that the outside resource holds no rights.

8. **Informal subcontractors**
   If the seller uses an informal arrangement with subcontractors where they may legally own rights to some of their work, then you may be in trouble on the purchase. You may have to get signed documents from every current and previous subcontractor assigning the rights to the seller.

9. **Person who did the work is no longer with the company**
   This introduces two issues: (a) intellectual property rights, and (b) the possibility that no one currently with the company knows enough about the previous work done to maintain it or enhance it.

10. **There is no (or not enough) budget for maintenance and replacement of equipment**
    In some cases the seller will strip "unnecessary" items like maintenance and equipment replacement out of the budget so that the P&L (profit and loss statement) will look better.

11. **No contingency or disaster plans**
    Like #10, in many cases the seller will try to operate without any contingency or disaster plans in an effort to minimize cost. This obviously introduces a high degree of risk for the business.

12. **Lost in technology translation**
    This is like #13 from the first list but it refers to the situation where a business claim is made by the seller that can't be supported by the technology the seller is using. For example, the seller claims 24×7 system availability, but only has one server with no fault tolerance.

2026 Commentary

Aside from acquisitions, this list is relevant during discussions with a vendor about a major software purchase, or a major outsourcing contract.

100. In with the New—Out with the Old. A Conservation Approach to IT

Originally published on SEPTEMBER 16, 2009 in #IT, #ALIGNMENT, #MANAGEMENT, #PROJECTS, #STRATEGY

Many years ago my sister Mary taught me a valuable lesson. At the time she and her husband lived in a very small apartment. But each time I visited her I was amazed by how neat everything was. There were no overflowing bookshelves, no overstuffed closets and pantries, and none of the usual clutter of everyday living. It didn't look like my house at all. Instead there just seemed to be total order, with everything in its place.

I asked her whether she had some offsite storage location somewhere for all of the "stuff" that we all seem to accumulate during our lifetime. She said that everything she and her husband have is in the apartment, but that they carefully control the number of things they buy. Their simple rule is "for anything that comes into the apartment, something else has to leave." If they buy a book, then they sell or give away a book. If they buy a lamp, then they get rid of a lamp that's already there. They buy food only when other food has been consumed or when they decide to throw something out. It's the ultimate conservation of "stuff": the amount of stuff in the apartment always stays the same.

It's a simple concept, but one that's very difficult to practice for most of us. We tend to hang on to things because we think we'll use them later, because we attach some sentimental meaning to them, or because we feel like we've spent money on them that would go to waste if we discard them. Clothes might come back into style, or might fit after we lose weight. We'll reread that book one day. We'll find a use for that old gadget. Keeping things seems to be the "safe" approach.

But there's a downside to being "safe." As we get more books, we get more bookshelves. As we get more clothes, we expand our closet or get our closet redesigned to hold more stuff. Our house becomes cluttered and uncomfortable. Our refusal to deal with things from our past costs us in space, maintenance, complexity, and ultimately in time. It's just too much effort to have to manage and track all of that stuff.

Peter Drucker

The other day I was reading a book about Peter Drucker, *Inside Drucker's Brain* by Jeffrey A. Krames. A quote from the book reminded me of Mary's approach:

"The first step in a growth policy [for a business] is not to decide where and how to grow. It is to decide what to abandon. In order to grow, a business must have a systematic policy to get rid of the outgrown, the obsolete, [and] the unproductive."

Drucker is essentially advocating the same approach for business that my sister Mary used to keep her apartment under control. And this exact same approach can be applied to IT as well.

Progress Hurts the Business Perception of IT

In my book, *Boiling the IT Frog*, I talk about the fact that maintenance and infrastructure typically start as small parts of an IT budget, with new projects making up the largest part of the budget. Then over time as new projects get completed, the percentage of the IT budget dedicated to maintenance and infrastructure rises, and the percentage of the budget dedicated to new projects declines. Everything new gets converted into something that consumes resources and has to be maintained. It's the cost of progress.

But from the business perspective, it appears that the IT organization is getting less done. There's a decline in the amount of money allocated to new projects, or the IT budget has to increase. Either way it's not good for the business perception of IT. And this perception is something that we all have to struggle with on a day-to-day basis.

A Conservation Approach to IT

I'm not against progress. But I think that many IT organizations tend to follow the same approach that their people follow at home. They accumulate systems the way some people accumulate clothes. A new system is purchased or developed, and the old system stays in place. The organization starts using a new operating system or a new development platform, but the software using the old approach is still supported. That old system probably even has a maintenance budget, and year after year you spend money on enhancing something that's destined for the junk pile.

So here's my message: Try to apply the rule "for anything that comes in, something else has to leave." And take it one step further by including in the new project plan the cost of migrating and closing down the old system. Don't just accumulate old applications, hardware and operating systems the way some people hang on to old clothes and old books. Clean out the old as you move to the new.

Your budget will improve. Your business reputation will improve. And you'll find that your job is a lot less hassle without the burden of all those old systems. Simpler is better, but things don't get simple all by themselves.

2026 Commentary

Although I think this is a great idea, I still have problems following this advice in my personal life. You should try to be better at this than I am.

101. What Ever Happened to the "Truth"?

Originally published on SEPTEMBER 23, 2009 in #CAREERS, #LEADERSHIP, #MANAGEMENT

I grew up in the 1950s and 1960s when there were just three TV channels, two local daily newspapers, a few local radio stations, and no cable or satellite TV. There were no personal computers—let alone the Internet—and so our news sources were pretty limited. We each picked our standard of truth for current events, but the choices were limited. Some defined truth by Walter Cronkite, some by Huntley-Brinkley, some by the position advocated by the local newspaper or radio station.

There were people with radical views, of course: Socialists, communists, right-wing militants, and extreme environmentalists. But these people had their own ways of communicating: low cost pamphlets, local meetings, quiet get-togethers at the homes of their members.

We're taught as children that there's a difference between a lie and the truth. But it's described by our parents and teachers as such a black-and-white difference that we attach an emotional significance to the words. "That's a lie" is not just a statement of fact—it's an accusation. Liars are despised, vilified and punished. But what is the truth? And what is a lie?

Absolute Truth?

There are few, if any, cases of absolute truth. If I drop something, it will fall—except in space or in a rapidly descending airplane. That balloon is red—or is it orange or magenta? Babies come from a man and a woman—or a test tube. Two plus two equals four—except in certain chemical reactions where it might equal eight or even eight hundred.

We internalize our standard of truth because it simplifies life. We can't question everything around us—we wouldn't be able to react fast enough to deal with what life throws at us. So we accept certain things as truth, and we build on these truths to put together our own philosophy, our own moral code, our own way of explaining the universe.

We do this all subconsciously—we do it without even being aware that we're doing it. We build a standard for truth based on our parents, our environment as children, what we're taught in school, and based on the "truth" we see reported in the press.

Finding Truth used to be Easier

When I was growing up there was little variation in what we saw in the regular press. Big daily newspapers and the three television networks all reported pretty much the same thing. They looked at an event and interpreted it for us. We could disagree of course, but if we disagreed we knew that we were disagreeing with the "mainstream media," and that made us view our disagreement in a different light. We might disagree, but we knew that we were in the minority. We knew that we had to do something to persuade others to join us. We didn't just accept our version of the truth as the standard. We knew that we were opposing the standard, and that action would have to be taken to change the standard.

I never associated with any radical organizations when I was growing up, but I imagine that this would be the way that they would think. They might think they were right and that the majority of people were wrong. But they dedicated themselves to changing the minds of the majority. They knew they were in the minority and they took steps to make their views known.

Today, Truth is Harder to Find

Today we have hundreds—maybe thousands—of cable and satellite TV channels. Many of these channels offer news programs that offer interpretations of current events from their own perspective. But that's just the beginning, because there are millions of Internet web sites that offer their own interpretations of current events and their own versions of reality.

Choosing your version of truth isn't as easy now as it once was. There is no longer a single interpretation of events that we can use as a reference point. There is no Walter Cronkite with whom we can agree or disagree. Now there's Nancy Grace, Thomas Friedman, Rush Limbaugh, Al Gore, Glenn Beck, Michael Moore, Jon Stewart, and many, many more. And that's just in the United States. Consider the opinions of Pope Benedict XVI, Gordon Brown, Hu Jintao, Desmond Tutu, Kim Jong-il, Ayatollah Ali Khamenei, and Osama bin Laden. Each of these people defines the truth for some part of the world's population.

Each person offers a different view of the truth. A logical person might listen to some of these people, get an understanding of their motivations, investigate the facts and make an informed judgment on who to trust. But who has the time to be logical?

Few of us investigate the facts. After all, it's difficult, since we never see the facts—we just see *interpretations* of the facts. Once in a while we'll see a web site or television show that attempts to pull together evidence for something important—like global warming—but most of us lack the specialized education that's required to understand the difference between normal weather variation and a pronounced trend. Even for something that ought to be easy to interpret—like whether a health care bill will or won't advocate euthanasia of the elderly—the words get in the way and an already suspicious public reads more into a paragraph than was ever intended.

Conclusion

"The truth" is a funny concept. We act like there's an absolute truth—one that's irrefutable and correct—and human nature compels us to build our personal philosophy on top of that absolute truth. But in fact truth is not an absolute concept—it's a relative one. What's true to me may not be true to you, and our behavior may be different because of our interpretations of the truth.

When I was growing up there was more consistency in our definition of truth. There were still differences; I'm sure the U.S. version of the truth was vastly different from the version in the Middle East or in the U.S.S.R. But within a single country there was more consistency. And I attribute that consistency to a clearer differentiation between the mainstream media and the fringe groups.

There is no mainstream media any more. Millions of Internet sites and hundreds of cable channels have leveled the playing field so that any communications medium is as mainstream as any other. No longer is there such a thing as a fringe group—all groups are both mainstream and fringe.

This puts the responsibility for determining the standard of truth squarely on the shoulders of the individual. Never before in history has this been the case. Never before in history has there been equal access to so many divergent views. Never before in history has an individual been put in a situation where they are pummeled with propaganda from so many directions.

And frankly, I don't think we're ready for this.

2026 Commentary

There are a number of dated references in this article from 2009, and there is no mention of social media, which complicates things even more. But the problem is still the same: I don’t think human beings are built to handle this much confusion and disinformation. See the next article.

102. 10 Ways to Find the Truth

Originally published on SEPTEMBER 30, 2009 in #CAREERS, #LEADERSHIP, #MANAGEMENT

In my previous article I talked about the problem of determining the truth in current events (and in other areas) when we're faced with conflicting views from thousands of media and Internet sources. In this post I'll offer some advice for dealing with the problem:

1. Become more conscious of the assumptions that you've been making about truth. Most of us bury our truth assumptions so deep in our subconscious that we don't even realize what we're assuming (see my article, "25. Don't Get Stuck in a Learning Stage"). We accept things as the truth because people around us believe them or because we were "brought up that way." That doesn't mean those things are really true. They *might* be true, but it's important for us to decide for ourselves.

2. Accept the fact that different people will have different views on many subjects. Those different views aren't necessarily right or wrong—they're just different. Everyone is entitled to their own view, even if they disagree with you (see my article, "94. 4 Reasons We Disagree, and What to Do About It"). Try to be open-minded and understand where the other views are coming from.

3. Don't commit yourself prematurely to a particular point of view. It's perfectly OK—even a good thing—for you to see and understand the different sides of an issue and to understand how different people can believe different things. As F. Scott Fitzgerald said, "The test of a first-rate intelligence is the ability to hold two opposed ideas in the mind at the same time, and still retain the ability to function."

4. To understand a view that's different from your own, try to focus on similarities rather than differences. What are you both agreeing on? What assumptions are you both making? Are you starting out with some of the same objectives but then differing in your methods? Are you considering the same methods but disagreeing on how you assign importance to different decision factors? Once you sort out similarities and differences then the other view becomes more logical. You may still disagree, but at least you'll understand the basis for your disagreement.

5. Remember that language and emphasis are at the root of many disagreements. Often you may find yourself "in violent agreement" with someone. You may use different words, or you may use the same words with different meanings. But if you really clarify what you're both saying, you may find that you're trying to say the same thing.

6. For important issues, ask yourself these questions about the views of the speaker or writer:

- Does this make sense? Is it likely? The more unlikely something is, the more skeptical you should be. People make outrageous statements because that kind of statement gets them attention or gets them in the news. Outrageous statements are also more likely to be repeated to others—without validation of course.
- Is it hurtful to others? If so, then you owe it to those people—even if you don't know them—to verify the statement before you repeat it to others.
- Is the statement based on facts? How do you know? What is the source of the alleged facts? Is the source trustworthy? Where did the source get the information? Trace the facts back to their origin to confirm their validity.
- If the statement isn't based on facts, then the speaker or writer is asking you to trust them. Should you? Do you have a basis

for that trust? Is there any reason the speaker/writer might want to deceive you or to "spin" the facts in a self-serving way?

- Watch for loaded words (e.g., racist, terrorist, meltdown, bailout, etc.) that add emotional meaning to an alleged factual statement. Replace the loaded words with unemotional words and see if the statement still makes sense.
- Look for exaggerations that distort the importance of something that's otherwise commonplace (e.g., "A recent study shows that half of the people in the world are below average.")
- Don't let statistics fool you (see my article, “114. How to Create Misleading Statistics in 6 Easy Steps”). Adding numbers to something that's not true shouldn't make it more credible, especially if the numbers are made up or if the numbers come from a biased survey or study. Verify the source of the statistics, and if the statistics are based on a survey then make sure that the survey questions aren't misleading (a common way to bias an answer) and that the sample used for the survey fairly represents the relevant population.
- Watch out for confusion between correlation and cause/effect. Just because some factor (e.g., poverty) is correlated with another factor (e.g., obesity) doesn't mean that the first factor causes the second factor, or that the second factor causes the first factor. Both factors may be caused by something else. But advocates of a particular point of view may use statistics to make it appear that there is a cause/effect relationship between two things when in fact there is only a correlation.

7. Build a base of sources you trust, and make sure it's a solid base. For media, political and religious sources, look at how other people view your potential source. If a large number of people view your source as extreme, then reconsider whether the source really ought to be in your base.

8. Use your base of sources as a "norm," and evaluate other sources of information based on their variation from your norm. Any source of information that varies a great deal from your norm ought to be subjected to additional scrutiny before you accept that new source as credible.

9. Don't be too volatile in your views. If a news story suddenly goes off in a crazy direction (e.g., a massive conspiracy theory), then question it. Let your response be incremental, in stages, changing your view slowly as you get more and more information to verify the news. Don't overreact.

10. Be an extremist only if you want to be. If your base of sources is considered to be extreme, then you'll be an extremist. But recognize that extremist views are quite different from the norm, that those views are considered by many to be inappropriate, and that extremist views are seen by most people as lacking credibility. You might want to consider getting some of your news and information from more moderate sources so you'll understand mainstream views better.

Conclusion

Everyone has beliefs. The beliefs may be different from your own, but that doesn't make those people wrong, and it certainly doesn't make those people bad for having those beliefs. It just makes them individuals. If the individuals aren't interfering in the lives of others, and if the beliefs aren't leading to hurtful or unlawful actions, then let people believe what they want. Respect the beliefs of others, and maybe they'll respect your beliefs as well.

2026 Commentary

This topic is much more important today than it was in 2009. Conspiracy theories are being repeated everywhere, and some of our traditional sources have become untrustworthy. It's more and more difficult to find the truth.

103. Who's Better at IT? Younger Digital Natives or Older Digital Immigrants?

Originally published on OCTOBER 14, 2009 in #CAREERS, #IT, #MANAGEMENT

I was asked this question at a recent speaking engagement in Utah, and I've thought about it a bit more since then. "Digital natives" are people who grew up using digital technology; they used computers as children and so they never lived in a non-computer world. "Digital immigrants" grew up in a world that didn't have computers, and they made the leap to the digital world as foreigners.

Each group brings its own advantages to IT.

Digital natives:

- Are more intuitive about new technology. They more easily understand how to use digital devices and systems in their everyday lives.
- Easily understand modality in user interfaces. It's natural for them to see the same control used for different things in different contexts.
- Naturally relate to parallel processing, since they already multitask in their everyday lives.
- Easily understand virtual metaphors like computer mice, virtual documents and avatars. The controls for video games like the Wii have taken virtual metaphors to a whole new level.
- Have an inherent understanding of the advantages of mobile computing and being connected. Digital natives are never out of touch.

- Want quick results. They like projects which offer early benefit.
- Use devices and apps as natural tools the same way that some people might use a hammer or a screwdriver. Pick it up, use it, put it down. No big deal.
- Are better with interactive systems since all of their games, devices and systems are interactive.

Digital immigrants:

- Understand the non-technology ways of doing things, and have a better appreciation for how and why things in our current business got to where they are.
- Better relate to processes that contain a mix of manual and automated tasks, and have a better feel for the advantages and disadvantages of automating a process.
- Can be better at applying non-digital models to digital systems because of the digital immigrants' richer and more varied experience.
- Are more detail focused, and are more likely to take a craftsman approach to systems design and building.
- Have longer attention spans. They look for long-term solutions to their problems.
- View systems and processes holistically—it's much more than the devices and apps themselves. Systems are designed to be comfortable for constant use—not necessarily for quick in-and-out efficiency.
- Are better with non-interactive systems like large-scale data manipulation or transaction processing, or systems which require great reliability or scalability.

Together, the two groups can achieve great things. But is it really fair to ask which group is better? You can ask this about any two generations that grew up on different sides of a major paradigm shift. For example, were people who grew up without electricity better at mechanical design than "electrical natives?" Were people who grew up with a car better at social skills than "automotive immigrants?"

Ultimately, I don't think one group is better than the other. The world changes, and each generation adapts to the needs of the new world. Inevitability, some people are slow to adapt and others adapt quickly. But it's not necessarily younger people who adapt faster—it's a function of the individual.

I believe that each younger generation will develop the skills they need to resolve the problems of the future. The approach may be different than the prior generation, but "better" isn't the right word to use. We learn, we grow, we evolve, we adapt to the new world in which we live. The natives of one world become immigrants to the next.

In the next article: A Reaction to the Reader Responses I've Received on this Article

2026 Commentary

It's amazing how fast the world is evolving. If I were writing this article today, I would probably be comparing AI natives to AI immigrants.

104. We're All Biased—Learn from It

Originally published on OCTOBER 21, 2009 in #CAREERS, #IT, #MANAGEMENT

Last week I posted an article about whether younger "digital natives" or older "digital immigrants" are better at IT. In responses I saw on Reddit or that I received directly, I noticed a pattern:

1. A lot of people were disappointed (to put it mildly) that I didn't draw a conclusion in favor of one group or the other. They didn't like the fact that I saw the issue as shades of gray rather than black and white.

2. Some readers thought that the article was biased toward one group or the other. Interestingly, the younger readers thought that the article was biased toward the digital natives. And the older readers thought that the article was biased toward the digital immigrants.

One of the readers in particular thought that I had sold out to my bosses, saying

"Don't bother reading. It's basically 'I can't pick between my peers and the fat cats that give me my paycheck.'"

In fact I run my own business, so I've got no bosses to answer to. And I'm flattered that this reader seemed to think that I'm a digital native, although in fact I'm a digital immigrant (an early one—I was an immigrant on the digital equivalent of the *Mayflower*).

Quite honestly, when I first had the idea for the article, my gut feel said that older digital immigrants are better at IT. We've got more experience under our belts, we understand the challenges of trying to change a business process that's been used so much that it's practically carved in stone, and we understand the complexities of large-scale fault-tolerant systems. But I tried to look at the issue from the perspective of the things that the older generation is lacking, and that's when I started seeing all of the advantages that digital natives bring to the table.

As I started listing the advantages that digital natives offer, I began to realize that there are distinct advantages to a fresh viewpoint, and that maybe this new way to look at things will produce better systems and better business processes. That's when I decided to change the conclusion of my article to point to a balance between the two points of view, and I ultimately ended the article on a positive note (so I thought) by optimistically pointing to new solutions coming from the younger generation.

But going back to the two patterns I observed in responses to the article, let me make a few observations:

1. Not Drawing a Conclusion

- I'm concerned that the world is viewing everything as too black and white (see my article, "101. What Ever Happened to the 'Truth'"). To some degree I think this is a result of the current "sound bite" mentality in which things are always summed up in just a few words. When we're overwhelmed with information, it's easier for us to reduce all of the detail to a high-level summary (see my article, "35. Information Overload: Why You Won't Read This Newsletter"). But I think we do this too much, and as a result we lose track of the subtlety that the world offers. It's like the pictures from the early digital cameras—the ones that only had a few pixels and very little ability to distinguish one color from another. The photos from those cameras were a representation of reality, but it was an oversimplified reality, like looking at the world through half-closed eyes. Oversimplification of information—boiling it down to sound bites—is an insult to natural human intelligence. And it discards so much valuable data that it often gives us incorrect results.

2. A Biased Perception of Bias

Everyone is biased in one way or another. We each see the world through the lens of our environment and our upbringing. We're taught about the world in a biased way by our biased parents and teachers. We don't see the bias, of course, since the bias is taught as the "truth." We gradually see the word "bias" as meaning "the way that *other* people see things." That's not honest, but it's natural. We have to use something as a reference point by which other points of view are measured, and our natural bias becomes that reference point.

So it's not surprising that to a young digital native it's more important that a software application be mobile, interactive and intuitive than it is for the application to be fault-tolerant, scalable and consistent with good business processes. And so in their eyes the skills to create mobile, interactive and intuitive applications must be more important to IT than the more mundane skills held by older IT veterans.

And it's also not surprising that to an older digital immigrant, a different set of skills is more valued. Again, the natural bias comes into play—both bias in upbringing and the bias that has developed from years of making mistakes, learning, and then doing things better.

Conclusion

To the younger digital natives who read this article: Please recognize that there's a lot more to IT than mobility and sexy user interfaces. Don't so readily discount the experience of older IT workers. They may not have your intuitive grasp of computer technology, and they may not relate to video games the way you do, but the digital immigrants have attained a lot of practical business systems experience that will take you years to understand. See their value, and learn from them.

To the older digital immigrants who read this article: Please try to remember your own early days using computers. Remember the excitement you once felt when you got a new technology to work for the first time (however basic that technology might look today). Remember the thrill of putting a system in place that actually improved the lives of business users. Recognize that you were once headstrong and abrupt, and that you learned from your experience, just as these younger people will learn. Where you see mistakes being made, try to teach. But recognize the difference between a mistake and a different approach. Not all differences are mistakes.

To both generations: Be more open-minded, and try to understand the reason (and inherent bias) behind an unusual behavior. We're in the middle of a knowledge transfer between digital immigrants and digital natives. And that knowledge transfer is bi-directional, so each group needs to learn from the other. Neither group is "right" and neither group is "wrong"—we're both biased, and we can each learn by seeing a point of view that's different from our own.

2026 Commentary

The advice in this article is specific to IT, but it applies equally well to everything in life.

105. Have You "Pulled Your Goalie" in IT?

Originally published on OCTOBER 28, 2009 in #IT, #ALIGNMENT, #MANAGEMENT, #STRATEGY

I lived in Boston during the height of the Bobby Orr days, and I got caught up in the enthusiasm that Boston felt for their Bruins. I had never seen ice hockey before I moved to Boston, and I learned the game by watching the Bruins win the Stanley Cup.

One of the intriguing tactics used in ice hockey (just called hockey in Boston—the ice is implied) is the practice of "pulling your goalie." The game is played with six players per side on the ice at any point in time (minus any in the penalty box). Normally one of those six players is a goalie who stays in front of the goal net and stops any shots on goal from the opposing team. But in certain situations when a team is losing a game and desperate for a score, the team may "pull its goalie" and replace the goalie with an attacker to improve the team's chances of getting a score on the opponent's goal.

Pulling your goalie is trading a defensive player for an offensive player. It intensifies your attack but it can critically weaken your defense. Any random shot by the opposing team in the direction of your goal can score a goal for your opponent because of your open net. It's a risk that you might take when you are sure that you're going to lose the game without desperate measures. It's a last ditch effort to win something that otherwise might be lost.

Why talk about hockey?

In our current economic situation, we have a lot of companies who are cutting back on expenses in extraordinary ways. Millions of people have lost their jobs, and thousands of companies have closed their doors. But not all businesses are equally desperate. Most companies have suffered losses in revenue, but many businesses have solid foundations that can withstand temporary losses and then recover when times get better.

The economic turndown has affected IT organizations along with the rest of the business, and we've had our share of layoffs, project cancellations and budget cuts. IT has been under extreme pressure to cut costs wherever we can, and some very difficult decisions have been made.

Some of those decisions are equivalent to the tactic of pulling the goalie. Many of us have cut the budget in areas that fundamentally weaken the IT and business infrastructure—cuts that expose our business to tremendous risk. We have "pulled the goalie" in IT, maybe without even thinking of the consequences, and in most cases without making business executives aware of the risk we're taking.

Let's take a few examples. In your IT organization, have you:

- Stopped updating virus prevention software?
- Fallen behind on implementing operating system and network software updates?
- Cancelled offsite backup services or disaster recovery site contracts?
- Discontinued contracts with "white hat" Internet security audit firms who look for Internet security vulnerabilities?
- Postponed outside audits of business processes and systems?
- Delayed replacement or upgrade of servers which offer inadequate response times to your customers?
- Let your infrastructure deteriorate to the point where customers are complaining?
- Laid-off the only people who really understand how to maintain and support some of your current custom systems?

If you answered yes to any of these questions, then you've pulled your goalie. You've taken resource away from a crucial defensive area, presumably with the intent of using the money for a more critical need. You've acted out of desperation in a way that could cripple your business. And in most cases your CEO doesn't even know. That's because most CEOs don't understand the implications of some of the technical decisions we make in IT, and business executives will never understand those implications unless you tell them.

Conclusion

In hockey, pulling the goalie is a last ditch effort—a team only does it when they know that they're going to lose the game unless they take this drastic action. But in hockey there's always another game and another season. The score resets with each new game, and so battles are fought on a game-by-game basis.

In business, it's one continuous battle. We don't restart each year with a fresh score and new equipment. In business, a pull-the-goalie decision is a bet-the-company decision. A loss can cost you the entire business. And pulling the goalie in IT is a hidden decision—the CEO and stockholders won't know that you're betting the company unless you tell them.

Pulling the goalie is an exciting part of hockey, and it's great fun for the spectators. But pulling the goalie in business is a desperate move, and you shouldn't do it unless you're truly desperate, and unless your CEO agrees.

2026 Commentary

It occurs to me that the current U.S. government has "pulled their goalie" in a number of different government agencies. Some of these cutbacks have been in technology areas, and some critical people have been let go. Time will tell whether these decisions are the right ones.

106. 8 Reasons Businesses Can't Connect Business Strategy and IT Strategy

Originally published on NOVEMBER 4, 2009 in #IT, #ALIGNMENT, #LEADERSHIP, #STRATEGY

Last week I met with a client to discuss a presentation I'm going to do for his company. The client company has a good process in place for business strategy, and they have the beginnings of an IT strategy. But they're having difficulty connecting the business strategy and the IT strategy, and they want me to help them.

There are a lot of reasons companies have difficulty connecting the business strategy and the IT strategy. Here are eight:

1. Strategy is a Secret

- Business executives often keep the business strategy a secret, and don't share it except on a need-to-know basis. So the important information that can guide the direction of every employee in the company is kept under wraps. After all, we wouldn't want our competitors to find out.

The situation reminds me of a Dilbert cartoon I once saw. A hacker breaks into the company's system and discovers the company's business strategy. Dilbert asks, "Did he post it on the Internet? I'd like to read it."

2. No CIO Participation in Strategy

In companies where the CIO doesn't report to the CEO, the CIO usually doesn't participate in the creation of the business strategy.

3. Technology isn't Important Strategically

In many companies, the business executives don't see how CIO experience and knowledge could contribute to the creation of a business strategy. "After all," they naively think, "why would technology impact the future of the company?"

4. IT Doesn't Need to Know

Executives don't see why IT should need to know where the company is headed. The executives know that large IT projects—like large building construction projects—often take a long time to accomplish, but somehow they don't translate that knowledge into a desire to give the IT organization more time to get ready for a future large IT project. Or worse, the executives don't understand that business assumptions are at the heart of IT systems, and that when the business strategy changes, those assumptions and systems may have to change as well.

5. Strategy = Vague

Business strategy is often written in very high-level terms that really don't convey much useful information (to business *or* IT people). Even if the IT organization sees the strategy, there isn't much that the IT organization (or anyone else) can learn from it.

6. Strategy = Objectives

There's a tendency to confuse strategy and objectives. Many business "strategies" contain a list of objectives like "increase sales 20%" or "expand into Country X" which don't really describe your strategy for accomplishing those objectives. *How* are you going to increase sales? Are you going to add more salespeople? Are you going to implement a new marketing campaign? *How* are you going to expand into Country X? Are you going to acquire a business in Country X? Are you going to open an office in Country X from scratch? Without knowing the strategy behind those objectives, the information is pretty useless.

And yes, I realize that the actual strategy for accomplishing those objectives is then worked out in subsequent meetings. But why isn't the result of those subsequent meetings considered important enough to include in the strategy document (or at least in *some* strategy document)? Instead, those "details" are often never documented at all.

7. Strategy Omits the Omissions

Most strategy documents don't include one of the most important parts of a strategy: the list of things that the business *won't* do. In many cases this list can provide guidance to IT (and other parts of the business) on areas where shortcuts can be taken. For example, if you have no intention of expanding or selling internationally then IT doesn't have to worry about multi-currency orders and accounting. If you have no plans for selling direct to consumers—you'll always use distributors—then that implies a different type of systems emphasis.

8. Strategy = Evolutionary

Many processes for the creation of business strategy treat the strategy as an evolutionary step from the previous strategy rather than a revolutionary response to changing times. Relatively few business strategy sessions start out by brainstorming answers to the questions "What has changed in the world since we last had this conversation? How have people's views of their environment changed? What new tools and technologies exist that might impact our customers' perception of our products and services? What new tools and technologies are now reliable and affordable enough for us to apply in improving our products, our services, our processes, or our way of communicating with our customers?"

Conclusion

The presentation I'm going to do for my client is going to highlight some of these issues, and illustrate the impact of business strategy on IT by using a few examples. In the presentation, I want to make these points:

- Business strategy has to be specific enough to be actionable and unambiguous.
- Giving IT visibility into the business strategy can make your IT organization more effective in their role of implementing parts of that strategy.
- IT information about viable cost-effective technology can influence the direction that your company is taking in certain strategic areas. For that reason (and to facilitate better communication), IT ought to be represented in the business strategic planning process.

Try to get this message across in your own organization. And let me know if you want my help.

2026 Commentary

I don't think much progress has been made in this area since 2009.

107. The 7 Biggest Challenges of a Manager

Originally published on NOVEMBER 11, 2009 in #CAREERS, #LEADERSHIP, #MANAGEMENT

I've previously written articles on "85. Why Do You Want to Be a Manager?" and "87. How to Become a Manager—13 Skills You'll Need." This article explains the seven biggest challenges faced by a manager.

1. Achieving a Stretch Goal

The organization you're managing is responsible for something—whether it's performing a business process, supporting some other organization, developing a new product, or getting new customers. There are goals associated with your objectives, and if your organization is aggressive then those goals require more than the typical amount of effort. It's going to take some careful planning for you to figure out how to apply your organization's people and resources to achieve an aggressive goal. You're going to have to motivate people, remove roadblocks from their path, and focus them on the things that are most important. It's a stretch goal, but you can achieve it—maybe even surpass it.

2. Bringing Out the Best in Your Employees

All employees have good days and bad days. Some of the causes are out of your control. But it's important that you take steps to make as many days as possible "good days." Here are some of the things that you can do:

- Treat every employee with respect. If you have praise for the employee, give the praise in front of coworkers. If you have criticism for the employee, give it in private. For all but the worst underperforming employees, make sure that the praise happens much more often than the criticism.

- Help employees align their personal goals with their work goals. Talk with each employee about their personal goals: what they want to get out of life, where they want their career to go. To the extent possible, use this information to help you allocate work assignments.

- Provide a work environment that is appropriate for the work and conducive to employee well-being. A comfortable work environment makes your employees more productive.

- Encourage employee communication and cooperation. For example, in one of my management jobs, I held a monthly lunch for my employees. During the lunch I updated them on any company news I'd heard, and I had some of the employees describe their recent work and some of their challenges. We also had a series of awards.
  But these were not your typical awards. Each award was given by the previous award recipient to someone who exemplified the spirit of the particular award. There was a "Gumby" award (a Gumby character) given to the most flexible employee, and other awards for things like putting the team ahead of yourself, most creative outside-the-box idea, and unluckiest employee. Employees sometimes even created their own one-time awards when something special or unusual happened. Over time the number of awards grew, and the interchange of enthusiasm and ideas made the organization a happy and fun place to work.

3. Dealing with Underperforming Employees

Not all of your employees will do their best. Some will have personal issues that interfere with their work. Technically it's not your problem, but in reality any issue that contributes to an underperforming employee *is* your problem. You'll help employees cope with personal issues, you'll provide motivation and counsel, maybe steer them to appropriate resources inside or outside your company. You'll "carry" your underperforming employees to a point, and then beyond that point you'll have to ease them out of your organization. You'll be humane, but you have to balance the needs of the organization with the needs of the employees.

4. Dealing with Outstanding Employees

Some of your employees obviously outperform the others. That's good news for your organization, but it presents its own set of challenges. Outstanding employees need special treatment. You want them to keep doing an exceptional job but that usually means that you'll have to pay them special attention. They need recognition for their talents and efforts. They need encouragement, training and guidance. And above all they need to know that they have a career path in your company, even if that career path takes them out of your organization.

You'll be tempted to hold on to your outstanding employees and keep them from being promoted out of your organization. You shouldn't do that. When an employee star outgrows your organization, the best thing for your company is to make sure that the employee finds a home in another part of your company where they can continue to contribute. And ultimately, you'll be rewarded for your good deed of helping the employee achieve their potential. Your reputation as a "team player" and good manager will grow, and your own career will be enhanced.

5. Hiring the Right People

No matter how happy your employees are, you'll get occasional turnover. And if your organization is successful then you'll often find that your budget and headcount will grow as you are assigned more and more responsibility. Either way, you'll need to hire. Hiring is easy, but hiring the right person is extremely difficult.

The trick in hiring is to get an understanding of how an employee will actually perform the work—not just how the employee does in interviews. Interviews are seldom a good predictor of work and work habits, so going beyond the interview is crucial.

I've sometimes used unconventional interviewing techniques. I've done the traditional interview, but then I've tried some things that gave me a better feel for how the interviewee will perform in an actual work situation. For example, for some programmer positions, I had the prospective employee spend some time with their future coworkers, going through a task that the current employee was doing. Getting feedback from the current employee (who had a vested interest in finding someone who would carry a part of the workload) made a big difference in our choice for some potential new hires. And the process also increased our acceptance rate for job offers, since the job applicants had a better feel for the environment into which they were being hired.

6. Responding to a Crisis

No matter how much planning you do, things will go wrong. An employee will get sick at a critical time. A weather disaster will hit your facility and disrupt your plans. A crime will be committed—maybe a theft or even something that harms an employee.

Planning is a part of managing, but perhaps more important is a manager's ability to change plans on the fly in response to changing conditions. When a crisis hits, you have to be able to deal with it—calmly, quietly and without being overwhelmed by stress.

7. Continuous Improvement

No matter how good your organization gets, it can do better. There's always some type of improvement that can be made: a change in a process, a better working environment, better employee motivation, more focus on the essentials. If you ever get to the point where you honestly have no idea how to improve things further, then you should either (a) seek outside advice, or (b) look for another job. There's always a better way, and you have to keep looking for it.

Conclusion

Management is complicated. It requires skill and motivation. But most of all it requires commitment—the commitment needed to rise to these seven challenges.

2026 Commentary

This article deals with manager challenges—not *project* manager challenges, which I address in different articles. A manager achieves success through their people, and most of these challenges are directly related to people skills.

108. Getting Ready to Move using Push and Pull

Originally published on NOVEMBER 18, 2009 in #MANAGEMENT, #PROJECTS

My wife and I have decided to move. We originally picked our home location because it was equidistant between my work and my wife's work. But I work from home now and my wife is retiring, so there's no longer a good reason to stay here. Instead, we're going to be looking for a house that's closer to our grandkids.

We've been living in our current house for sixteen years—that's the longest we've ever lived in one place during our entire lives. But there's a real downside to living in a single place that long: During the next few months we're going to have to go through sixteen years of accumulated stuff and figure out what we really need.

I'm a process person by nature, and so naturally my thoughts have gone to how I can set up a process to make this weeding-out challenge easier. I've mentally categorized my stuff, and I've made some initial conclusions about how to deal with each category. But in thinking about all this, it struck me that the main decision I have to make is whether I'm going to use a "push" process, a "pull" process, or a combination of the two.

Push versus Pull

A push process starts with the stuff I have, and then pushes each item through a conceptual sieve, eliminating the things that aren't as important and allowing only the truly important things to pass through to the new house. A pull process, on the other hand, starts with the new house and defines all of the things that I will need. Then the needed things are moved ("pulled") to the new house, and everything else is sold off, given away or thrown out.

A Pull Process Example: Cleaning Out a Kitchen Utensil Drawer

The pull process is a little harder to understand, so let me give you an example. All of you probably have a kitchen utensil drawer that has filled up with spatulas, whisks, pizza cutters, lemon zesters, vegetable peelers, shrimp deveiners and other random utensils. If you're like most people, your drawer is so full that you occasionally have trouble opening and closing it. Here's how to clean out that kitchen utensil drawer using a pull process:

1. Take everything out of the drawer and put it in a box. Put the box someplace that's easy to access but out of sight—maybe under the sink.
2. When you need a utensil, look in the drawer first. If the utensil is not in the drawer, then take it out of the box and use it. But after you've cleaned the utensil, put it in the drawer instead of the box. Initially you'll go to the box quite a bit for the utensils you need. But after the first few days you won't go to the box very often at all.
3. After several weeks, take the box and move it to someplace that's harder to access.
4. After several months or more, get rid of all the stuff in the box. You obviously don't need it.

Now it's easy enough to apply this approach to a kitchen utensil drawer, but it's harder to do this with the entire contents of your house. And it's especially difficult when some of the things are seasonal—we probably won't need our swimsuits during the next few months before the move, but that doesn't mean we should throw them away. So that means that I'm having to do "thought experiments" along with my pull process. I have to make a list of our seasonal activities, and then I have to imagine us doing those things so that I can see what stuff we need.

Conclusion

I'm early in the getting-ready-to-move process, and I will undoubtedly learn more as I go along. But ideally I would like to achieve the goal set forth by the English designer, William Morris: "Have nothing in your house that you do not know to be useful or believe to be beautiful." I've got a long way to go.

Sidebar: Applying Push/Pull to IT and Management

Some Examples:

- Apply the push/pull thought process to a facility move or a data center move
- Use a push or pull process for systems design. For example, when moving from an old system to a new system, you can use the push approach and figure out what old system capabilities have to be migrated to the new system. Or you can take a "clean slate" approach and define the business processes required in the new system, then figure out what system processes in the old system should be migrated to meet those requirements.
- Use a push or pull process for staffing a new business location. You can use a push approach and figure out which current people you want to move to the new location. Or you can use a pull approach and define the hypothetical staffing requirements for the new location, then determine which current people best fit those requirements.
- Use a push or pull process for migrating from an old PC to a new one. Push the stuff you'll know you need. Then hold back on the rest until you have a specific need on the new PC. When you do, "pull" the software or data from the old PC. (This assumes that you either hold on to the old PC or keep a full backup of the old PC's hard disk where you can get to it.)
- If your data center is overwhelmed by creating hundreds of business reports, then use the pull approach to cut back. Delay

selected reports by a few days and see whether anyone complains. For reports that get no complaints, gradually delay them more and more. If you still get no complaints, then eliminate the report—it's obviously not being used. Warning: make sure the reports aren't required by SEC or Sarbanes-Oxley regulations.

- Use push and pull for connecting business strategy and IT strategy. Start with the business strategy and pull the projects you need into an IT strategy. Then start with the technology you already have, and think about how that technology can be better used to build enhanced business benefit. Push those ideas into the business strategy.
- Use push and pull for business cost cutting. You can push your existing costs through an evaluation process, or you can do "zero based budgeting" and pull your essential services into your new budget based on what you really need in the business.

2026 Commentary

I had forgotten about this article. What I don't mention in the article is that the main reason my wife and I were moving closer to the grandkids is that my wife was diagnosed with ALS, a terminal disease. We knew she didn't have much time left, and she wanted to see her grandkids more before she died.

That adds an emotional dimension to this push/pull process: we wanted to move as fast as possible, and yet we had to deal with all of this stuff we had accumulated. Ultimately, we ended up giving away a lot of stuff. In the long run, stuff really doesn't matter.

109. 6 Things I'd Like to Tell Your Boss about IT

Originally published on NOVEMBER 25, 2009 in #IT, #ALIGNMENT, #LEADERSHIP, #MANAGEMENT

I wrote a book on this subject, *Boiling the IT Frog*, but it focused on the management and leadership aspects of IT—not the technical stuff. Now I'm looking at doing a two-day class on IT for business executives. I'll include the stuff from my book, but I'm trying to identify the technical topics that I should also include in the class.

What technical topics do you feel are important for business executives to understand? What should your CEO know about IT? Here are my initial thoughts:

1. Technology trends that are changing the nature of best IT practices

- Cloud computing, Software as a Service (SaaS), open systems software licensing, virtual software licensing
- Virtualization of storage, computers and services
- Data de-duplication
- Business analytics and data warehousing
- Access to the Internet from anywhere, and thus access to your systems from anywhere

2. Technology trends that are driving down costs and making more things possible—for companies on a tight budget, and for consumers

- Mobile computing with increasing power and bandwidth, enabling remote video and increasing amounts of data transfer
- Increasingly more affordable location tracking through inexpensive GPS devices

- Netbooks and tablet PCs making traditional PC use more portable
- Increasingly inexpensive data storage that is changing the rules on how much data you can affordably keep
- Inexpensive high-volume data communication, making outsourcing and offshoring increasingly attractive, and enabling real-time connections between multiple business locations
- Videoconferencing capabilities that are almost as good as being in the same room

3. Social trends that are taking advantage of technology trends

- A change in the way we get news about the world around us
- Everyone is a publisher—no barriers to entry, no cost
- Everyone looks like an expert, so who do you trust?
- The shift from an observer mentality to a participant mentality
- An increase in real-time interaction with "the experts"
- Social network sites like Facebook, MySpace, YouTube, Twitter, and the tools that make their use ubiquitous
- Group interaction tools like Wikis, Second Life and multi-player games
- Tools for easier collaboration and sharing of business information
- Telecommuting and virtual employees
- Huge amounts of data being collected about consumers, their purchases and their behavior
- The difference between data and information, between information and knowledge, between knowledge and wisdom

4. Major purchasing and systems design decisions that will be crucial in positioning your company to take advantage of new technology

- When to use thin client, browser-based applications and when to use thick client (my opinion: use thin client unless it's absolutely impractical)
- When to enable mobile access (my opinion: mobile access to almost everything)
- When to use generic off-the-shelf software and when to use proprietary software you write yourself (my opinion: use proprietary software only when it offers a significant competitive advantage for your company)
- How to select systems products
- How to make your systems resistant to disasters

5. Data security reality and myths

- Discrediting some of the incredible technical tricks you see people use in movies and on TV. Movies make it seem effortless to break into systems and get the information you need. Reality: a) it's very difficult to break in, b) even if you do, you'll have trouble finding the data you want, and c) the data you find is probably inconsistent, incomplete and inaccurate.
- The truth about phishing, social engineering, hacking, spoofing, and exploiting software vulnerabilities
- Tools that can help you fight the security battle: Virtual Private Networks (VPNs), two-factor identification, audits, various products and services

6. The truth about "IT doesn't matter"

- How IT enables and empowers competition, and how to take maximum advantage of IT

- The stages of IT use in a business: from automating routine processes to empowering business executives and their customers
- When to use "leading edge" technology and when to use "trailing edge"
- Where you ought to focus your IT effort (it depends on the current state of your IT)
- Should you use centralized or decentralized IT
- What to expect from IT
- How to measure IT success

Conclusion

What important topics have I missed? What suggestions do you have?

2026 Commentary

The list was probably appropriate for 2009. Since that time, mobile computing has become ubiquitous, with everyone having their own smartphone. That has both good and bad consequences: Remote access is easier, but data security is much more difficult. And social media algorithms are giving people more of what they want, but this increases the division among people because they only hear their own point of view. AI video, audio and social media posts are blurring the line between reality and fiction, so no one knows who or what to trust.

110. How to Deal with a Bad Boss—3 Approaches

Originally published on DECEMBER 2, 2009 in #CAREERS, #MANAGEMENT

You've got a bad boss. Maybe it was a surprise—they seemed nice during the interview. Or maybe it was a gift from higher-up in the organization—they were brought in to replace your previous boss. Whatever the reason, now you're stuck with a bad boss, and you have to do something.

There are basically three approaches to dealing with a bad boss: leave, get rid of the boss, or make the situation better. Here's more information about each approach:

1. Leave

This seems like the most logical answer, but it's actually the last alternative that you should consider. The unfortunate fact is that there are a lot of bad bosses out there, and you're just as likely to leave one and end up working for another one. And it's a shame to give up your knowledge and seniority in your position just because of this idiot you work for. And then there's the current economy—jobs aren't that easy to find these days.

Of course there's more than one way to leave your current job. You can try to find another position in your current company—maybe something in another area that reports to someone else. That way you can broaden your experience base, continue with your company, and get away from your boss all at the same time.

And if your boss is just difficult to be around, then there are ways of "leaving" while staying in your current job. Telecommute as much as you can. Travel if it's justified for your job. Spend time with customers. Do whatever it takes to decrease the amount of time that you actually have to spend face-to-face with your boss.

2. Get rid of the boss

This is the least common alternative, but I've had it work for me twice. There are two ways to get rid of the boss—well, two legal ones anyway. One is to get your boss a different job, and the other is to get your boss fired.

Getting your boss fired is the more difficult way unless your boss is doing something illegal. You have to somehow make your boss look really bad while making sure that the problem can't be blamed on the employees. I don't recommend this approach—it tends to backfire on you.

Getting your boss a different job is a bit counterintuitive. You have to make your bad boss look so good to upper management or an outside company that your boss is promoted or hired away. This is a much safer approach than getting your boss fired, and it can have huge benefits for you as well.

Think of it this way: In spite of your boss's failures in their relationship with you, there's probably something they do well. Maybe they were promoted into management from a "doer" position, and they were really best at being a super doer. Maybe they excel at independent thinking and strategy, but just have trouble with people skills. In either case you need to convince your boss that they need to focus on what they do best, and that they need to build on their true strengths. Work with your boss on a plan to highlight their skills and get them promoted or moved to a different position.

And if your boss doesn't want to work with you on this, then do it anyway. Play up your boss's true strengths every chance you get. Mention them to higher management whenever the occasion arises. Praise the things that your boss does well, and make them a good candidate for promotion.

Why does finding your boss a better job have huge benefits for you? Because you've turned your bad boss into someone who owes you a favor. If the boss doesn't immediately get a new position then you've improved your relationship with your boss. And even if you're caught trying this tactic then what can they say? "You shouldn't try to make your boss look good?"

3. Make the situation better

The last alternative is more practical in most situations. First you have to understand that the bad relationship between you and your boss may have something to do with *you*. A relationship has two sides, and your behavior may be aggravating the problems between you and your boss.

The best way to improve your relationship with your boss is to try to put yourself in their shoes. Try to see things from their perspective: How are they being measured? What pressures are they under? What goals are they trying to achieve? If your organization isn't achieving its goals, then their being rough on you may be their way of just trying to do their job.

Maybe their method of coping isn't the best one, but to improve the situation you've got to ignore their methods for the moment and focus on the result they're trying to achieve. If you help your boss achieve their goals, then you will change your position in their eyes from an adversary to an ally. And if you're an ally then maybe you'll be treated better.

So the best way to change your boss's behavior toward you is to do the things that will help make your boss successful, and to make sure that your boss knows it. It will be hard for your boss to be harsh to someone who's helping them, even if they're a bad person at heart.

There's also a bonus benefit to approach #3: If you help your boss become successful, then there's a good chance that approach #3 will turn into approach #2.

2026 Commentary

Bosses haven't changed since 2009. And I think the advice still applies. To better understand what a boss ought to look like, see "51. 8 Attributes of an Ideal Boss."

111. What Penguins Know about ERP Success

Originally published on DECEMBER 21, 2009 in #IT, #MANAGEMENT, #PROJECTS, #STRATEGY

Have you ever watched a nature documentary showing penguins entering the Antarctic ocean? They gather at the edge of the water, hesitating until more penguins arrive. They look at each other as if saying, "Do you want to go first?" or "Are we ready yet?" Then finally the hesitation will end and a large number of penguins will dive into the water together.

The penguins' hesitation makes sense when you learn about penguin predators. Seals hunt in those waters, and a single penguin swimming alone has a good chance of becoming seal food. But by waiting until there are significant numbers of penguins before entering the water, the penguins increase their chance of survival. There's safety in numbers, or at least there's an increased likelihood that the seal will go after someone else.

ERP

Last week I had dinner with some IT executives who are considering replacing their old systems with a new ERP (Enterprise Resource Planning) system. They seemed to be hesitating before making their move, and from the discussion it became clear that they hadn't yet reached the point where they felt they had enough executive support. Like the penguins, they knew it didn't make sense to dive into the ERP waters alone; to do so would prematurely end their careers.

The next day I met a CIO who had successfully implemented major parts of SAP, a popular ERP system, in his company in only nine months. He attributed his success to (1) implementation of an off-the-shelf industry version of the software [see secret 10 in *Boiling the IT Frog*], (2) help from a third-party consulting group who had extensive experience in implementing this particular industry version of SAP, and (3) enough executive support to require everyone in the company to use the off-the-shelf processes "as is"—no matter how attached the employees were to their previous processes and systems.

The successful CIO didn't dive in alone. And the penguins would be pleased.

2026 Commentary

It occurs to me that those of you who haven't done any sort of implementation of a major software product won't see the problem. What typically happens in a large company is that people are so wedded to their current processes that they refuse to implement a software product "as is." Instead, they insist that the product be bastardized to follow the same business processes they currently use. If they succeed in their lobbying, then these people will force a product which works well to be used in a way in which it was never intended to work. The product implementation will fail, and the lobbyists will blame the product and the IT organization.

Secret 10 from *Boiling the IT Frog* is "If you're going to use off-the-shelf software, then use the business processes that come with it. Otherwise you're just paving the cow paths." The "paving the cow paths" reference is to the legend of Boston's initial street design, which literally paved the cow paths that already existed, rather than design streets in an intelligent manner.

112. The Bad Economy is an Excuse for Layoffs, But They Should Have Happened Earlier

Originally published on DECEMBER 31, 2009 in #CAREERS, #ALIGNMENT, #MANAGEMENT

I'll let you in on a secret: Most companies have used the bad economy as an excuse for laying off people who the company wanted to get rid of anyway. Now I'm not saying that these companies haven't had financial issues—most companies have experienced a loss of revenue as a result of a decrease in consumer spending. But the truth is that businesses have made enormous productivity gains over the last ten years. Technology—and particularly information technology—has allowed most businesses to do more with less. The number of employees required to run the business and handle day-to-day operations has declined for most businesses. But even though fewer people are needed, most businesses hadn't taken steps to reduce their labor force until the recession gave them an excuse. There are a lot of reasons that companies were holding on to unnecessary employees:

- It's nice to have extra employees around if you anticipate growth.
- Many managers see the number of employees reporting to them as an ego thing. The more employees you have, the more important your job.
- It's nice to have labor in reserve to get you through peaks in demand and temporary shortages: sick leave, maternity leave, vacation, holidays.
- The annual budgeting process ensures that if you cut your budget, then you'll have a hard time increasing it back again later. Cutting an employee is cutting your budget.

- HR rules make it tough to fire an employee even if the employee isn't doing their job.
- Some managers are kind hearted, and don't want to cost people their jobs.

Put all these reasons/excuses together, and repeat this situation in department after department in company after company, and you can explain why technology productivity gains hadn't translated into a significant savings on the bottom line. The productivity gains were there—there just wasn't enough motivation to actually realize them.

So along comes the recession. Now all of a sudden everyone is in lifeboat mode—the ship is sinking and we have to throw all of the unnecessary stuff over the side. Executive management says "cut your budgets" and all those managers who were squirreling away spare employees suddenly have to get rid of them. The managers start with the slackers of course—now you don't have to go through all that HR bureaucracy to get rid of a bad employee. Then gradually the managers work their way up to the adequate employees who were actually superfluous to the day-to-day process. In some cases the managers have to cut good employees who are contributing to actual productivity. But let's face it: Productivity has risen to such a point that we can get by with a lot fewer employees that we had before. And here's the important part of higher productivity: *we don't ever need to hire them back!*

The Jobs Aren't Coming Back

That last statement is the key to understanding the slow economic recovery right now. Most companies were operating with more employees than they actually needed. And now that they've become accustomed to getting along with a smaller number, it isn't very likely that those jobs will come back. Instead, we need to look in different directions for new jobs. New jobs will come from new types of businesses, from growth of existing businesses into entirely new areas, and from an increasing number of freelance employees who will be working as supplemental employees.

So if you've wondered why it is that the stock market is recovering but the job market isn't, this is why. Business profitability is up because the businesses are producing the same products and services with fewer people—they've gotten rid of the jobs they didn't need. The old jobs are gone, and will never return. And the job market will only recover when totally new jobs are created to satisfy totally new needs.

Happy New Year! And welcome to the new economic reality.

2026 Commentary

This article obviously only makes sense when considered from the perspective of the economic downturn going on during 2009. But it seems like we're very close to repeating that downturn, and Artificial Intelligence (AI) may be the latest technology advance that leads us to once again eliminate superfluous employees.

113. There are Only Two Reasons for Strategy

Originally published on JANUARY 20, 2010 in #ALIGNMENT, #LEADERSHIP, #MANAGEMENT, #STRATEGY

A lot of people think that the creation of business strategy is a mysterious process—something that's secretly practiced behind closed doors in the boardroom. There's a lot of mysticism around strategy setting, and so we avoid it for fear that we'll do it wrong.

But there's no magic in strategy—just as there's no magic in IT. There are only two reasons for strategy: focus and communication. Let's look at each reason in more detail.

Focus

In management, focus is the process of concentrating your effort on the things that are important while stopping work on the things that are unimportant. You don't need focus if you have unlimited resources and you can pursue all possible objectives simultaneously. But most of us don't have unlimited resources, and so it's necessary for us to pick and choose where those resources will be used.

As a child you may have moved a magnifying glass back and forth to focus the sun's rays on a piece of paper. The magnifying glass was able to concentrate the sun's energy on a single point, generating enough heat to start a fire.

Focus in strategy works the same way. We have a limited amount of resource, but we pick the small number of things that are most important for us to accomplish, and we're able to use our strategy to focus our resource on those important things.

Communication

If you've ever seen the *Blue Angels* or the *Thunderbirds*, then you may have marveled at how these pilots can fly their jets at over 600 mph with such precision. The planes pass over our heads at an air show just 18 inches apart. How can the pilots achieve such control?

Part of the answer is, of course, amazing training and lots of practice. But one of the secrets to being able to fly in formation with the planes so close together is the way that the pilots steer their planes.

In any formation, one of the pilots flies the lead plane. This lead pilot sets overall direction for the formation, picking the altitude and direction. The other pilots don't look out of their front cockpit window or even at their instruments. Instead, they steer their plane based on their relative distance to the other planes in the formation. For example, if a pilot knows that their right wing tip is supposed to be a certain distance from a certain part of the lead plane, then they steer their plane to maintain that distance. Then when it's time for the formation to break up, the pilot of the lead plane gives commands over the radio, and the pilots move apart just as they practiced.

Strategy in a large organization works the same way. It isn't possible for everyone in the organization to steer by "looking out the cockpit window" at market conditions, competition, the economy, shifting customer demand, and all of the things that drive a business strategy. There just isn't enough time for everyone to do this, and it would be impossible to coordinate everyone if this approach was taken.

So instead, certain people in the organization are designated to act as the "lead pilot." These people—often the key executives of the business—decide which way the business will go and what things will be done. They write this down in a strategy, and the strategy is used to communicate this overall direction to the rest of the organization.

Lower-level executives steer their own organizations by "lining up" with the overall strategy. Lower-level strategies may be created to provide more detail on how each organization will fulfill the requirements of the higher-level strategy. Then these lower-level strategies are used to align the actions of the still-lower-level departments and employees.

If all goes well, then the large organization acts just like the *Blue Angels* in formation. Everyone does their job, the different parts of the business move together in formation, and the business achieves its objectives. The strategy has become the basis for communicating direction and alignment.

Dealing with Human Nature

Human nature tends to oppose the use of a business strategy. People have their own individual objectives, and want to spend their time on *those* things rather than on the things the strategy tells us are important. And people keep wanting to look out the cockpit window—they see the world changing around them and want to respond directly rather than just focus on doing the tasks that have been defined for them by the strategy.

How do you deal with these weaknesses of human nature which threaten to sabotage our use of strategy in business? It's not enough to just develop a strategy, write it down and disseminate it to your employees. You have to give the employees frequent status reports. They need to know whether or not the strategy is working and what you're doing about it. They have to celebrate progress, mourn failure, and see the response—if any—to changing business conditions. They need a way to provide feedback on the strategy so that they can apply their own experience to make it better. But they also need constant assurance and encouragement to keep them focused on the things that matter, and to discourage their pursuit of non-strategic goals. Basically, you have to inspire trust in your strategy—enough trust that your employees are willing to follow the strategy instead of their own short-term interests and distractions.

Conclusion

It's not enough to set strategy—you have to keep your organization "connected" too. Focus and communication are the only two reasons for strategy. But for strategy to succeed, you've got to keep everyone in the know.

Use your strategy to focus your limited resources. Use your strategy to communicate where you're going and how you're going to get there—to your business units, to your employees, and even (in subset) to your customers, suppliers and partners. If everyone believes in the strategy, understands their individual contribution to the effort, and sees where the business is going, then it's a beautiful thing. Just like seeing the *Blue Angels* scream across the sky in a tight formation.

2026 Commentary

It's easy to write about strategy this way, but it's much more difficult to actually implement the strategy. I've seen so many organizations use a different "bottom up" approach where the corporate strategy is assembled from a collection of contributions from lower down in the corporation. This doesn't work. You can't direct a corporation "in formation" by summing up the individual goals of a bunch of different people. Without a focal point, strategy is ineffective.

114. How to Create Misleading Statistics in 6 Easy Steps

Originally published on MARCH 25, 2010 in #ALIGNMENT, #MANAGEMENT, #PROJECTS

Want to deceive people? Here's how the professionals do it:

1. Do a survey and use a biased sample population

People focus on the survey result and seldom pay attention to information about your sample population. So feel free to bias your result by surveying people you know will answer the way you want. Want a survey result that shows a majority of people favoring smoking in public? Just ask smokers. Want a strong statistic showing approval for your new system? Just ask people who you know are on your side.

2. Use biased questions in your survey

You can further strengthen your survey results by asking biased questions which make the survey respondent more likely to answer the question the way you want. Political organizations are masters of this. Want to show approval for guns? Just ask a question like, "In these days of increasingly frequent home invasions, do you feel that having a gun in the house would help your personal defense?" (and don't forget step 1 and use gun owners as your survey population). You'll get an amazingly strong show of support for guns.

3. Ask the wrong question

You can often support your cause by asking the wrong question (the one that's not quite as relevant to your desired result), and then answering it in a way that implies something you want. For example, if you want to show public support for a war, then ask a survey question about supporting the troops (which is, after all, an entirely different thing). If you want to hide the fact that you've got occasional critical system downtime, then show *overall* system uptime—those system failures will be lost in the statistical noise. If you want to show high consumer interest in product A, then ask consumers about A versus C, where C is a not-so-popular alternative. But don't ask consumers about A versus B, where B is A's closest competitor.

4. Use misleading graphs

If you want to show high variability in a statistic, then chop off the bottom of the graph and just show the top (for example, using a y axis of 5,000 to 5,001 instead of 1 to 5,000). This magnifies the area you're interested in, and makes small variations look like huge peaks and valleys. On the other hand, if you want to show stability in a statistic, then make sure you show the full x axis (e.g., from 1 to 5,000), and make the graph as small as possible to minimize the appearance of variation.

5. Imply cause and effect when you only have correlation

Two factors can often be correlated without any cause and effect relationship. Both factors can be caused by a totally different factor, or the cause and effect can even be backwards from what's being implied or stated (i.e., B causes A instead of A causing B). For example, I once saw a claim that older people who walk are more likely to live longer. That may be true, but it's also just as likely that older people who are sick and can't walk are more likely to die sooner. So maybe the walking isn't the cause—maybe the ability to walk is just an indication of continued good health.

6. Make your results more precise

A number with higher precision gives the impression of being more accurate. If I say that 91.3% of system users are happy, it sounds more accurate than 90% of users, even if both numbers are made up. In the old advertising campaign, Ivory soap was touted as "99 44/100% pure" (pure *what*?). People trust precise numbers because they sound more scientific, but that doesn't mean the numbers are correct.

Conclusion

In reality, I'm not advocating your use of any of these six techniques, but I wanted to show you how easy it is to mislead people with statistics. So to avoid being fooled:

1. When you see survey results, look at the survey methodology. How was the survey sample determined? Is the sample large enough to be meaningful? Is the sample biased?

2. Where possible, look at the original questions in the survey to see if the questions themselves are biased. Don't trust a survey that uses biased questions.

3. Make sure that the statistic being cited is one that matters. Are the right questions being asked? What is the analysis leaving out? What is being deliberately omitted?

4. Where graphs are used, look carefully at the x axis to see if the graph has been enlarged. Has the graph been manipulated to bias your impression?

5. Look carefully at cause and effect. Does the evidence support the cause and effect that's stated? Is the opposite cause and effect possible? Is it possible that both factors are being caused by something else entirely?

6. Don't be deceived by precision in numbers—extra decimal places for the sake of looking good. There's often little justification for a more precise number, and more precision doesn't imply more accuracy.

In general, look at the source of the statistic and think about what that source wants you to believe. Look carefully at the case the source has made. Be skeptical of anything that isn't consistent with your common sense. Trust your gut on these things—radically new views on things happen only rarely, so if the statistic seems surprising, it has a good chance of being wrong.

2026 Commentary

There has been no decrease in the use of misleading statistics since I wrote the article. If anything, the lies are increasing.

115. How to Deal with Complaints—7 Steps

Originally published on APRIL 22, 2010 in #ALIGNMENT, #MANAGEMENT

How do you deal with someone who complains—whether it's a customer of your company, a customer of your department, or even an employee or family member? How do you turn the complainer into a supporter? Here are some steps to take:

1. Listen

If you're going to make the complaint go away, then first you have to listen to what the complainer has to say. Take notes so the complainer will see that you're serious. If there is a factual basis to the complaint then get those facts using the old reporter's list of questions: who, what, where, why, when, how. If there's no factual basis—it's just the way the customer "feels"—then try to determine why the customer feels that way.

2. Take Responsibility

If it's a problem you've caused (or your department has caused), then take responsibility for the problem. Even if there's the slightest chance that your department has caused the problem, you still need to take at least partial responsibility. Many times the customer will be much happier just having someone acknowledge that the problem exists, even if there's nothing you can do about it.

3. Apologize

Acknowledge the customer's feelings. Say something like "I'm sorry you feel that way" or "I'm sorry you're having difficulties" even if your department didn't cause the problem. Showing empathy puts you on the same team as the customer and lets you work together to solve the problem.

4. Offer a Solution If a Solution Exists

Some problems have immediate solutions—or at least workarounds. If a solution or workaround exists, then offer it.

5. Offer a Future Solution

If a solution is coming in the future, then tell the customer. But be careful not to build false hopes or commit to something that can't be delivered. Don't be specific about a date unless you're absolutely sure, and even then you should use qualifying terms. For example, you could say, "We expect this problem to be solved in the v3.0 release which is currently scheduled for next June. But if difficulties are encountered then the solution may not appear until the software release after that."

6. Deal with Disappointment

Many times a customer will be unhappy with news of a future solution. After all, the customer has the problem *now* and wants it solved immediately. Let the customer know if there's something that they can do to influence the priority of the solution. Would a letter or email to the right person help? Could the solution be implemented sooner if other tasks were delayed? Help a disappointed customer deal with the priority issues.

7. *Follow Up Later*

Add the customer to a list of follow-ups related to the particular problem the customer is having. When a solution is implemented, let the customer know. If the solution is moved up in priority (or down in priority), let the customer know as well. Keeping the customer informed will go a long way toward turning an irate customer into a supporter.

When the Complaint Isn't Something in Your Area

Many non-work-related complaints fall into this category. It's not your fault that the government is doing this and not that. It's not your fault that human nature causes people to make mistakes—sometimes *huge* mistakes. Here's what I think about people who complain about this sort of stuff:

- They see people and organizations doing things that they fundamentally disagree with (or just don't understand), and they don't like it.
- They feel helpless and out of control. Many times they're afraid.
- They talk about their concerns because it makes them less afraid. Talking about a fear—sharing the fear with others—makes them feel like they're not alone in having the fear.
- These people are not looking to solve a problem. They're not looking for the truth. They're just looking for a companion in fear—someone who can acknowledge the fear and make them feel better.

Fear is the difference between a solution-seeker and a complainer. Sometimes you have to recognize the fear, validate it, and separate it from the problem before you can begin to actually seek a solution.

2026 Commentary

Human nature hasn't changed, and this approach still works. For a related article, see "32. IT Lessons from a Waitress."

116. How to Deal with Complainers—2 Approaches

Originally published on APRIL 27, 2010 in #CAREERS, #MANAGEMENT

People who complain fall into two categories: those who complain because they want help in resolving a problem, and those who complain because they want sympathy. Often the complainers themselves don't understand why they're complaining, so it's up to you to figure it out for yourself.

The Complainer as Problem Solver

The first category of complainer is easier to deal with. They have a problem and they sincerely want the problem solved. The problem may be simple—"my computer keeps crashing"—or very complex—"children are starving in Africa." But this first category of complainer isn't just looking for sympathy—they honestly want help in solving the problem. For simple problems you might be able to provide help yourself (see the previous article), or you might know a good resource to provide the assistance the complainer needs. Once you provide the help or the resource, the complainer stops complaining on *that* subject, but may move on to complain about something else.

For complex problems it's more difficult to provide assistance. The problem won't be solved with a simple solution, but you might be able to offer some general guidance on first steps toward a solution. Moving in the right direction often provides relief for the complainer—they feel like they're on the way toward a solution.

The Complainer as Sympathy Seeker

The second category of complainer is more difficult to deal with. These people aren't looking for solutions—they're looking for sympathy. This type of person may feel overwhelmed by certain circumstances and possibly by life in general. They want to feel that they're not alone in their perception of things being the way they are. They're looking for the psychological equivalent of a hug —a sense of sharing that enables them to face the world knowing that no matter how bad the world might be, we're all in this together.

Sometimes this second category of complainer can get out of hand. They complain so much and so loudly that their friends and co-workers begin to tune them out—to ignore their complaints and discount their importance. The problem is compounded if the complaints are repeated over and over. The complainer gets a reputation as a negative person, and people begin to avoid the complainer. If things go this far then intervention may be required. The complainer must be confronted and told that they need to consciously edit their words or risk losing the respect of their coworkers and friends. Constant complainers are annoying.

The Non-Complainer

There's another type of person who never complains at all. Some people are naturally optimistic, and they see the good in most situations. They seldom complain because they see nothing to complain about.

Other non-complainers have learned to work within the system. They don't complain—they act instead. If they have a problem then they actively try to solve it. If they want help then they ask for it instead of complaining. This is usually a more productive route toward problem resolution, and it's much less annoying to coworkers and friends.

For this type of person, complaints turn into "war stories"—anecdotes shared over dinner or drinks about problems they encountered and how they solved or avoided them. The war-stories approach gains the same sympathy and comradeship that complainers seek, but without the annoying negative side effects. The key difference between a complaint and a war story is that a war story has a resolution while a complaint is open-ended and leaves the audience with unresolved tension and stress. War stories are appreciated when shared among a group, especially if they're told in a funny way. But care must be taken to avoid telling the same war stories over and over; repetition destroys the value of the story.

A Memorable Complainer

Many years ago I had an employee, Steve, who constantly complained about everything. Some of the complaints were problems that could be solved. But most of the complaints were general comments on his work environment that weren't problems to anyone else. Steve had an overly sensitive sense of "the way things ought to be," and anything that didn't live up to his standards resulted in an immediate complaint to me as his boss as well as to his co-workers.

I tolerated Steve because he was good at his work, but he was definitely high maintenance. I tried to help Steve see that his volume of complaints was inappropriate, and I counseled him on a combination of acceptance and quiet problem-solving. This seemed to help, but not enough.

Steve eventually got fed up with what he considered an unacceptable work environment, and he left to join another company. But he soon discovered that the other company was much worse, and his new boss was less tolerant of Steve's complaining behavior. Steve sheepishly returned to me asking for his old job back, and I consented. If Steve had been less good at his job then I would have never taken him back. You can only deal with so many high maintenance employees at a time.

Conclusion

Everyone is different. All of us have good personality traits and bad ones. Complaining is a behavior that can come across as annoying to most people, so if you're a complainer then you should make an effort to change. Changing doesn't mean that you have any fewer problems—it just means that you try to be more positive, that you try to make *requests* instead of complaints, and that you're selective about the problems you take to other people.

And if you're on the receiving end of constant complaints, then try to refocus the complainer on a more appropriate behavior. Show the complainer this article, and maybe they will get the message. But above all, try to be sympathetic. Sympathy is what most complainers are really trying to get, so give it to them once in a while.

2026 Commentary

In hindsight, this article sounds a little harsh. I think it comes across as too insensitive, but remember that I was talking about chronic complainers—not just people who complained once in a while.

117. The Difference between Success and Happiness

Originally published on JUNE 1, 2010 in #CAREERS, #LEADERSHIP, #MANAGEMENT

A lot of people tend to confuse these two words. We work hard, focus on our goals, and figure that happiness will come once we achieve success. Don't fall for this deception.

Happiness

Happiness is a feeling. It can be triggered by external factors—a friend, a lover, a place, an event, a food or a drink—but it comes from inside ourselves. We can *choose* to be happy, even if things around us are chaotic and we're all alone.

There's a tendency for some of us to rely on others for our happiness. It's more likely if we were raised in an environment of rewards and punishments. It's more likely if we were not given unconditional love as a child, or if we were constantly criticized by those around us. In a hostile environment like this, we can unconsciously give away our ability to find happiness in ourselves. We become dependent for our happiness on external forces. We set goals and tell ourselves, "I'll be happy when I achieve that goal." Then often we'll find that happiness continues to elude us when the goal is achieved, and so we set yet another goal, continuing to strive for that ultimate happiness that never seems to arrive. We fall into the trap of equating happiness with success.

Success

On the simplest level, we succeed when we achieve an objective. That's technically what the word "succeed" means, and success is what comes from succeeding. But for many of us the word "success" means so much more. It's more personal. Success is not something we do—it's something we become: we strive to become "a success." And so being successful becomes our all consuming goal, the thing that justifies all of the frustration and pain that we endure on a day by day basis.

Ironically, our personal definition of success keeps enlarging as we get closer to our objective, so we never quite reach it. According to a survey I once read, most people say someone is "well off" if they make 50% more than you, and a "rich" person makes twice as much as you do. Your definition of wealth changes as your income increases—the more you make, the more you think you need —and the same thing is true with your definition of "being a success." It's like a dog race where the dogs can never catch the automated "rabbit" that runs in front of them—the target always stays a little bit ahead of you.

Happiness Does Not Come from Success

This is the biggest mistake many people make. They think they can postpone happiness until they achieve success. But happiness does not come from success. Instead, success comes from happiness. Ultimately your success in life will come from the amount of happiness you feel and the amount of happiness you share. Someone who enjoys life and makes life a joy for others will have a far greater impact on the world than someone who achieves an executive position or becomes rich.

And here's the real irony: someone who enjoys life and makes life a joy for others is more likely to achieve an executive position or become rich! That's because people who find their passion—the thing that makes them happiest—are more likely to be successful at it. When you really enjoy doing something—whether it's art or design or building or math—you get so immersed in the process that you become truly great at doing it. And that greatness will position you for whatever kind of success comes from your field.

We've Got It Backwards

So many of us have been looking at things the wrong way. We have assumed that we have to postpone happiness in order to achieve success, and that the happiness we're missing now will come to us once success is achieved.

But that's backwards. We need to focus on the things that make us happy, find our passion, and discover the strengths that will maximize our happiness. And through that happiness we'll achieve success.

I remember a cartoon in the *New Yorker* many years ago. It showed a songbird talking to a human. The songbird was saying, "I don't sing because I'm happy. I am happy because I sing." The bird discovered its passion, and so can you.

2026 Commentary

I think I wrote this article to attempt to convince myself that this is true. I have had trouble finding happiness my whole life, looking more to the future than to the present. I believe what I wrote in the article, but believing something and internalizing it are two different things. I'm still fighting to internalize this message.

For a related article, see "119. Which Do You Have? A Job, a Profession, a Career, or a Passion?"

118. Should You Use a Business/IT Liaison Person? No!

Originally published on JUNE 23, 2010 in #CAREERS, #IT, #ALIGNMENT, #LEADERSHIP, #MANAGEMENT, #PROJECTS, #STRATEGY

The question about the pros and cons of using a business/IT liaison person came up at a meeting I attended last week. I've got to admit some bias on this issue. Long ago I tried using a business/IT liaison person for one of my software development groups, and I wasn't happy with the result.

The person typically assigned to a liaison job seldom has enough technical savvy or in-depth application experience to fairly represent IT to the business. As a result, the liaison person will often make inappropriate technical assumptions that create unrealistic expectations for the business. That's what happened when I tried using a liaison, and I ended up getting rid of the position.

Here's the Problem

In theory using a liaison person sounds like a good idea. We know that many IT people have difficulty communicating with business people, and so it seems reasonable to put a "translator" in place. But while language translators like those in the United Nations mostly translate what's being said in an unbiased way without having their own agenda, the same cannot be said for liaison people. Invariably I've found that over time the liaison people gravitate toward either the business or IT camps, and begin to take sides in disagreements. Often their job turns into more of an expeditor function than a translation function—they badger the IT people to get the business people what they've asked for, and usually do little to clarify requirements or to help negotiate a smaller set of requirements that can be delivered faster.

Eventually a two-way disagreement between business and IT evolves into a two-way disagreement through an intermediary. This magnifies the problems rather than solving them, since all of the interactions take longer, and it's much less likely that personal relationships will develop between the business people and the IT people. Often the business people will feel better using a liaison person, but the IT organization will lose touch with what's really going on in the business. The net long-term result will drive a wedge between business and IT.

Why Do Companies Use a Liaison?

So if liaison people are so likely to cause problems, why do so many companies use them? There are a number of scenarios:

- The business forces the position on the IT organization. This is a sure sign of imminent IT organization failure, and usually precedes the search for a new CIO. The business is essentially saying, "I've given up on dealing with you—give me someone new to talk to."

- The IT organization is so caught up in constant interruptions that they create the liaison position to keep the business people off their back—kind of like setting up a complaint department. Again, this is not a good sign. Why are the interruptions occurring? You should *solve* the problem instead of trying to cover it up.

- The IT organization has done a poor job of hiring project leaders and analysts with business communication skills, and the liaison position is created in an attempt to work around the hiring problem.

- The business organization has abdicated their own responsibility for designing and improving business processes, and the liaison position is created to fill the gap. In some cases this means that the liaison person tries to design systems (usually with limited success due to lack of experience). In the worst case (I've seen this happen), the liaison person can get so frustrated with the IT group that the liaison person tries to outsource the systems design or even purchase an outside system directly without going through the IT organization at all. But guess who ends up inheriting the support and integration of the outsourced system? The IT organization, of course.

Is There a Solution?

If you've already fallen into the liaison trap—or if you're seriously considering it, then here's my advice:

1. First, **get the support of your CEO** (or lower-level senior executive if you're trying to solve this problem at a lower organization level) for integrating process and system responsibility into the business organization. I'm not talking about the technical responsibility—that stays with IT. I'm just saying that business processes (how a business operation is carried out) are a part of the business—not something that should be handled by IT.

2. A business should be responsible for the processes it uses—you can't delegate processes to another organization like IT. Bring in a process consultant if necessary, but some way or another, **get a basic understanding of the fundamental business processes that you use in your business**, and how the computer systems fit into those processes.

3. **Get an overall picture of the status of each of your business processes**. Are they working well? Do they need improvement? Are they broken? Can they handle the current and future volumes that are required by your business? Are they cost effective? Where deficiencies are noted, what plans exist for corrective action? Joint meetings between business and IT people are required to plan and implement the required corrective action.

4. **Eliminate the liaison position(s)**. Replace them with regularly scheduled get-togethers between business and IT people. Hold joint training sessions if necessary to improve communication between the two groups. It's not important for business people to understand all of the technical stuff in IT, but it *is* a requirement that IT people understand the fundamental aspects of the business processes they support, and are able to communicate with the business using business language. If your people can't do this, then find someone who can (as a replacement—not as the totally unnecessary liaison position).

5. **Prioritize your business needs** so that IT can focus its limited resources on the most important things. Many liaison positions are really just ways of coping with a lack of business priorities. When everyone in business wants everything at once, how do you expect *anyone* to successfully perform? Having a liaison doesn't solve the problem—it just gives you a full-time person to hear complaints.

6. **If you don't have an IT strategy to help you prioritize your IT effort, then get one**. See some of my other articles on strategy, or take a look at my book, *Boiling the IT Frog*.

Conclusion

There are situations in life where a liaison person makes sense. You typically see the liaison position successfully used when two independent organizations want to share information. For example, the liaison job is widely used to coordinate the sharing of information between different parts of the armed forces, or between different government agencies. The key to success in these situations is that there are no critical dependencies and few time-critical issues.

That's not the case between the business and IT organizations, where there are many dependencies and tons of time-critical issues. Using a liaison person forces you to do partial delegation—something that's almost impossible to control. To the IT organization, the liaison person is representing the needs of the business users, but you can't get user buy-in through a liaison, so a project failure is much more likely. To the business organization, the liaison person is representing the collective technical expertise of the IT organization, but the nuances of design choices often get lost in the translation. To both organizations, the liaison person creates a barrier to trust—how can you trust the members of the organization when you don't ever spend any time with them?

The net result is that using a liaison person is a short-term way to postpone dealing with a longer-term problem. You're better off facing the problem directly and helping your business and IT organizations work together to achieve a good day-to-day trusting relationship.

2026 Commentary

The article is very IT centric because I wrote it to address a very specific IT problem. But I feel like using a liaison person is equally bad for any two organizations where there are critical dependencies and time-critical issues. This is particularly true where there is an inherent personality difference between the members of the two organizations.

So, for example, I would think that a liaison person between an engineering organization and a marketing organization would be a bad idea, as would a liaison person between a scientific research organization and a manufacturing organization.

The bottom line is: Don’t try to solve personality disagreements by throwing a middle-person into the process.

119. Which Do You Have? A Job, a Profession, a Career, or a Passion?

Originally published on JUNE 28, 2010 in #CAREERS

A Job

A job is an exchange of work for money: for every hour you work, you get an hour of pay. Although you might derive some satisfaction from doing the job, and you might enjoy the people you work with, the reason you do the job is for the money. If you win the lottery, the first thing you'll do is quit your job, because, after all, you're only doing the job for money anyway.

There's usually very little advancement within a job. Maybe you'll get a seniority raise if you stay in the job long enough. But for real advancement you'll have to change jobs. You'll have to get "promoted" from one job to another higher-paying job.

Jobs can be tedious, frustrating and dull. But most people put the job behind them at the end of the day when they go home to their family and friends. A job is something you compartmentalize—there's the job, and then there's your life. The two are quite different, and often unrelated.

A Profession

For many people, a profession is a glorified job. That's true whether you're a lawyer, a doctor, an accountant, a policeman, a fireman, or an executive. You still get paid for each hour of work you do, although that fact is often disguised by a "salary" that is paid to you every few weeks. The salary makes it possible for your employer to ask you to work more than 40 hours a week without any extra pay. Successful professional people rarely leave their work at the office at the end of the day. The work is constantly on their minds, and their personal lives may be interrupted by work issues at any time of the day or night.

There are three primary distinctions between a job and a profession. First, a profession usually requires more training, and likely an advanced degree or certification. Second, a profession has more prestige than a job. People who have a profession are more likely to be regarded as being successful. Third, a profession usually reflects a personal choice in the type of work you like to do, where a job is more associated with finding the highest-paying available position.

A Career

When people use the word "career" they are usually referring to a progression of professional positions, starting with an entry level position and then moving up to higher-level positions with more responsibility, more prestige and higher pay. A career may reflect a conscious plan for advancement, but more than likely a large part of a career is determined by accident. You're in a certain position and another position becomes available. You jump to the new position, and your career "advances."

People often refer to a career in hindsight rather than looking forward. I seldom hear a young person talking about a career choice (which is a shame, since people seldom achieve what they really want without a plan). You're more likely to hear an older person at a retirement dinner talking about their career, with that career being viewed through the rear-view mirror. A career is often a reflection of what we did during our life, viewed from the endpoint.

A Passion

What are you passionate about? What gets you excited? What do you enjoy doing that you can do for hours without noticing any time passing? That's your passion. All of us have one, but many of us ignore our passion to focus on a job, a profession or a career. Why should this be? Does it make any sense for you to work 40 hours a week at a job while somewhere in the back of your head you're longing to go somewhere else and do what you're passionate about?

If you look at truly successful people (defined in just about any way you like: money, power, happiness, acclaim), you'll find that almost all of them have figured out how to turn their passion into an income. Their time is focused on doing their passion, and doing it really well. And because they live and breathe their passion, they become really, really good at it. They become the best in their field, and that's how they achieve their success.

To these people, their work isn't a job. It isn't a profession (except by someone else's definition). And it isn't even a career. It's just spending their time doing what they truly enjoy.

Conclusion

Do you have a job? A profession? A career? Or are you pursuing your passion and getting income from it?

And if you're not pursuing your passion, then why not? What's holding you back? And what are you going to do to change things?

2026 Commentary

I believe this is still true, although I don’t think most people consider it. The people who pursue their passion are probably in the minority, and I think that’s sad.

120. IT Governance is Like Pushing a Rope

Originally published on JULY 13, 2010 in #IT, #ALIGNMENT, #LEADERSHIP, #MANAGEMENT, #PROJECTS, #STRATEGY

IT governance has been getting a lot of attention lately in the press. Frankly, it bothers me, since I think that a focus on IT governance is misguided. Four years ago I wrote an article about the difference between management and leadership. In the article I said that:

"... management is like pushing a rope; it works only if you make continuous small adjustments to keep the rope in line. Leadership is like pulling a rope; the parts of the rope that are out of line will automatically align themselves with the direction of motion. When you push a rope you have to focus on the rope; when you pull a rope you focus on the direction in which it's being pulled."

In my opinion, the focus on IT governance is a focus on management at the expense of leadership. A lot of IT governance is concerned with project decisions: trying to figure out the best way to allocate scarce IT resource. There are many ways to do this. You can use steering committees, you can use review boards, and you can come up with complex "impartial" formulas for project benefit evaluation (see my article, “20. Why ROI Isn't Working” to find out why I put "impartial" in quotation marks). But all of these different approaches are pushing the rope—they're ways of forcing certain projects ahead of other projects without really looking at what's most important strategically for the business. The optimum benefit for the business shouldn't be determined by a vote. It shouldn't be determined by some complicated formula. It should be determined by "pulling the rope”—by setting a strategic direction for the business and then figuring out what IT projects provide the most benefit and support for that strategic direction.

I made the comment in a meeting a few weeks ago that all of the companies who are focused on IT governance instead of strategy are in deep trouble (actually I used another word other than "trouble" but the meaning is the same). Some of the people at the meeting understood what I meant, but I suspect a large number of the attendees were clueless. We tend to become absorbed by current projects—not strategy. And we're too busy fighting day-to-day battles to recognize our omission.

Allocation of a scarce resource in a business is a political issue, and the larger and more complicated the business, the more political it is. Politics is about optimizing the personal agendas of many different people. And that's what IT governance is all about—it's a way of choosing the "least bad" set of solutions from a minefield of conflicting personal agendas. IT governance is like a multi-dimensional traffic light: some projects get the green light, and some projects get the red light until it's their turn to go. But projects that are more strategic aren't typically given higher priority, and in fact there's often no consensus on which projects are considered to be strategic.

In many of my articles I've talked about focus—the need to put your scarce resources to work on the most important things. It's my experience that strategy can provide focus, but IT governance cannot. Instead, IT governance just forces various projects through IT for all of the wrong reasons.

Is your business "pushing the rope" and *managing* your IT workload? Or are you "pulling the rope" and *leading* the business and IT to its strategic objectives? In the long run we'll all know because of your business success or lack thereof. Pushing the rope is the best approach for business mediocrity. Pulling the rope can give you superb IT and an outstanding business.

2026 Commentary

This article isn't just about IT projects; the same thing is true for *all* projects in all organizations. "Pulling the rope" means that the projects you do are dictated by the overall strategy of the company. "Pushing the rope" means that the projects are determined by some other means that don't optimize your projects to fulfill your corporate strategy.

121. Why We're in this BP Gulf Oil Mess, and What We Should Do About It

Originally published on JULY 26, 2010 in #LEADERSHIP, #MANAGEMENT, #PROJECTS

I've talked a lot about human nature in my articles. I believe that human nature is the biggest challenge to most successful management, and especially the biggest challenge for IT managers.

Information technology is all very logical. Software does exactly what you tell it to do. Computers—for the most part—behave the way we expect them to behave. But people are on the opposite end of the behavior spectrum from software and hardware, and people behavior follows a different set of rules.

Let's take the BP Gulf Deepwater Horizon oil spill as an example. Putting an oil well in the ocean is an inherently dangerous process. Actually, putting an oil well *anywhere* is an inherently dangerous process, but putting a well thousands of feet below the surface of the ocean is much more complicated and therefore has a much higher risk.

No one—not BP, not BP workers, not the government and not the population of the planet—wants to see a leak in an oil well. In all of the criticism of BP no one has even suggested that the leak was deliberate. Instead, the discussion about blame and fault has centered around the safeguards that were employed: Were they safe enough? Were they properly maintained? Were warning signs ignored?

In my book, *Boiling the IT Frog*, I talk about the unfortunate use of magic in IT—the suspension of reality that occurs when a technological process is so sophisticated that people develop a blind trust in its capabilities. I believe that this same magic plays a large role in the BP Gulf oil disaster. Here's how:

1. Because of the inherent risk in drilling so deep in the ocean, a number of safeguards were installed to automatically stop the oil flow in case of disaster.
2. There is no such thing as a 100% reliable system, especially one that includes hardware that is exposed to the elements. The ocean is a hostile environment to most machinery. Thus any system installed in such an environment must be regularly inspected and maintained, and certain parts that are subject to wear and corrosion must be regularly replaced.
3. Because of assurances from engineers, safety experts and scientists, the BP employees working on the oil rig developed a sense of magic around the safeguards. They trusted the safeguards to protect them from any disaster. And because there were multiple safeguards (primary systems as well as backup systems), the workers developed an especially unreasonable trust. This is, unfortunately, human nature: when you have to live day-to-day in a risky environment, you suppress your fear by developing an unreasonable trust in things going right. That trust persists right up until the point when things go wrong.
4. When the safeguards began to malfunction, there was probably some concern. But because there were multiple safeguards, the concern was not severe. Each worker on the oil rig theoretically had the ability to stop operations if there were safety concerns. But as long as there were multiple backup systems, the workers reasoned that things weren't that dangerous. No one wanted to be the "bad guy" who stopped production. Thus the multiple safeguards—which are good for reliability—are *bad* for human nature: multiple safeguards lead to unreasonable trust, which leads to a lack of concern when one of the safeguards fails.
5. One by one the safeguards malfunctioned, went out of maintenance (in some cases waiting on parts), or developed small

problems that would later snowball into something big. But the "magic" persisted and no one recognized how close they were coming to a disaster. Finally a relatively small problem triggered a sequence of failures which caused the rig to explode and the well to spew oil into the Gulf. It wasn't one thing that caused the disaster—it was a series of things that were ignored because not one of those things was considered significant enough to cause alarm.

This Isn't the Only Case

There are many examples of this kind of thing: a series of cascading small issues that roll together into a single big disaster. Chernobyl followed the same pattern, as did Three Mile Island, the Challenger shuttle failure, the Twin Towers (not the planes, obviously, but the building collapse), and even the Titanic sinking.

All of these disasters could have been avoided, and *should* have been avoided, but human nature got in the way. It's human nature to develop a false trust in safeguards (especially multiple safeguards) when we're told by "experts" that the safeguards are sufficient. It's human nature to avoid being the "bad guy" who makes a big deal out of something when everyone else assures you it's OK. And now, in the aftermath of the BP Gulf oil disaster, it's human nature to believe that all disasters have a single cause and so we look to find a single person to blame.

What Can We Do?

I've thought about this a lot over the last couple of months. I wish there was a simple answer—a "silver bullet" solution that would prevent this sort of thing in the future. Some people call for more regulation, and maybe that would help, but I'm afraid that the inspectors and regulators would also fall victim to the same unreasonable trust that we saw on the oil rig. And personal agendas and politics have a way of twisting regulations and enforcement to provide personal rewards to the occasional person who is corrupt enough to look the other way for a price.

One possible solution is to learn from some of the things that have been done in commercial flight training, in NASA, and in some parts of the military. When you're learning how to fly a large jetliner you don't just learn how to do everyday things like take-offs and landings. You spend a huge amount of time learning how to deal with various malfunctions and flight issues. There's a protocol that you learn on what to do when certain things happen. You learn that if a certain instrument gives a certain result, you're supposed to do a certain thing. There's no judgment—except by the people who originally develop the protocol—you learn the protocol so well that you do it automatically when the actual event happens. Flight simulators make the process realistic—you go through the protocol in a simulated cockpit that looks and feels just like the real one.

Commercial aviation also relies on checklists. Before a pilot ever moves a plane away from the gate the flight crew has gone through an extensive checklist designed to make sure that conscious attention is paid to every relevant instrument and flight indicator. The process isn't perfect, but that checklist—in association with a predetermined protocol of how to deal with each abnormal checklist item—has prevented many potential disasters.

I don't know enough about the processes and procedures on the oil rig, but I get the impression that:

- Clear protocols were not in place to force workers to take certain actions when certain problems occurred. These protocols would have removed the need for a personal decision (and eliminated the "I don't want to be the bad guy" problem), since the protocols would have dictated the required behavior.
- There had been no "flight simulation" type training that took the oil rig workers through simulated problems until the solution was part of the workers' unconscious thought.

On the other hand, I know that flight-simulation-like techniques have been used for workers on nuclear reactors, and we still had the Chernobyl and Three Mile Island disasters, even though I suspect that many other potential nuclear disasters have been avoided as a result of the simulator training. I don't know what happened at Chernobyl or Three Mile Island: maybe the problems were outside the training, maybe the simulation training wasn't used in those facilities, or maybe the workers ignored their training and caused the problems.

Conclusion

I believe that the solution to the BP Gulf oil type of problems lies in:

1. Getting a better understanding of the safeguards that are in place in any risky situation, with specific focus on whether they're safe enough. I know that 100% safety is not possible, but we ought to be able to approach that number.
2. Determining the risk when one or more of the safeguards is not working, and making sure that protocols are in place to deal with repair or replacement of the broken safeguard in accordance with the risk. High risk should dictate an expedited repair or even a temporary shutdown.
3. Putting checklists in place for regular inspection and maintenance of the safeguards, and putting protocols in place to deal with any area that doesn't pass the checklist requirement.
4. Training workers to deal with exceptions, with specific instruction on what steps to take in each situation, and who is supposed to take those steps. This seemed to be missing on the BP oil rig, and so everyone just assumed that everyone else was dealing with the problems.

5. Using simulation training where it makes sense, so that workers get an actual exposure to what would happen in certain unlikely situations, and so that they learn how to deal with those situations.

And for those of you who don't work in the oil industry (most of you, I'm sure), think about what types of disasters might befall your own business. Are you prepared? Or will your employees fall into the human nature trap and avoid the problems rather than acting? What can *you* learn from the BP oil disaster? And what can you do to put human nature to work *for you* instead of having it work against you?

2026 Commentary

Most readers of this article in 2026 won't even remember the Deepwater Horizon incident, but I think you can get enough of a sense of what happened by reading the article. In my book, *Boiling the IT Frog,* I use a quote from Arthur C. Clarke, "Any sufficiently advanced technology is indistinguishable from magic." When the technology becomes so complicated that people don't understand it, there is a very high probability that people will begin to take the "magic" for granted. And when that technology is dangerous, the danger is compounded by the artificial trust placed in that magic.

122. A Model of Job Performance

Originally published on AUGUST 10, 2010 in #MANAGEMENT, #CAREERS

Back in 1979 I put together a model of job performance to help with some process improvements we were doing at Digital Equipment Corporation. Here's the model:

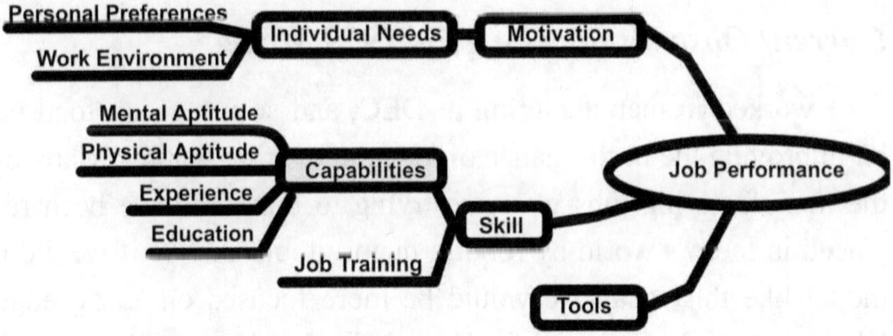

I ran across the model when was going through some old papers, and I thought you'd like to see it. Here's my explanation of the model that accompanied the diagram:

The model shows that an individual's performance is a function of three primary elements: Motivation, Skill and Tools.

An individual's Motivation depends on their personal needs and the ways in which the work environment satisfies those needs.

Skill is a function of an individual's Capabilities and the Job Training that the individual is given. Capabilities, in turn, depend on mental and physical aptitude, experience and education.

The Tools used for a particular job enable an individual to multiply the effect of their Skill in order to increase Job Performance. [The tools depend on the job, but computer systems are often used as tools for knowledge workers.]

The point of doing the model was to look at the various variables involved in achieving superior job performance, and then to go through the variables one by one to see where improvements could be made. Some of the variables are determined by the hiring process: picking people who are capable and who are motivated to work in your environment. Other variables are changeable after hiring; you can always give an employee better tools, better training or a better working environment.

Current Observations on This 1979 Model

I worked in manufacturing at DEC, and so my group focused on improving the performance of a manufacturing worker. Many of the manufacturing jobs we were trying to improve have been replaced in today's world by robotic factory tools, and so if we did a model like this today we would be more focused on Knowledge Workers. I think the model for Knowledge Workers would be similar, but a few clarifications are needed:

- The term "Job Training" is a bit misleading in today's world. There are many kinds of training, including professional training (how to carry out the day-to-day activities of your profession), tool training (how to use a computer system, application or some other kind of tool), simulation training (using a simulator like a flight simulator to see how you would do job tasks in an environment that tolerates failure), and shared best practices (learning from the way other people do the job).

- The word "Tools" also seems a bit dated, although conceptually the category is still valid. Most of us hear the word "tools" and think of hand tools like a hammer or screwdriver. But an iPhone or BlackBerry is a tool. A spreadsheet is a tool. A browser-based computer application is a tool. Anything that helps you enhance your job performance is a tool.

- In today's world of autonomous decision-making, context is an important—and often omitted—part of the thought process used by a knowledge worker. Context is provided by the leadership of the business through statements of strategy and direction. It's not enough to tell an employee what to do—you have to provide a set of principles for the employee to follow when making decisions.

Overall Conclusions

The world has changed drastically since 1979, and yet human beings are pretty much the same. I still believe that motivation, skill and tools are the three keys to job performance. Giving people what they need is still the best way to motivate. And training is still the best way to turn capabilities into skills.

What do you think? What factors are missing from this model? How would your model be different?

2026 Commentary

I think the model still applies. Bear in mind that "Individual Needs" is very broad, and includes everything from salary to the type of management needed by the individual. Most importantly, I think, it includes the type of work environment needed by the individual. If that work environment isn't provided, then low motivation is the result.

123. Why Cloud Computing is Good for Your Financials

Originally published on AUGUST 17, 2010 in #IT, #MANAGEMENT, #PROJECTS, #STRATEGY

Here's the way most people justify automation of a manual process: they replace a high variable labor cost with a relatively fixed system cost. I'll illustrate using graphs, then show how cloud computing fits into the picture.

Before: A Manual Labor-Intensive Process

Before any automation, the financials for a manual process look like this:

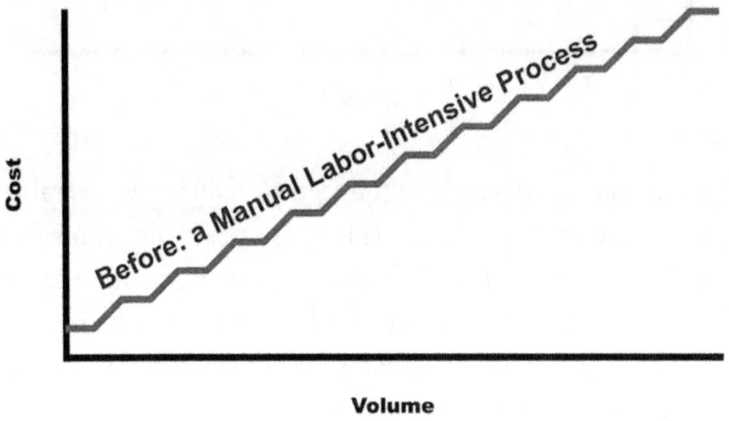

With a manual process, there's usually a direct relationship between volume and cost. If your volume goes up, your existing people can handle the increased volume for a little while. But if your volume continues to increase, then you need to add people, which raises your cost.

On the other hand, if your volume goes down, then you'll probably hold on to the people you have for a while to make sure the decrease will persist. Then when you're sure the volume decrease is going to be with you for a while, then you'll reduce the number of people through attrition or layoffs.

Traditional Process Automation

The financials for traditional process automation look like this:

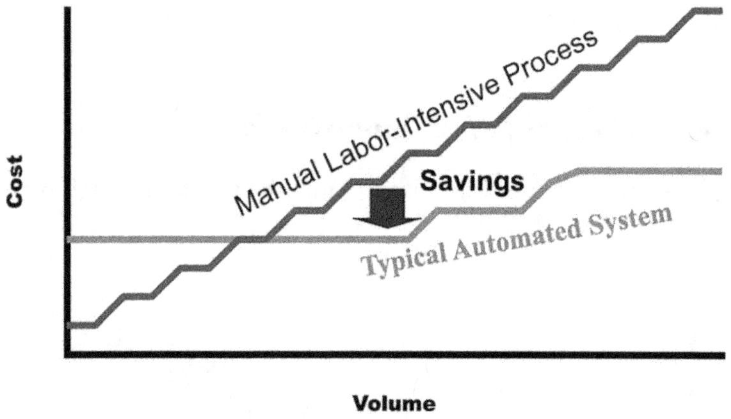

The automated system is introduced to reduce the overall cost as well as the cost per volume. This gives the business a savings, and lets the business expand its volume without adding more people. The cost of the automated system is flat over a certain range of volume, and then has a few increases at higher volumes due to the need for additional hardware and software.

The Problem with Traditional Process Automation

There are a couple of big problems with this automation approach. The business has replaced a flexible approach with a rigid automated process, and the business has traded a widely available and incremental cost—people—for a fixed cost that doesn't respond to change nearly as quickly.

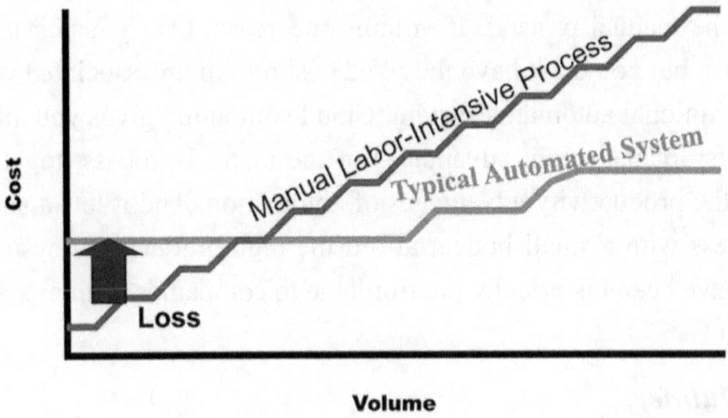

So when volumes decline below the expected minimum levels—as they have in the past few years of recession—the business ends up with a higher cost for the automated system than the previous cost for a manual process.

That's where cloud computing offers a huge advantage.

The Cost Advantage of Cloud Computing

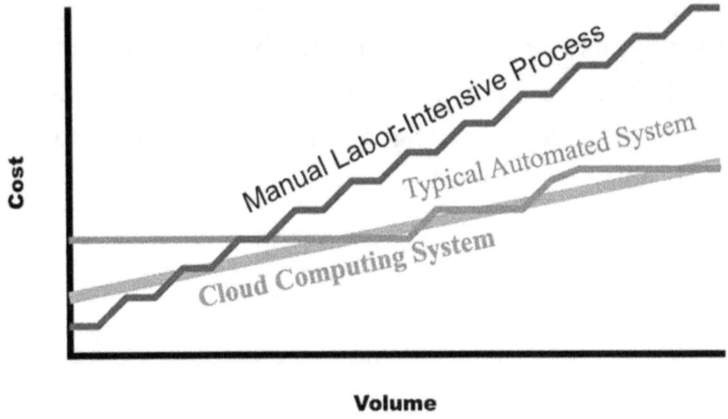

Cloud computing offers all of the advantages of an automated system with the more linear cost model of the old manual process. If volumes increase, then your cost increases, but at a lower rate than the manual process. If volumes decrease, then your cost decreases, but you don't have the fixed cost minimum associated with the traditional automated system. Cloud computing gives you all of the easy management advantages of the manual process, together with the productivity advantages of automation. And it lets a small business with a small budget afford the high productivity systems that have been historically unaffordable to companies of their size.

Disclaimer:

This article looks only at the financial advantages of cloud computing. For an overall look at cloud computing, see my article, "93. What is Cloud Computing? And Why Should You Care?." And remember that unless you're using a private cloud, cloud computing is a form of outsourcing. For an article on the advantages and disadvantages of outsourcing, see my article, "14. When to Outsource and When to Offshore."

2026 Commentary

Sixteen years later, this is probably obvious to many readers. But Cloud Computing was in its infancy in 2010, and it was more common for companies to bring expensive computer hardware in-house to handle systems needs.

124. Why IT Magic is Never Good

Originally published on SEPTEMBER 14, 2010 in #IT, #ALIGNMENT, #LEADERSHIP, #MANAGEMENT, #STRATEGY

I recently spoke at a conference attended by senior business and IT people from accounting firms. I described my usual view on IT Magic: that when Information Technology gets too complex, it's perceived as magic; then there are lots of problems for IT because business people develop unrealistic expectations. In the Q&A with attendees, one of the audience members asked an interesting question, "Is there ever a situation when IT magic is good?"

In the conversation that followed, it became apparent that this particular questioner—the head of IT for an accounting firm—is so trusted by his business partners that they blindly turn all of the IT direction over to him. He doesn't bother to explain to them why certain technologies are chosen, why he uses the level of infrastructure security that he does, or how his software and outsourcing choices contribute to the success of the firm. He is apparently given carte blanche to do whatever he thinks needs to be done, and he lets his IT work become magical to his business partners. Now he was describing his situation and asking me, "What's wrong with that?" And you may feel the same way.

Trust is Good

Let me separate out two different things that are going on with this IT leader. First, he has been highly successful in getting the trust of his business partners. That is undoubtedly a good thing. Trust means that he is respected by the business leaders of his firm, and they know that he has their best interests at heart. It probably also means—at least I hope so—that he participates in high-level strategy discussions on the direction of the firm, the firm's weaknesses and its plans to strengthen them, and the firm's strengths and its plans to leverage them. All this high-level communication puts him in a good position to put together a complete and appropriate IT strategy for the business, and to align IT direction with that strategy.

So far, so good. I applaud his ability to gain and earn business trust. I've been in that position myself a few times over the years, and I know how nice it is to be trusted enough that you don't have to plead for project money.

But Magic is Bad

On the other hand, I'm very concerned about his emphasis on magic and even wizardry (the perception that IT people are wizards because of the magic they can provide). I feel that with trust comes obligation—not just the obligation to act in a trustworthy manner, but also the obligation to be transparent.

Transparency is kicked around a lot lately in politics, but to me transparency means that people have a certain level of visibility into what you're doing with your time and with the money that you've been allocated. It doesn't mean that they have to watch over your actions on a day-by-day basis—it just means that they can see traceable evidence of your progress, and they constantly get positive reassurance that you're doing the right things.

Transparency of Complexity is Important Too

But there's another aspect of information technology transparency that people tend to forget. Business processes that involve information are complex, and information systems are often put in place to simplify things for the business user. But the original complexity of the business process doesn't disappear—it's just passed from the business user to the IT organization. This follows Larry Tesler's Law of Conservation of Complexity:

"Every application must have an inherent amount of irreducible complexity. The only question is who will have to deal with it. "

You may remember that Canon used a similar slogan a few years ago in advertising their cameras. They said that a Canon camera was "so advanced, it's simple." In essence, Canon was claiming that because the Canon engineers successfully dealt with the complexity of taking digital photographs, the camera user shouldn't have to deal with that complexity.

Often the business forgets how much inherent complexity has been passed from the business user to IT. And if the IT organization isn't careful, that forgotten complexity turns into the perception of magic, including the feeling that the business user is getting something for nothing. The ongoing effort required in IT to support the hidden complexity is forgotten and taken for granted, and business people begin to wonder why the Information Technology organization is spending so much.

Magic versus Transparency

To me, magic and transparency are opposites. Magic means that things happen without a good explanation. Transparency means that things happen, but that you're aware of how they're being done, and you can appreciate both the things themselves and the work that went into them. Transparency shows effort, progress and result. Magic shows only result, and makes you wonder how it was accomplished.

Now again you may be wondering, "Why is it bad to hide the effort and progress? Why is transparency a good idea?"

Why Transparency is Good

Here's why: Because sooner or later, the business is going to experience financial obstacles which cause it to reevaluate its expenditures. Or maybe the business will be acquired, and new management will question whether to keep the old IT or replace it with IT from the parent company. In those situations, all areas of business expense will be reevaluated. Budget cut-backs may ensue, or whole departments may be eliminated, combined or outsourced. If your department has relied on magic for its good reputation, then you've got a good chance that your department will be cut back. Senior management will be looking for cuts which will maximize their savings-to-loss ratio (see my article, "88. What's the Opposite of ROI?—How to Compare Cost-Cutting Opportunities"). And if your IT department hasn't been transparent about where business investment has gone, then it will be very difficult for you to explain how budget cuts will affect the business.

On the other hand, if you've been transparent, then it ought to be easy to say something like this:

"Yes, we can cut the IT budget 10%. It will have the following impact on the business: These projects will be delayed [and you name them]. The support levels for these applications will decline [and you give specifics]. These key business systems will run more slowly because of decreased network capacity [and you name the systems and give your estimate of slower performance]. The probability of a system outage or data security breach will climb from x% to y% [and you give specifics and provide the logic behind your calculations]. And the impact of all this on the business will be [and you give specifics on how these changes will affect the business]."

With transparency, you can do that. With transparency, you can easily show the impact of budget changes (either down or up) because you are accounting for the resource you've been allocated.

And Why Magic is Bad

But if you've chosen to use magic instead—if you've deliberately hidden your hard work and taken credit as a wizard who can do anything—then woe be unto you. Because wizards who can create something great out of minimal resources are expected to be able to reduce their resources without impacting their great results.

And that's what I told the questioner who asked about good magic: IT magic is never good, and if you insist on using magic—on deliberately avoiding transparency in your IT work—then sooner or later it will come back and bite you in the butt. Because sooner or later the perception of magic won't be enough; you'll need to actually be able to tie specific parts of the IT budget to specific deliverables and services. And if you can't, then your days as a successful IT magician or wizard will be over.

2026 Commentary

Because this is based on human nature, it’s still absolutely true. And it doesn’t just apply to IT—it applies to any organization that is not well understood. Examples include Research & Development, Engineering, and even such mundane areas as Facilities Maintenance.

125. One Situation Where IT Magic Can Be Valuable: Requirements Definition

Originally published on SEPTEMBER 17, 2010 in #IT, #LEADERSHIP, #MANAGEMENT, #PROJECTS

In my last article I talked about why IT magic is never good. Well, I guess I should have known better than to use the word "never." In his "Thoughts by Techxplorer" blog, one of my readers came up with a pretty good exception: a situation where the *thought* of IT magic—but not the actual use—can be valuable.

Here's the situation: Imagine you're in the early part of a new project doing the initial requirements definition. This particular project is trying to improve the way that a certain business process is being done, so naturally you talk to the people who are carrying out the existing process to get their ideas on improvements. But now you run into "the ditch digger syndrome": If you ask uneducated ditch diggers what they need to do a better job, they'll more than likely say "a bigger shovel." They won't know to ask for a backhoe if they've never seen one.

This is a common problem when we try to get people to talk about requirements: people tend to be limited by the way they've always done things. And if they're not encouraged to do otherwise, people tend to think small. They'll tell you that a certain field should be moved to another place on a screen, or they'll ask for a new report. But they won't ask for big things because they won't think of them. As Steve Jobs once said, "It's really hard to design products by focus groups. A lot of times, people don't know what they want until you show it to them."

Unfortunately, the result of people thinking small during the requirements process is that you end up specifying a system that looks very much like the one they already have. And that's usually a tremendous waste of money (why bother replacing something with itself?), and it gives you only single digit percentage increases in productivity.

Now, as Techxplorer points out, the business users you're talking to may just be trying to be reasonable. They are trying to make your job easier by asking for small things to minimize the effort required. So how do you shake them up? How do you get them to open themselves up to whole new ways of viewing the process you're trying to improve?

That's where Techxplorer suggests you use the idea of IT magic. Take the business users through a hypothetical situation: "Suppose you could get anything you want. And suppose that we could use magic to make this process ten times easier for you. How would you change the process in *that* situation?"

I've also heard this called "blue-sky brainstorming" (the sky is the limit). And frankly I've never used the word magic when I've done this—I've just relied on the phrase "make the process ten times easier." Most people have to think big when they improve a process by a factor of ten.

But no matter what you call it, and no matter how you explain the concept to business users, I think this is an important tool to employ during the early stages of requirements definition, and I'm glad Techxplorer pointed it out. You need to shake up your users and make them think about what *could be*, because even if it's something you can't deliver—even if it's impossible—it gives you valuable information about radical new ways to improve business processes. A totally ridiculous idea may spark a great idea that's very practical to implement.

A Caveat

I only have one concern with this hypothetical use of IT magic. If you use this method, then you need to be very careful to bring your business users back to reality before you leave them. It's all well and good to tell them, "suppose we could use magic," but before you leave you need to remove the magic spell. You should remind them that there is no magic, and make sure they realize that this was an exercise, and that the only way this kind of thing can be implemented is through hard work using real day-to-day tools.

Or, if you prefer, you can avoid use of the m-word altogether (or save it in reserve for extreme cases) and just use the "ten times easier" approach to broaden their thinking. I've found that it works just as well.

Remember that my argument against IT magic is that it creates unreasonable expectations. So be careful that during your requirements gathering you don't create those unreasonable expectations yourself.

2026 Commentary

Another way to communicate this idea during requirements gathering is to ask system users how they would design the process if they had to do it all over again from scratch. Or you could ask them whether they've ever seen a better way to do the process, maybe in another company or in a different industry altogether.

126. A Manager's Biggest Burden, and 5 Ways to Deal with It

Originally published on OCTOBER 25, 2010 in #MANAGEMENT, #PROJECTS

One of the biggest surprises to new managers is the intense pressure to keep people working productively. This is especially true in a project environment like IT where employees aren't doing the same thing day after day. Managing an organization is like being in a taxi with the meter running and only a few dollars in your pocket, except that the meter is burning up hundreds of dollars per hour.

Let's do a quick, back-of the envelope calculation. If you have ten people working for you, and if they have an average salary of $70,000 (not an unreasonable number for today's IT organizations), then you've got a payroll of $700,000 per year. Add another 30% for benefits and expenses, and you've got a people budget of $910,000. The average worker in the U.S. has at least 4 weeks of holiday, vacation and sick days, so you've got no more than 48 weeks worth of actual work time. Divide $910,000 by 48, and your organization has a "burn rate" of about $19,000 per week, about $3800 per day, about $475 per hour, or about $8 per minute.

So think of it this way: Every minute that you hesitate in telling your organization what to do is costing your business $8. Every day of giving your people the wrong work is costing your company $3800. And every week of rework is costing your business $19,000.

Want even more pressure? Consider the next level up. If a second-level manager has five organizations your size reporting to them, then the meter is running five times faster. The second level manager's organization is burning $95,000 per week, or about $40 per minute. Go up more management levels, and the burn rate multiplies again and again, until you easily get rates of millions of dollars per week and thousands of dollars per minute. Doing the wrong thing can be very expensive, and the pressure to properly use your burn rate is intense.

5 Ways to Deal with the Problem

So how do you minimize the wasted time and money? Here are some tips:

1. Don't give people just *one* thing to do—give them a view of what comes next.

If you give people one thing to do, then they are likely to do it and then "take a break" before asking you what to do next. If they see a stream of work instead of an individual task, then they're more likely to finish one task and then start right up with the next one.

Even better, if people know the sequence of things they have to do, then they'll take advantage of natural "waits" in their current task (waiting on someone else, waiting on resources, waiting on a scheduled event) to get some preliminary work done on their next task. This will increase their overall productivity.

2. Make sure that you properly prioritize the work that is to be done.

When given multiple task assignments, it's human nature to spend time on the tasks that are easier or more fun, putting off the other tasks until later. But if employees know the priorities associated with the tasks, and if you insist on them working on higher priority tasks before lower priority tasks, then they'll stay focused on the work you want them to do.

3. Keep the number of tasks in front of each person to a reasonable number.

While you might know all of the things that a person will do for months in advance, it will only confuse and distract a person if you assign all of those tasks at the same time. There's no best number for how many tasks to give a person, but you generally want to stick to single digits (i.e., 2 to 9 tasks), and lower numbers (2 to 5) are better. The actual number should depend on how much wait time will be associated with doing each task.

4. Have a "standby list" of tasks that can be assigned when there's nothing else to do.

Some of these tasks may be non-critical maintenance tasks that can be done whenever resource is available. This standby list is important because you never know when an entire project might be put on hold, and then you can assign the standby list while you're figuring out what your organization is going to do next.

5. Use "pull management" instead of "push management."

In pull management you assign a task objective, tell your worker how success will be measured, and then turn your worker loose on achieving the objective, essentially "pulling" the worker toward the objective. In push management you micromanage, keep the employee in the dark as to the true objective, and just give the employee enough task instruction to tell them specifically what to do, "pushing" toward the objective.

With push management the employee will keep coming back to you for detail after detail, and productivity will be wasted while the employee is waiting for your feedback. With pull management you'll need to do a little more communication up front when you're defining the objective and the measurement, but the employee will be largely self-sufficient during the performance of the task. Employee productivity is much higher using pull management, and business results are usually better.

Conclusion

There's nothing worse for an organization, for a business, or for a manager than to have employees who are idle because they don't have anything to do. The meter is running constantly for business organizations, and every hour lost is an hour that you'll never get back. Dealing with high organizational burn rates is a manager's biggest burden, but proper management can minimize wasted burn rate and maximize employee productivity.

2026 Commentary

The numbers in the example were from sixteen years ago, so the costs are obviously higher now. Do the calculation yourself with typical salaries for your industry. No matter what the result, there's a large amount of money that is spent on labor cost, and you don't want to waste that much money by having idle people.

127. Two Questions to Ask After Each Project is Completed

Originally published on NOVEMBER 8, 2010 in #LEADERSHIP, #MANAGEMENT, #PROJECTS

Know what the biggest difference is between an *adequate* project manager and a *great* project manager? The *great* project manager always learns from every project and applies that learning to the next project.

Here's a simple technique to help you learn from every project too. After each project completion, before you send the project team members on to other things, get all of the project principals together and have a postmortem review. You can kick around as many ideas as you like during the review, but make sure you answer these two questions:

1. **What did we do right that we want to make sure we do again in the next project?**
2. **What could we do better next time?**

In each question, the wording is critical.

What did we do right?

Most postmortem project reviews dwell on the negative, but in all of the projects I've seen—even the project disasters—there were always a few things that were done right. It's important to recognize those things because unless we specifically emphasize them on the next project, there's a strong likelihood that they won't be repeated. Overemphasis on negatives in a project review is counterproductive. Everyone walks away from the review with a bad attitude, and no one really learns anything. By stressing the positives, the project team can reinforce the good things that were done, and the team members can leave the review with an upbeat attitude.

What could we do better?

Of course it's important to learn from project mistakes as well, but it's too late for blame. Don't focus on what went wrong—instead focus on how things could be done differently to make them more successful. There's a subtle difference here. If you talk about things that went wrong, then people get defensive and even antagonistic. But if you recognize the things that went wrong only as a jumping off point to generating ideas for doing things better, then the defensiveness is defused, and team members honestly look for better ways to prevent that kind of mistake on the next project. Even though the bad things that happened on the project come up in the review, the review team doesn't dwell on them, and so the result is more upbeat and positive.

This works for meetings and events too

I've used the same two-question review after a long meeting and after a special offsite event. By asking the two questions at the very end of the meeting or event, we were able to generate a lot of ideas for future improvement, and we ended on an up note. Everyone likes improvement—we just don't want it to come at our own personal expense. Approaching improvement in this way makes the process much easier to accept, and everyone walks away encouraged.

And that's how progress is made: reinforce the good things, and figure out ways to improve upon the bad things.

2026 Commentary

I highly recommend this process. It also works for non-work-related things like team sports (after each game), social events, family get-togethers, and even dating.

128. Are You a Manager or a Leader? Who Cares! Just Do What's Needed!

Originally published on NOVEMBER 29, 2010 in #LEADERSHIP, #MANAGEMENT

Business literature is full of distinctions that some very smart people make between a manager and a leader:

- *"Management is doing things right; leadership is doing the right things." - Peter Drucker*
- *"...Leaders are concerned with what things mean to people. Managers are concerned about how things get done." - Abraham Zaleznik*
- *"Leaders are the architects...Managers are the builders." - John Mariotti*
- *"Management is getting people to do what needs to be done. Leadership is getting people to **want** to do what needs to be done. Managers push. Leaders pull. Managers command. Leaders communicate." - Warren Bennis*
- *"Management is coping with complexity; leadership is coping with change" - John P. Kotter*
- *"Leadership focuses on the creation of a common vision ... Management is the design of work .. it's about controlling." - George Weathersby*
- *"Leadership has about it a kinesthetic feel, a sense of movement. Management is about 'handling' things, about maintaining order, about organization and control." - Kouzes and Posner*
- *"Leadership is communicating to people their worth and potential so clearly that they come to see it in themselves." - Stephen Covey*

- *"Management is nothing more than motivating other people." - Lee Iacocca*
- *"A great leader has two key qualities. He knows where he wants to go; he's able to persuade others to go with him." - Ted Turner*
- *"A good manager is best when people barely know that he exists. Not so good when people obey and acclaim him. Worse when they despise him." - Lao-Tzu*
- *"The first responsibility of a leader is to define reality. The last is to say thank you. In between, the leader is a servant." - Max DePree*
- *"The first duty of a manager is to translate vision into reality." - Peter Drucker*

But the bottom line is that the distinction between a manager and a leader is all semantics. No one hires you with the job title of Manager and then says, "Now don't try leading—that's not in your job description." And no one hires you with the job title of Leader and then says, "Just focus on leading—managing isn't something you should care about." Let's face it: All management jobs have a leadership component, and all leadership jobs have a management component. And frankly, it doesn't really matter what you call it; what matters is that you get the job done.

Sure, higher-level management jobs and executive positions probably deal more with change than with maintaining the status quo. And lower-level management jobs often focus on fine-tuning a process that already exists. But that doesn't mean that the jobs have to stay that way. A job is a blank slate, and it's up to you to use what you've got to make things better for the business. If you're hired to lead things in a new direction and you discover that the existing direction is plenty good enough, then by all means speak up and convince your organization that radical change isn't necessary. And if you're hired to just keep things running the way they are and you discover that the way things are is in drastic need of a change, then it's up to you to lead the organization in a different direction.

Job descriptions are guidelines—not rigid laws. What's most important is to understand your constituency—to understand who your customer is and where your organization should be heading. All direction setting should be defined in terms of what's best for that constituency, and it's not enough for you to placidly accept what you've been told and follow orders in the face of huge amounts of evidence to the contrary.

Are you a manager or a leader? Yes! You're both. And the sooner you realize that, the sooner you'll focus on what's important for your business and begin to put your resources to work to move things in the right direction.

Tom Peters said, "Never, ever, ever, ever, ever, ever, under any circumstances, accept an assignment as given. Shred it, expand it, twist it, turn it, until you create something that you're red-hot-coal passionate about." I agree: don't get caught up in semantics—just do what's needed for your business.

2026 Commentary

If you work for someone who keeps restricting your ability to do what's really needed, then find a way to work around that person, or find a different job. It's not fair to you or the business to give you an objective and then stand in your way of achieving that objective.

129. IT isn't about Analysis and Programming Anymore

Originally published on JANUARY 6, 2011 in #CAREERS, #IT, #ALIGNMENT, #LEADERSHIP, #MANAGEMENT, #PROJECTS, #STRATEGY

I've talked about the changing nature of IT in a previous article ("18. IT is Moving Toward Property Management"), but it's amazing to me how fast some of the changes are taking place. Ten or twenty years ago the key skills for someone in IT were systems analysis and programming—mostly technical skills. But more recently the need for those skills in a typical IT shop has declined drastically. Now the key skills are project management and system integration—skills which are not nearly so technical.

Changes in the Way We Deal with Food

There's a parallel to be drawn between IT and the way we get our food in the United States. A couple of hundred years ago we were a nation of farmers. Most people participated in the growing of their own food. They prepared the ground, they planted the seed, they tended the growing plants, and they harvested their food. Only then did they think about cooking and eating the food.

Now the farming process is performed by a small minority of people, mostly operating under the control of a few large corporations. We don't participate in the growing any more—instead our focus is on shopping and cooking. It's no longer important for the average person to know how to best prepare a field for a certain crop, to know how to correctly place seeds in the ground and then tend to them to best encourage their growth, or to know how to harvest the resulting crop to avoid damaging the food. Those skills have been outsourced.

Changes in the Way We Deal with Software

Forty years ago there was virtually no packaged software. The IT world was all about systems analysis and programming because everyone had to develop their own software. The IT organization determined business requirements, designed systems, and then built those systems from the ground up, writing every line of code. Since the software was all home grown, it was up to the IT organization to do systems integration as the systems were built. Part of every business requirements document was the detail on how the system needed to interface with other business systems.

Today most software comes from a few large companies. And just as the need for detailed information on farming has become less commonplace, the need for systems design and programming has moved from the everyday IT shop to those large companies like Google, Microsoft, Oracle, SAP and Apple which specialize in designing and building software. Your average IT shop does very little systems design and programming. In fact, I'm guessing that the majority of IT shops in small and medium-sized businesses do no programming at all—they just configure newly acquired software, adjust parameters to optimize the software to their operating environments, and use high-level systems tools to integrate systems into something that resembles a cohesive whole. You'll still find IT systems design and programming in larger companies, but it's mostly where those companies use home-grown software to provide a specialty product, service or process that differentiates them from their competition.

So just as we've gone from a nation of farmers to a nation of shoppers and cooks, our IT organizations have evolved from designers and programmers into a group of project managers and system integrators, with a few support people included to help resolve day-to-day issues.

What's the Impact of This Change?

Looking back at the impact of the change in the food process on our lives, we see the following:

- High volume farming is more efficient but the compromises made in high-volume farming lower the quality of the food. Because the time from crop to plate is longer, the high-volume farmers optimize their seed and harvesting around that longer time. You won't find anyone who will tell you that corn bought at a store tastes as good as corn you just picked. Yes, the corn from the store is OK: it generally tastes like corn and it provides most of the nutrition that fresh corn provides. But you can't beat fresh food grown without the compromises made on a high-volume farm. That's why there's a demand for local organic food—it brings back some of the advantages of growing your own: fresher food and fewer preservatives.
- In an attempt to satisfy the majority of the market, the high-volume farmers focus on a more generic range of tastes. If you grow your own tomatoes, you can pick the seed that gives you exactly what you want: big tomato or small, just the right amount of acidity, the perfect texture. Your range of choices in store-bought tomatoes is much smaller.
- With meats, the high-volume farms optimize their production in a way that is often inhumane. Chickens are confined to tiny cages for their entire lives. Cows and pigs are mistreated and sometimes tortured. All of this is hidden from us because we only see the finished product—not the process used in its production.

Now let's look at these same issues in the evolution of an IT shop:

Large software companies are more efficient at creating products like inventory systems, customer relationship management systems, payroll systems, and accounting systems—essentially any software that's large and complicated. A standard software product can be developed for any business process that's relatively consistent across an industry. With software you don't have the issue of preservatives, but you do have the problem of freshness. One of the compromises you'll make in using off-the-shelf software is that any improvements or bug fixes will take longer to implement. You don't have the option of choosing to use your IT resources to make an immediate improvement to the way that a particular screen works or the way that a particular process step is handled.

Trade-offs are made in developing a product that has to satisfy so many companies in so many different industries. An off-the-shelf software product is never as good for your individual company needs as a product that's been custom built. Sure, it's less expensive—it would cost a small fortune to develop a custom-built product, and most small and medium-sized companies can't afford it. But you definitely have to compromise when you buy an off-the-shelf solution, even if you configure it for your company, and even if you tailor it (actually change the software code—not something I would recommend).

Is there a backlash? Is there a demand for "local organic software"? I haven't seen it yet. Maybe I'm wrong, but U.S. businesses don't seem to care about where their software comes from. And as for the compromises required to tailor the software to your exact needs, most companies don't see enough cost/benefit to justify going the custom route. And the freshness issue is one that companies seem to live with.

And how about inhumanity? Much of the off-the-shelf software being used today was developed using inexpensive developer labor in other countries. Those developer jobs you no longer need in your business don't just move to the software development companies—they move offshore. So while I haven't heard stories about programmers being confined to tiny cages their entire lives .... No, wait a minute. Do cubicles qualify?

2026 Commentary

Artificial Intelligence (AI) is entering the software development picture, just as additives and artificial sweeteners came into food production. So far, I see AI being used to help generate software code, or to help document existing software code, or to automate testing of software. But it won't be long before AI helps you configure off-the-shelf software to optimize its use for your business.

I have no problem with AI being used this way, but I worry about people making assumptions about AI software being better. Remember the Arthur C. Clarke quote, "Any sufficiently advanced technology is indistinguishable from magic." If AI software is given undeserved trust, then quality problems will ensue. Don't treat AI like magic.

130. Future IT = Cloud + Mobile + Enterprise App Store

Originally published on JANUARY 31, 2011 in #CAREERS, #IT, #ALIGNMENT, #LEADERSHIP, #MANAGEMENT, #PROJECTS, #STRATEGY

There's been a lot of talk about cloud computing, and mobility has been in the news for years. But apps and an enterprise app store are going to bring it all together to remake the face of IT.

Cloud Computing

Cloud computing is a method for delivering computing resource. Its principal attributes are outsourced management, ubiquitous access and elasticity.

- Outsourced management allows us to obtain the use of software without having to deal with installing it, maintaining it, or managing the associated databases.
- Ubiquitous access means we can use the Internet to get to our software and data. Anywhere you can get access to the Internet, you can get access to your cloud computing application.
- Elasticity means we don't have to worry about technology sizing. We don't have to buy a server that will accommodate our peak demand, which maybe occurs only when the last Friday of a quarter corresponds to a full moon. Instead, we just use the resource we need, and we pay only for the resource we use. This is a very fair approach, and CFOs love it.

Mobility

Mobility is one of those things that has snuck up on us. It started with mobile phones, then texting and mobile email, and now it's gotten to the point where you can do anything just about anywhere. And if you think about it, that's where things started in the first place. Before we got to the point of using computer systems, we relied on what we carried around with us to get our jobs done. We could move around freely to meet with customers and co-workers in the most appropriate environment for the job we needed to do. The idea that we should have to retreat to a single "corporate" location to find relevant data is an artificial creation of a time when computers were too bulky to carry around. The new reliance on portable computers and hand-held devices is just a return to where we started. We've thrown off the shackles of our behemoth computing captors and returned to a world where location is determined by our task—not by our limited access to data.

Apps

There's a huge difference between an app and a traditional software application—particularly a Windows application:

- Apps can be self-installed in a few minutes by a novice. Traditional applications require lots of steps, the answers to a number of unanswerable technical questions, and often a restart of your computer system.

- Apps are intuitively easy to use. Applications require training and rote learning of arcane codes. That's often because applications are more sophisticated than apps, but who came up with the stupid idea that an application has to be a Swiss army knife and do hundreds of things when only a few things are really needed? Apps just do what's needed; applications try to be adequate at everything and often end up being good at nothing.

- Apps automatically know how to get data from the principal data sources on your device. For example, apps which need address book or calendar information know how to get it. Traditional applications, on the other hand, require extensive configuration. You'll have to be specific about telling your application exactly where to get data, even down to the directory where you've chosen to store it or the server where the data can be found.

- Apps can be installed without hurting other apps. Applications often interfere with other applications due to DLL overlaps and conflicting registry entries.

- Apps can be deinstalled without incident. Applications often leave behind old files and even software which will continue to haunt you when it conflicts with other applications you'll later install.

Basically, apps are the new applications. They're what applications should have been if we had been working in an environment which made apps possible.

Enterprise App Store

Anyone with a smart phone is familiar with the concept of an app store. You search for an app for a particular need, then buy it, download it and install it all in one easy step. An enterprise app store is just like the app store for a smart phone, except:

- The apps have gone through a selection and weeding-out process so that only the best and most appropriate business apps are available for your use.

- The store contains a mixture of generic and proprietary apps. The proprietary apps were designed, tailored, written or customized by people in your company to provide for the specific needs of your business. Those proprietary apps know how to access your corporate information and the cloud databases they use. And so when you use one of these proprietary apps, you're actually updating the same corporate database that your company relies on for its day-to-day business.

- Access to the enterprise app store is secure, and only authorized employees are allowed to see certain apps. The apps you see in the store are restricted to those apps that make sense for your job and for your level of authorization within the company.

- There may be different versions of the same app. The version of the app you download will be determined by your job and level of authorization. For example, your app may let you look at your department, while your division vice-president's app may let them look at the entire division.

Bringing It All Together

If you put all these things together—if you use cloud computing, mobility, apps and an enterprise app store—you can totally change the face of IT in your company. For starters, you have no need for a data center, since all of your data is stored in the cloud, and all of your applications are delivered through the Internet. Then you can drastically cut the size of your help desk, since you no longer have to worry about PC configuration issues for applications, and since apps are so much easier to use than traditional applications.

Of course, someone still has to develop the proprietary apps, but you'll probably outsource most of that work. So the primary work of the IT organization will be:

- Strategic Technology Leadership—determining what ought to be done

- Systems Architecture and Integration—determining how corporate systems should look, breaking the requirements down into individual products, databases and apps, and then developing an overall approach for implementing what's needed

- Project Management—leading projects for major technology initiatives

And yes, there will still be a need for some help desk support, but with properly designed apps, the amount will be much less than what is required today.

Conclusion

Each of these factors—cloud computing, mobility and intuitive apps—has had an impact on IT. But together these three factors, in conjunction with an enterprise app store, are going to totally reinvent IT. You can lead the change in your business, or you can drag your feet and get left behind. The choice is up to you.

2026 Commentary

I see this happening faster in some businesses than others. I guess that makes sense, since some businesses derive more benefit from this approach, and some system needs are more appropriate for this kind of system architecture. In some cases, the "app" is implemented as a "web app" which is accessed through a browser. But the net result is the same.

131. No Surprises, No Rushing

Originally published on MARCH 21, 2011 in #MANAGEMENT, #PROJECTS, #STRATEGY

People who have worked with me know that two of my biggest project principles are "No Surprises" and "No Rushing."

No Surprises

Surprises are a sure sign of inadequate planning. When you do a project you have to anticipate what might go wrong as well as what might go right. Some of the things that can happen have a low probability, but you still have to develop at least a rough idea of what you'll do when they occur. By planning for these surprises, you'll eliminate—or at least reduce—the panic that's human nature when these crazy things happen. Instead, when a "surprise" happens, you'll just put in motion the contingency plan that you developed in advance.

Can you anticipate every possible surprise? No. But you should do enough contingency planning that true surprises are rare. When you do get a surprise, deal with it, then afterward ask yourself, "Could this have been anticipated? How can my planning process be changed to include the possibility of this sort of thing in the future?" Planning is an adaptive process: it gets better as you refine your technique to include the possibility of surprises.

Of course, it's also possible to "overplan." There's a trade-off here: you need to do enough planning to prepare for all of the likely contingencies and most of the less likely ones, with maybe just a rough sketch of how you'll deal with some of the unlikely contingencies. Then monitor the possible contingencies, and if some of the less likely ones become more likely, then beef up your plan for those areas. But overplanning can lead to overcomplexity, which in turn leads to a probable runaway project.

No Rushing

To me, rushing isn't about speed—it's about moving faster than your thinking. I'm all for speed: speed in decision-making, speed in planning, speed in implementation. But action without at least a little bit of thought is just reckless.

When you drive a car at night, you have to be careful not to overdrive your headlights—you have to stay at a speed low enough for you to stop your car in the distance your headlights illuminate ahead. I think there's a parallel problem in projects: you have to move at a speed that allows you to make course corrections and to overcome obstacles, or you're going to get a runaway project. Just as you drive slower at night on a curvy road, you should move slower in a project when there are a lot of unknowns and you're feeling your way through. When you get through the slow patch, and you're dealing with stable, known technology and a clear business direction, then you can move faster.

One of the biggest causes of project failure is doing the wrong project: putting a whole bunch of work toward a badly-defined objective. Rushing doesn't get you a result faster if you have to throw things out and start over. So move at a slower pace when you're getting things defined, pick up the pace when you know exactly where you're going, and slow down again as you encounter roadblocks or obstacles.

Conclusion

Surprises and rushing are the sign of an inexperienced project manager. They can also come from an over-eager upper manager who is more interested in the appearance of action than in a good result. If you want to be successful as a project manager, then don't let yourself be pushed into skipping planning steps and moving faster than you should on a project. Proper planning can save time and money on a project, even if the planning process seems overly time-consuming at the start.

2026 Commentary

I talk about surprises and rushing in projects, but these two issues are equally important in most things in life. For example, when planning your career, you need to slow down at critical decision junctures and think about what surprises might occur if you pursue different paths.

132. 8 Ways to Deal with Employee Personal Problems

Originally published on MAY 3, 2011 in #MANAGEMENT

There's a naive belief among many new managers that employee personal problems should be irrelevant to job performance and therefore something that managers can ignore. We like to believe that when employees walk through the office door, all of their personal problems are left behind. To managers with this belief, employees are like robots: treat them all the same way and just focus on the work.

But the fact is that people are not like robots. And as much as we would like to believe that personal life doesn't impact a person's work, it very much does. Exhausted new parents suffering from lack of sleep due to a crying infant aren't able to be as creative as they'd like. Workers dealing with problems at home often find their minds wandering, and don't do their best work. Employees who are in pain—either physical or emotional—don't operate at peak levels.

It's our job as managers to get the best work out of our employees. But it's also part of our job to keep our employees motivated and happy so that they will continue to be a contributing part of our organization for the long term. And to do that well we have to know the employees as individuals, and to help them through some of the personal issues that interfere with their ability to do their best work.

Here are some tips on how to help employees with their personal problems:

1. Listen

Often it's enough to just listen to the employee, be sympathetic, and offer kind words of encouragement. But pay close attention to the difference between listening to an employee and solving the employee's problem. Remember that it's up to the employee to solve their own problem. You can listen, but you shouldn't try to take charge of getting a solution.

2. Refer employees to appropriate resources

As much as you would like to give personal advice, it's seldom the best course of action. A better approach is to steer employees toward professional counselors or clergy, or at least refer them to understanding and sympathetic family and friends.

If you have personal experience with a similar problem, then you can tell your own story and describe your solution. But be careful to keep the focus on the employee—it's not about you, it's about them. And don't try to solve the employee's problem yourself or try to force your own solution on the employee—just gently push the employee toward finding their own personal solution.

3. Accommodate short-term needs

Be as flexible as you can in making temporary accommodations to help the employee through a crisis. But make sure you communicate that these changes are in fact just temporary. Work out a date when the accommodations will end, or at least a date when you will get together with the employee to reassess the situation.

4. Be flexible in working hours or working location

This is usually one of the easiest accommodations to make. Give short-term time off if it's needed (use vacation or sick time if it's available). Or if the problem will continue for a while, then maybe work can continue as is but with a revised working schedule (e.g., come in an hour later, but leave an hour later to make up for it). Or maybe the employee can work from home on certain days. Offer alternatives which meet the needs of the employee.

5. Temporarily assign an employee to different work that is better suited for the employee's current state of mind

This is a more extreme accommodation but it may be required in certain situations. For example, you may temporarily assign a project manager to other work, putting someone else in place as temporary project manager. Or for an employee who typically travels a lot for the company, you may temporarily assign the employee to a job that requires little or no travel.

6. Make it clear that these are short-term accommodations

It's important to make it clear to the employee that long-term continuation of these accommodations may ultimately result in a change in job title and/or reduction in pay. Don't make this a threat —just let the employee know that you're willing to let things slide for a while, but ultimately the employee will be expected to deliver the same level of performance that was previously provided.

7. Keep in touch with the employee during the crisis

Monitor the situation to ensure that the employee is in fact taking steps to resolve the situation. Provide encouragement and positive reinforcement along the way.

8. When the crisis has passed, refocus the employee on the steps needed for success and career advancement

Congratulate the employee on making it through a difficult situation. Then provide assistance to help the employee get back the focus that they once had.

Conclusion

Every employee is an individual with their own personal strengths and issues. You can't expect to take advantage of an employee's strengths without occasionally running up against some of the issues. Management is about achieving business results by focusing resources, and to properly focus human resources you'll need to deal with employee personal problems.

2026 Commentary

I'm reading this fifteen years later, and it sounds a bit cold hearted. I think that the action you take depends a lot on the relationship you have with the employee. If the relationship is strictly employer/employee, then the article's advice is fairly sound. It's basically saying to not go down the rabbit hole of turning into some sort of psychological counselor. If you get personally involved in the employee's problem, then it may be difficult to figure out how to separate your feelings for the employee as an employee versus your feeling for the employee as a friend. And it's not fair to other employees for you to give too much special consideration to the employee who is having a problem. It's up to you to figure out where to draw the line.

133. Are You Listening with Confidence or Listening with Arrogance?

Originally published on JUNE 1, 2011 in #CAREERS, #LEADERSHIP, #MANAGEMENT

Listening is one of the most important traits of a good manager. Good managers spend most of their time listening: listening to their employees describe the problems they've encountered, listening to what their bosses tell them to do, listening to what customers have to say about products and services.

Beginning managers listen

It's easy to listen when you're early in your career. After all, you don't know very much, and the people around you can provide you with useful information. You learn about how things have always been done, and if you push a little, you'll even learn why. Dig a little deeper, and you can learn whether the reasons are still valid.

You'll learn about your customers, what they like, what they don't like, how their lives can be made better through use of your products and services, and how your products and services sometimes make them miserable. You'll learn that customers enjoy being listened to, and you'll learn that very few companies listen enough.

You'll learn about your employees, who they are when they're away from the office, what they want out of life, what they enjoy and what they worry about at night. You'll learn that they're individuals—not just robots who take direction—and that they respond to different kinds of motivation depending on their nature.

You'll learn that your bosses are individuals too. They have doubts and fears just like you do, and they're often driven more by individual aspirations than by the perceived good of the company.

Higher managers tend to stop listening

As you move higher in the management ranks, you start filtering what you hear based on your experience. It's not a conscious thing. It's natural for your brain to apply pattern recognition to the new information you receive. It's the brain's way of making sense of all of the increasing amount of data. When you hear something the first time, you listen carefully. But by the 100th repetition of the same information, you tend to filter it out. You zone out during the repetition and try to skip to the bottom line.

Experienced managers think they've heard it all before. They listen to a message, categorize it based on their experience, and apply the standard response that they've developed for this particular message. That can be good up to a point, but you have to be very careful.

Good managers listen with confidence

Listening with confidence means that you apply all of your knowledge and experience, but that you still listen very carefully and attentively. You listen to the message, but also to the nuance. You listen to the words, but also to the emotion. You listen to the meaning, but also to the implication and hidden subtext. You listen to what's familiar, but you carefully explore subtle shadings that make this telling of the message slightly different. And you always remember that even though you've heard this message (or a similar one) a hundred times, it's probably the first time that this particular speaker has given the message to you. It may seem boring to you, but it's an emotional event for the speaker. And every version of a message is slightly different depending on the situation and the speaker.

Bad managers listen with arrogance

Listening with arrogance is hardly listening at all. It's hearing what someone says, but rapidly discarding the words, and then pigeonholing the event based on prior experience. It's listening while doing something else—maybe checking email or looking out the window. It's cutting off the speaker in mid-sentence, not allowing the speaker to finish. It's listening passively—not asking questions to clarify or get a better understanding of intent. It's being impatient and showing that impatience to the speaker. It's knowing the answers without asking the questions. It's applying your experience in a way that only perpetuates the past, instead of learning that things can indeed change.

As a speaker, how do you tell the difference?

How can you tell if someone is listening to you with confidence or listening with arrogance? Good listeners pay attention. They aren't distracted—they're focused on you and your message. They try to truly understand you, whether or not they agree. They ask questions to clarify areas of your message that are unclear or ambiguous. They probe for intent, and push for clarity. They ask you for your opinion on how to resolve any problem you describe. They ask for pros and cons of various approaches. They seem interested in you and in what you have to say. You leave the conversation feeling good about yourself and about the communication.

Arrogance sneaks up on you

I don't know of any manager who chooses to be arrogant. It's the kind of thing that sneaks up on you. There's a progression from timidity to confidence that continues into arrogance if you don't keep your ego in check. But arrogance is one of the worst traits a manager can have, because arrogant managers resist change. They build their organizations to work a certain way and then use their arrogance to set their beliefs and their processes in concrete. Arrogance leads to bureaucracy, and bureaucracy leads to failure in a world that's rapidly changing.

Are YOU an arrogant listener?

If you see yourself in any of these descriptions of arrogant listening, then it's not too late, since a truly arrogant listener probably wouldn't have read this far in the article. Take steps to return to confident listening, and you'll get a better organization with better morale. And you'll probably find that you'll enjoy your job a lot more too.

2026 Commentary

I've also heard confident listening described as "being present" or "being in the moment." It means that you're focusing 100% of your energy on actually listening to a person, and not half-heartedly listening while thinking about something else. It's easy to do once in a while and difficult to do consistently. But consistency is what you need in this area.

134. Should You Let People Go, or Keep People and Reduce Salaries?

Originally published on JULY 1, 2011 in #LEADERSHIP, #MANAGEMENT, #STRATEGY

My friend Derek Cheshire made an interesting observation yesterday:

Tell me if I'm being stupid but after reading about the Greek austerity measures I do wonder why we have to try and make hundreds of public sector employees redundant. Why not just trim pay by say 10%? At least there would be more people with money to help keep the rest of us afloat.

Derek's question got me thinking. In general if you're with a business or government agency and you need to cut labor cost, should you let people go or should you try to keep people employed by cutting everyone's pay?

First, some perspective

Most businesses and government agencies have gotten more productive over the last twenty years. For a long time people wondered whether technology really makes a difference, but we've come to realize that it indeed makes a huge difference in productivity. It's not uncommon for a business to operate with a small fraction of the labor required in the past, and in many cases the quality and customer service is actually better with more technology and fewer employees.

In spite of productivity increases, most businesses didn't cut their number of employees as productivity grew. In cases where business volume increased, the productivity gain enabled the business to absorb the increased volume without having to add as many people. And in cases where business volume remained relatively stable, there wasn't a big economic incentive to get rid of people. It was easier to just hold on to the employees you already had than to go through the heartache of reducing your labor force.

Then the economic downturn hit. Profitability disappeared, and companies had to make hard choices about all of their expenses. Projects got put on hold. Travel and training budgets got hit hard. And eventually the cuts got deeper and layoffs became more common. Companies realized that they really don't need all these people since technology has made the remaining employees more productive. Jobs are being cut that will never be needed again.

Here is the way most businesses look at the problem of reducing labor cost.

Businesses look at a workforce reduction logically and unemotionally:

1. All employees are not equal. In a given job, the difference in output between your best employee and your worst employee is significant. The actual difference depends a lot on the job, but I would expect a minimum difference of 20%, and I've seen differences in programmer productivity of around 1000% (the best programmer can easily have 10 times the output of the worst programmer).

2. If you reduce your work force, you typically get rid of the lower productivity people first (unless labor or union constraints force you to take a different approach). It makes sense; if you have to get rid of people, you want to make sure you keep the best ones and get rid of the worst ones.

3. And for those of you who factor salary or pay rate into the equation, look at it this way: For any given job category, mea-

sure output per pay dollar, and get rid of the people who have the lowest output per pay dollar.

4. So the result of a workforce reduction is not just a reduction in overall labor cost; it's also an increase in employee productivity. If you get rid of people who contribute less to overall business output, then your overall business output per employee pay dollar will go up.

Now consider the alternative scenario where you keep all of your employees but cut their pay proportionally to achieve the same reduction in overall labor cost.

1. As before, all employees are not equal.
2. When faced with a reduction in pay, a certain percentage of employees will look for employment opportunities in other businesses. But the higher productivity ("better") employees are more likely to be able to find jobs in other companies. So you are more likely to lose higher productivity employees than low productivity employees.
3. So the result of reducing pay across the board is a decrease in overall employee productivity. If employees leave who contribute more to overall business output, then your overall business output per employee pay dollar will go down.

So from the business viewpoint, it generally makes more sense to reduce the workforce than to keep the workforce and reduce employee salaries.

Important Exceptions

As with any rules, there are exceptions:

1. If you expect that business volume will bounce back soon then you might want to consider a short-term pay reduction which is specifically announced as short-term to keep employees from

jumping ship. This is especially important if your workers require lots of proprietary training and so there is a high cost for a new hire.

2. If your business is the primary employer in a geographic area and so your layoffs will have a large impact on the economy of that geographic area, then you might want to do whatever it takes to keep most of your workforce employed. In this situation you have such a high degree of dependency on the economy of your geographic area that large layoffs will backfire; the economic disaster you create will be worse for your company than the money you save from reduced labor costs.

3. If the whole culture of your company is based on the team rather than the individual, then doing layoffs might upset the delicate balance of your company. If labor cost reductions are required, then you should work with the teams themselves to come up with creative solutions. You may end up with a hybrid approach that eliminates poorly performing workers (even teams recognize deadbeats) and reduces pay for everyone else in return for future benefits to be paid when the business becomes profitable again (stock options are one choice, but there are other possibilities). Using the team to solve your economic problem bridges the management-versus-labor divide and instead includes worker participation in solving the company's problems.

Conclusion

When faced with the choice between layoffs and reducing salaries across the board, most businesses and governments follow the logic I've outlined above, and they go with layoffs. It's the easy choice, but as I've noted in my list of three exceptions, it's not always the best choice.

And as for the situation in Greece that sparked Derek Cheshire's original comment, I honestly don't know. But I would be willing to bet that unrecognized technology productivity gains were at least part of the rationale behind the decision. Should Greece have considered exception #2? Maybe. Time will tell as we see the impact on the Greek economy.

2026 Commentary

In retrospect, this article seems a bit cold-hearted. But although I wrote about it in a fairly unemotional way, I know from experience that these decisions are heart-breaking to make, no matter what the result.

Most managers understand the impact that the decisions make on employee lives. But if the economy and the business get bad enough, something has to be done to reduce costs, or the whole business will fail, and *everyone* will be out of a job. You can't just hide from this sort of decision, so you have to approach it in a pretty unemotional way, no matter how much it tears you up inside.

135. Grief and Disentangling

Originally published on AUGUST 29, 2011 in #CAREERS, #PERSONAL

My wife Sharon passed away July 31st, and I would like to describe some of the thought process I've been going through for the last month. My wife's death wasn't sudden. She was diagnosed with ALS two years ago, and she's been through a progressive loss of muscle control over various parts of her body. First she had difficulty walking, then she started to slur her speech and she gradually lost the ability to speak altogether. She had trouble swallowing and ended up getting a feeding tube through the stomach. Soon she needed a wheelchair, and then she couldn't hold her head up. And toward the end she lost the ability to use her arms and hands, and eventually the ability to breathe.

But this article isn't about Sharon—at least not in that way. It's about the way that we tend to tangle our lives up with the lives of others, until we find that we sometimes have a hard time figuring out what our lives are like as individuals.

Sharon

Sharon and I had been married for seventeen years when she began to show symptoms of ALS. We started our marriage as individuals, but as two people live together and love each other you find that sometimes it's hard to tell where one person stops and the other one starts. You make plans based on a combined set of goals. Sometimes you remember whose goals are whose but often you deal with an amalgamation, a blending of goals and desires that develops organically through the process of living together. You unconsciously divide tasks between the two of you based on ability, interest and desire: one person mows the lawn, one person vacuums the house, one person makes decisions about those seemingly trivial home decorations that turn a house into a home. It gets to the point where you don't think about it. Things just happen, whether out of habit or repetition or unspoken mutual agreement.

As Sharon's health declined, more and more responsibilities shifted to me, until I was doing all of the day-to-day stuff plus taking care of her. The time commitment caused me to cut myself off from many of my outside activities. I cut back on working, I resigned my positions in various community and non-profit organizations. I focused all of my time on caring for Sharon and in trying to keep ahead of her progressive decline by looking for tools and techniques to make her life better and easier.

Before the diagnosis, my life was already pretty much intertwined with Sharon's. But after the diagnosis, Sharon's declining health forced me to dedicate virtually every waking hour to making Sharon's last few years on earth more enjoyable and loving. I think I succeeded, although even now I can look back and see things I could have done better—I guess that will always be the case.

So why am I writing about this on a business blog? For a couple of reasons. First, because in trying to deal with the grief of my wife's death, I've learned things that I think are applicable to other situations—I'll get to some of those things in a moment. And second, because I now recognize that there's a lot more to grief than just getting control of your emotions, and I think my readers could benefit from some of my thoughts along those lines.

Grief

In reading and thinking about grief, I've found that:

1. Everyone grieves differently because the relationship between the griever and the lost loved one varies greatly from situation to situation. Yet in spite of this, you'll undoubtedly have 100

people tell you, "I know how you feel," even though they don't, and you know that they don't.

2. Grief happens at different levels simultaneously. There's the emotional grief of never seeing a loved one again—the loss of future love and enjoyment. There's the grief of guilt—wishing that you could have done things differently. There's life disruption grief—all of a sudden your world is turned upside down and you have to do things differently. And there's the grief of lost goals and objectives—your perspective has suddenly changed and now you have to reevaluate your life goals to figure out what you want to do with the rest of your life. There may even be PTSD—post-traumatic stress disorder—due to the intense stress of the circumstances just prior to the death of the loved one.
3. We commonly use the word "grief" for situations relating to a death, but the term applies equally well to other non-death situations that are traumatic in nature and which result from a sudden break from one reality to another. We can grieve, for example, after a divorce (essentially the death of a marriage) or even after a job loss (the death of our relationship with our employer). All of the same symptoms of grief may apply in these situations, and indeed in many cases the grief may be just as severe as it is for the loss of a loved one.

Entanglement

In thinking about these different aspects of grief, I realized that in many ways grief is about sudden, forced disentangling. We've spent years wrapping our lives around others, whether it's life as a married couple or life as a member of a work team. The longer we live or work in one place, the more our lives get connected to others. That's a good thing—not something to be avoided. We're a social species, and connections make us happy and give us joy on a day-to-day basis.

Often our relationships change over time, but usually the change is slow and easy to accept. We get promoted and our job relationships change. We move from one location to another and our neighbors change. Some friendships fade, other friendships blossom. Change is constant, but usually it's not abrupt. We still stay in touch with our old neighbors. We maintain loose connections with old friends on Facebook. All the change is gradual.

But then, once in a while, the change is startlingly abrupt. We're laid off. Our company goes bankrupt. Our child dies. Our spouse dies.

This is harder to accept. Too much is changing all at once. The mind boggles at all of the disconnections and sudden changes in our lives. All of our mindless day-to-day activities which we have performed on autopilot now have to be revisited and reassessed. Our world turns upside down. We're strangled by our entanglements. All of a sudden we're grieving.

Grieving

This is where I am right now. It's a weird place to be. I'm perfectly functional and lucid—most of the time. I can do mundane things like buying groceries and doing laundry. I can even write articles like this one.

But I get flashes of emotion, sudden floods of thoughts or images or feelings that take me back to the days before Sharon's death. Some of them are "woulda, coulda, shoulda" thoughts: things that I wish I could have done differently, words I wish I'd said or that I wish I hadn't said. I guess I would classify these as guilt, maybe deserved and maybe undeserved. Guilt is a funny thing—you feel it whether or not you should, because you tend to measure yourself in hindsight against perfection. And not just any perfection, but the perfection that comes from knowing all you know now, even though you didn't have that knowledge at the time you made those past decisions. You think, "If only I had seen that coming and done this other thing instead." But you can't keep beating yourself up over this. What's past is past, and the only thing you can do is to file your knowledge away in case the situation ever arises in the future.

Other flashes of emotion are empathy pains. I keep imagining what it must have been like to be Sharon, slowly losing all control over her body, as if she were being sealed inside a tomb while still awake, her brain still functioning perfectly but her body not responding to her brain's commands. I keep thinking of the actual moment of Sharon's death, when she died in my arms. But somehow I've got to train myself to stop thinking this way. It's like being in a car wreck and remembering the events of the wreck itself over and over. It does you no good; it's just torture. I've got to instead focus on remembering the good times that Sharon and I had together, and let those happy memories drive out the thoughts of Sharon in distress.

Disentangling

Perhaps the hardest thing to do in grieving, whether you've lost a loved one or a job, is to rebuild your sense of self. It's a disentangling process, but it's focused on the future—not the past. You don't have to disentangle the past. The things you did together—the experiences and pleasurable moments—will always be a part of you, and you don't have to forget them. But now it's time to separate out *your* goals and dreams from the goals and dreams you shared with your loved one or your employer. Maybe those goals you held together will take on less importance now that it's just you again, or maybe not. The key is to recognize what you really want for yourself. Not things you think you want because other people told you they're important, but the things you truly want and would want even if no one else cared one way or the other.

The disentangling process can be a shock. In my case I've been focused on Sharon's needs for so long that I'm having trouble figuring out what I want to do now. Sometimes it's difficult to hear your inner voice when all you've been hearing is the voices of others. And you'll probably have no shortage of people telling you what you ought to do now. Listen to them if you wish, but listen to your heart more. Do what makes you happy—not what other people tell you to do.

Conclusion?

There is no end to this process so there's no conclusion I can draw. I'm not done with my grief, but I hope and expect that I'll get better and better at dealing with it. And maybe I'll also get better and better at handling those sudden flashes of emotion. I hope I'll eventually recognize them as gentle reminders of a wonderful woman whose life I shared for twenty years, a woman whose love continues on inside of me and inside all the people who knew her.

Sharon, I love you, always and forever. And maybe someday I can put all of this into a better perspective. But for now, I'm still disentangling.

2026 Commentary

It's been fifteen years since Sharon died. The grief hasn't gone away, nor do I expect it to ever go away. But it's more in the background of my life now, reappearing every now and then when something reminds me of Sharon or of something we did together. The disentangling process is mostly complete, or at least I no longer think about it. Whatever goals we had together have coalesced into a set of goals for me alone.

136. 8 Techniques for Dealing with Grief

Originally published on SEPTEMBER 26, 2011 in #CAREERS, #PERSONAL

In a previous article I talked about the loss of my wife and some of the things I'm going through. Since that time I've gotten a little better at dealing with my loss. In this article I'll share some of the techniques I've been using. I think they're applicable in dealing with any loss, whether it's a loved one, a friend, or even a job. Here are the eight techniques:

1. Let go of the guilt

You probably feel guilty about things you did or said, or things you didn't do or didn't say. Unfortunately, it's too late for that now: what's done is done and there's no going back. So forgive yourself and forgive those in your past. There's nothing to be gained by carrying that guilt, and there's no one imposing it on you but yourself. Let go of the guilt and move on. Focus on what will be—not what was or wasn't.

2. Put regret behind you

Regret is a lot like guilt, but maybe not as intense or destructive. Nevertheless there's no room for it now. Regret is sometimes described as "woulda, coulda, shoulda thinking" because you're constantly thinking about what *would* have happened if you had done something differently, what *could* have happened if you had taken a different path, or what *should* have happened if only you had acted in a different way. But it's focusing on the past, and on things that are unchangeable. Maybe in an alternate universe things *did* happen that way. But you're living in the now, in our current universe. Push the regret into a tiny corner of your mind and try not to think about it. It should gradually take a smaller and smaller role in your life.

3. Accept things as they are—you can't go back and change them

You are where you are. How you got there is not important. What you do now is paramount. If you like where you are, then learn to live with it. If you don't like where you are, then decide where you want to go instead, and then take steps to get there. Yes, it's disorienting to be where you are, and maybe it's not where you had hoped you would be. But you're not stuck where you are—it's just a starting point. And you should think of it that way: as a starting point—not as an ending point. The old cliché applies, "Today is the first day of the rest of your life." It's not the end—it's just a different beginning. Start over—there's still time.

4. Integrate the good elements of the absent person (or job) into your life

You may worry that you'll somehow forget the person (or job) that you lost. Don't worry—it will never happen. Your interaction with that person has fundamentally altered you in ways you can't imagine. You're not the same person that you would have been if the two of you had never met. And you're a better person for having known the person you lost. So don't worry about forgetting. You may go increasing amounts of time between thoughts of your departed, but underneath it all—inside of you—you'll always remember because of the ways you changed. So consciously integrate some of what you've learned into your life if you like, but unconsciously you've already done so.

5. Accept the memory flooding and flashbacks, but learn how to control them. Be able to say, "not now."

I still get sudden flashbacks to things my wife and I shared. They're triggered by something I see, a song I hear, a sudden thought, or even a gentle breeze. But I'm getting better at controlling my reaction to a flashback. If I have a moment to savor the thought, then I'll do so. But if I'm in the middle of doing something else, then I've learned to say, "not now." I'll note the thought, think "maybe I'll think about this later" and then go on with my previous activity. This is an acquired skill, but you'll get better at it with practice.

6. Focus on the good memories rather than the painful ones

Think back to when you were a child and you lost a baby tooth. Or maybe you've lost a tooth as an adult. Either way, you probably had the experience of having your tongue continually find its way into the gap in your teeth. It just seems like you had no control over it; no matter how you tried, you couldn't keep your tongue out of that gap.

I think it's the same way with memories of the person we've lost. No matter how much I try to remember the good things, my brain keeps taking my thoughts back to some of the bad things, like the actual time of Sharon's death. It's just like my tongue—I have no control over it. But you eventually get used to the idea of a gap in your teeth, and your tongue gradually stops finding the gap. And I think it works the same way with the bad memories: eventually you get better and better at focusing on the good memories, and then you don't think so much about the bad ones.

7. Turn grief into pride

When the memories do come flooding back, you'll find that it makes you sad and depressed. But it helps if you sometimes let the memories come but train yourself to consciously pursue a different outcome. I've found that when I get a memory that makes me sad, it helps if I try to invert my thinking about the memory. So, for example, instead of having negative thoughts about Sharon dealing with her disability, I think how proud I am that she was able to deal with it so well, and how proud I am to have known and loved a person who was so loving and patient. Turn grief into pride, and you can begin to put a positive spin on a negative thought.

8. Remember the nice memories without adding "and we'll never do that again"

There's a tendency to take a positive memory and turn it into something sad and depressing because you think to yourself, "and we'll never do that again." But be fair: there are probably a lot of things that you did when you were younger that you'll never do again—not because someone died but because they were things that you've outgrown, things that you no longer have the physical stamina to do, or things that were better off just being done once. Let it be enough that you were able to enjoy them when you did. Life changes, we change, our bodies change. Enjoy the memories for what they were: an experience you'll never forget. When you find yourself trying to add the words "and we'll never do that again" to a memory, cut yourself off. Don't let your mind go in that direction.

Conclusion

I'm still disentangling, and I'm still trying to get focused on what I want to do now. But I've found some of my lost focus, and

I've found it by doing some of the things I've listed here. Getting better is an evolving process, and like any process it has techniques that work and others that don't. If I find other techniques that help me, I'll let you know.

2026 Commentary

It's been good for me to revisit this list. I still do most of these things, but there were a few ideas that I had forgotten. I've shared these thoughts with other people who have lost a loved one, and they seem to have benefited from the list as well.

137. The Naive CIO

Originally published on OCTOBER 11, 2011 in #IT, #ALIGNMENT, #LEADERSHIP, #MANAGEMENT, #STRATEGY

The naive CIO believes all the articles telling you that it's your duty as a CIO to prevent devices like iPhones, Android devices and tablets of all types from coming into your workplace. The naive CIO may actually believe that there *is* a workplace these days, even though more and more business is being conducted away from the office and maybe even away from your home city or country. The naive CIO thinks that "enterprise IT" is in control of your devices and that no unapproved devices will ever use your network.

The realistic CIO knows that IT exists to serve the enterprise and its employees—not the other way around. The realistic CIO knows that you can't prevent people from buying their own smartphones, e-readers and tablets, any more than you can prevent people from buying their own watches (and who wears a watch these days?). The realistic CIO knows that it's your job as a CIO to figure out how to help people use the devices they want to use to do the work they're paid to do, and that it's your duty as a CIO to prevent those devices from causing harm to your corporate data and systems. Yes, that's a difficult job, but that's part of what you're paid to do.

Technology isolationism is a short-term approach that won't work in the long term.

So what's *your* approach? Are you being naive or realistic?

2026 Commentary

When this article was written, BYOD ("Bring Your Own Device") wasn't a common policy in most companies. But it was obvious to me that we would end up there.

There was a similar issue for Internet access. Some companies believed that Internet access at the office should be restricted to strictly business use. Although there may still be some companies that have that policy, it's certainly not common these days.

138. Does a Higher GPA Make You A Better Employee?

Originally published on DECEMBER 1, 2011 in #CAREERS

Someone tweeted me a question yesterday, "Is GPA [Grade Point Average] an accurate summary of how someone will be as an employee?" I checked the source of the tweet and—no surprise—it seems to be coming from a student. I don't know the person but my guess is that he is either struggling and trying to rationalize his lower grades, or he's got a high GPA and he's trying to justify his hard work. But the question seems to be asked in a way that wants a "no" answer, kind of like a kid asking, "Do I really have to do my homework?"

Well I'm going to surprise the questioner by answering "yes," but then I've got a major disclaimer to go along with my answer. That's because I believe that GPA does in fact highly correlate with how you'll act as an employee. Students with high grades have shown that they can:

- Focus their energy on following directions even when the directions are unrealistic and make no sense
- Perform as well in subjects that they can't stand as they do in subjects they love
- Treat emotional and social growth as secondary to memorization of dry facts and performance of mind-numbing repetitive activities
- Defer gratification by postponing what they really want so they can focus on what other people consider important right now
- Suppress their creativity to slog through subjects that have no relation to real life

Yes, a high GPA does make you a better employee. But only under the 1950s definition of employee: a clerk, a ditch digger, a mindless construction worker, a soldier in the infantry, a machine operator.

And that's the core of the problem: our current educational system is still focused on turning out the kinds of workers that were needed 50 or 60 years ago. We still need some of those people today, so it's not a total loss. But advances in technology have drastically reduced the number of jobs that require mindless focus on following tedious directions without any understanding of why things are being done or how they might be done better. And our economy has shifted from a labor force of industrialized workers to a more diverse set of creative jobs that require specialization in design or people skills or problem solving.

When I hire someone right out of school, I look at GPA, but higher isn't better. I try to get a sense of the person. Is the person bright and creative? Does the person understand how things work or do they just follow directions? Is the person curious? What motivates the person? What are they interested in?

GPA is part of the picture, but only to see if it fits into the whole. A lower GPA without an explanation is a red flag, but there are lots of valid explanations: focus on part-time jobs, taking care of a family, interest in student politics, etc. A lower GPA can mean that you don't care about anything—low motivation, which is a bad thing. Or it can just mean that you care about something else more than you care about academics, which can be perfectly fine.

A higher GPA also demands an explanation. Did they push for the high GPA for the right reasons? Was it because of a strong interest in the major? If so, what about it was so interesting? Was it because they sacrificed all other aspects of their life? If so, why was that so important?

GPA is kind of like body temperature. You don't use a thermometer to take a child's temperature because you want to know how hot or cool they are. You want to know the temperature because a high temperature is a warning sign of an infection in their body. Similarly, GPA doesn't have a value in itself—it's just an indication of a student's dedication to following the rules in their curriculum.

In today's world there is a need for a certain number of employees who can follow the rules. They probably make up 50% of the work force. Another 25% of the work force is made up of people who barely hold on to their jobs—they're unmotivated and just do enough to get by. But the true success stories come from the remaining 25%—the employees who not only know how to follow the rules but they know why, and they know when to break the rules to make things better. This 25% of employees gets promoted to better jobs, to higher pay and to higher recognition. Some become outstanding individual contributors: the designers and engineers of a new era. Some become managers or executives where they set direction and strategy for workers who continue to just "follow the rules."

Yes, GPA is an accurate summary of how someone will be as an employee—at least to start. But sometimes those students with a high GPA manage to overcome their too-focused-on-grades early years and become successful anyway.

Hi, I'm Harwell, and I'm a recovering high school valedictorian. Yes, a higher GPA can make you a better employee. But why be *just* an employee?

2026 Commentary

Let me first clarify that my 50/25/25 division of employees is purely hypothetical. I have no idea what the actual percentages are, but I do agree that a certain number of employees are unmotivated, a certain number of employees are the stars, and the majority of employees are somewhere in the middle.

When you're in school, it's easy to be persuaded that a high GPA is the most important thing to strive for. After all, it goes on your "permanent record" (whatever that is). The reality is very different.

I hope that someday the educational system will be improved to recognize the vastly different talents and skills that exist in students, and to help those students take maximum advantage of those talents and skills during their post-educational years.

139. A Crisis is the End of an Illusion

Originally published on MARCH 19, 2012 in #LEADERSHIP, #MANAGEMENT, #STRATEGY

Every morning I post a new quotation on Twitter, Facebook and Google+. The quotation on February 9th, 2012 was from Jerry Weinberg, one of my mentors in the IT industry. The quotation was:

"It may look like a crisis, but it's only the end of an illusion."

Many years ago, I copied this quotation to my list of favorites when I read one of Jerry's books, *The Secrets of Consulting: A Guide to Giving and Getting Advice Successfully.* And when I recently reread the quote taken out of the context of the book, it struck me at first as a kind of wisecrack—the kind of thing you'd say as a quick rejoinder to someone's Facebook post about a problem they're having.

But then I read it again and, even without going back to the book for context, I saw how profound it is.

One dictionary reference defines "crisis" as:

The point of time when it is to be decided whether any affair or course of action must go on, or be modified or terminate; the decisive moment; the turning point.

But this defines a crisis based on what's happening now. Instead, we need to consider the entire timeline over which a crisis develops. For example, the BP gulf oil spill didn't just start when the wellhead blew—the problem started many years earlier when incorrect assumptions were made and then subsequent warning signs were ignored.

The same thing happened with the Challenger shuttle disaster. The choice to use O-rings in the design of the shuttle was based on some critical assumptions, and the crisis only occurred when there was a decision to launch the shuttle even after some cold weather stretched the limits on those assumptions. The assumptions had been all but forgotten, and they had been replaced by an illusion that the shuttle was safe under those conditions.

More recently in the United States we've seen the debt crisis in the news. And, like the others, this crisis really isn't something sudden that's appeared out of nowhere. Instead, it's the natural outcome of a recession that cut government income, together with the effects of a tax cut, an increase in medical and retirement benefits, an aging population, and a change in accounting policy that revealed the true nature of the debt. The crisis came when we suddenly realized how bad things are (the end of the illusion), but the sources of the problem go back many, many years.

I've written before about the root of humor being a sudden shift in perspective, and it's ironic to me that the root of a crisis is the same. And yet that's human nature: we continue to see things a certain way, day after day, week after week, year after year. And then suddenly we see things differently—we see them for what they've been all along. A crisis is generally considered to be a bad thing, but we also need to remember that a crisis is just an opportunity to see things clearly. It's not the crisis that's the problem—the problem is the way that we deceive ourselves. We ease gradually into our world of illusion, taking things for granted, basing our lives on certain unspoken assumptions. And then, "BANG!" The illusions are shattered all at once, and we call the result a crisis.

So when you're faced with a crisis, don't blame the people who happen to be around when the crisis is made visible. Instead, look at the source of the illusion, look at the incorrect assumptions that were made, and get an understanding of how we managed to deceive ourselves for so long. You don't prevent a crisis by developing a specially trained team to deal with the crisis when it occurs. You prevent a crisis by keeping things out in the open, by voicing assumptions and keeping them visible, and by actively preventing the illusion.

If there's no illusion, then there can't be a crisis.

2026 Commentary

I still post a daily quotation on Facebook, LinkedIn, BlueSky and X (formerly known as Twitter). But the crisis/illusion reference is probably more true now than ever. There's a huge divide in thinking among different groups of people in the United States. And the illusion that there wasn't a divide has come to a head in our current political crisis.

140. Why I Never Hire the Best Person for the Job

Originally published on JUNE 21, 2012 in #CAREERS, #MANAGEMENT, #PROJECTS

I never hire the best person for the job, but I always try to hire the best person for my company's future. Here's my reasoning:

A job is a task-oriented view of the business. Jobs are constantly changing because business needs are constantly changing. The best person for a job is the person who can do *right now* exactly what needs to be done *right now*. But due to the rate of change in business, it's unlikely that what needs to be done right now is what will be needed tomorrow. Requirements will change, customers will change, business processes will change, the world will evolve.

So what I really need is not someone who can do today's job today—I need someone who can do tomorrow's job tomorrow, and then continue to evolve to be able to do day-after-tomorrow's job day-after-tomorrow, and the day after that, and the day after that.

What I'm talking about is a person who can grow with the growing needs of the business, and this adds an additional dimension to the hiring requirement. If I hire the best person for the job then I'm going to hire a one-trick pony—a person who does one thing but does it really well because they have done that one thing over and over.

On the other hand, if I hire the best person for my company's future then I'll look for a person who has a deeper perspective. The person will have some experience in a situation similar to the one I need, but more important, will have a broader experience in a number of related but very different situations as well. And most important, the person will have the perspective that's required to understand why and how a technique that works in one situation has to be altered to work in a slightly different situation.

This deeper perspective and broader experience will help me in three ways:

1. The person will be able to see when the job moves in a direction that requires a different approach, and will have the depth to alter the approach.

2. When the job is complete, the person will be better positioned to move into a different job. Thus the person will be more useful to my business in the longer term.

3. The person will do better at leading, managing and training other employees in how those employees can gain a broader perspective themselves.

Now, you may ask, "Is there *any* situation where hiring the best person for the job is a better approach? What about for a short-term project that's absolutely cast in stone?"

And my answer is that (a) stone has a habit of breaking, and (b) if you really have a requirement *that* specific then you're better off bringing in a contractor for the job instead of hiring. You can subcontract the best person for the job, but you should never hire such a person. One-trick ponies have no place as employees—they'll prevent your business from growing and evolving.

2026 Commentary

The "one-trick pony" reference may be new to some of you. I first heard the phrase from a Paul Simon song called "One-Trick Pony." The song describes a horse that does only one trick, "He's got one trick to last a lifetime, but that's all a pony needs."

I've worked with one-trick employees: people who did the same job over and over for their entire career. They don't fit well in a constantly changing business environment. As I told a one-trick interviewee, "You don't have ten years of experience; you have one year of experience ten times."

141. Never Stop Questioning

Originally published on JUNE 25, 2012 in #CAREERS, #LEADERSHIP, #STRATEGY

There's a certain age that kids go through when they seem to have an endless supply of questions:

"Why is the sky blue?"

"Why do cows make a moo sound?"

"Why don't planes fall out of the sky?"

"Why are traffic lights red, yellow and green and not purple, orange and pink?"

And then, as we get older, the questions decline until the point when we stop asking. Albert Einstein said, "The most important thing is to not stop questioning." But as Buckminster Fuller observed, "All children are born geniuses; 9,999 out of every 10,000 are swiftly, inadvertently degeniusized by grownups."

Why Do We Stop Questioning?

- We lose the wonder of childhood. We're not as interested in things that aren't part of our everyday routine. We stop caring so much about the world around us.
- We focus on the specific things that are important to us and stop paying attention to all of the other things in the world that we think are distracting us from our focus. Hyper-focus can be good in short intervals when we've picked the right focal point, but often we're so intent on focusing on something that we don't pick the right thing to focus on.
- We've been beaten down by people telling us our questions are stupid. We no longer want to risk the embarrassment. We stop taking risks. We stop playing to win and begin to focus on just avoiding a loss.

- We've seen the consequences of being honest, and we don't want to risk those consequences. I've seen executives ask a group of employees what's wrong with an organization and get complete silence as an answer, because previous experience has shown that those who speak out are ridiculed, denied advancement or even driven out of the company.
- We're so rushed that we no longer take the time to wonder about things. We become so intent on being efficient (doing things right) that we forget how to be effective (doing the right things).
- We hide our ethical, moral or religious beliefs because we feel uncomfortable discussing them. Even when we're asked to do something which goes against our beliefs, we give in because we don't want to be viewed as contrary or annoying, and because we don't want to be labeled, judged or demeaned.

Why Is This a Problem?

- Progress occurs because people question the world around them. If you don't question something you do, then how do you ever get better at it? By questioning, you open up the possibility that there's another way to do things, another explanation for the results you see, another way to approach a problem you're trying to solve. If you don't ask questions, then you're essentially giving up: You're accepting the world around you "as is," without the possibility of any improvement or change. You're condemning yourself to live in your current world forever, without any chance of that world getting better.
- Studies have shown that group solutions to a problem are better than the individual solution of the worst member of the group but never better than the individual solution of the best member of the group. A group solution to a problem essentially averages the expertise and intelligence of the members of the group, giving an average solution. But if an individual in the group asks questions, speaks out, and gains the trust of the

group with an above-average solution, then the overall result is superior.

- Major technology and process advancements are more likely in smaller companies because the culture of a smaller company encourages outlaw thinking that would be unacceptable in a larger corporate culture. The smaller company culture encourages questions rather than discourages them, encourages extreme opinions instead of ridiculing them, and supports extreme research that would be impossible in a do-it-by-the-book organization. It has gotten to the point where many large companies achieve significant technology advances only by acquiring the small companies that have actually succeeded in those advances. Then further technology advancement slows or stops when the small company culture is overwhelmed by the negative culture of the large acquiring company.

Recommendations

- Try to reestablish wonder in your life. Spend some time with a young child and try to see the world through the child's eyes. Marvel at things you've been taking for granted, ask questions about things that have always puzzled you, and use your adult knowledge to find the answers to those questions. Let the answers raise more questions, and dive deeper into the unknown.
- Find someone you totally trust, and every so often have an actual deep conversation with that person about values, beliefs, and the meaning of life. It will invigorate you, and maybe help you put things into perspective. Do this with other people who have different beliefs from yours, and try to understand how their beliefs are different and how that affects their thought process and how they live.
- Encourage others in your family and your business organization to ask questions. Be supportive of a questioning

mentality, where no questions are considered stupid, and where contrary beliefs are tolerated and even encouraged.

- When you ask a question, be prepared to accept an honest answer, even if you don't like the answer and even if you disagree with the answer. Without being discouraging or offensive, try to understand why the person believes the answer they gave. If you disagree, try to get to the root of the disagreement: Do you have different fundamental assumptions? Do they have information that you lack? Are they looking at things from a different point of view? But never do this kind of probing in a public or group setting—do it privately to avoid embarrassment and to encourage honesty.
- Promote a "truth, whole truth and nothing but the truth" attitude at home and at work. The "whole truth" part is where most of us fall short. We hold back on telling the whole truth because we're ashamed or because we know it can get us in trouble. So we don't tell our manager that we're having difficulty with a certain task until we're so close to the deadline that there's no way to prevent being late. Or we don't reveal our concerns about a new potential vendor because we don't want to offend anyone. Telling the whole truth means actively sharing information instead of just waiting for someone to ask us. Telling the whole truth means not holding back.

Conclusion

We're full of questions as children, and then we gradually stop asking. It's not that we've learned everything there is to know. It's more that we've been sufficiently discouraged in our curiosity and questioning that we've found it easier not to ask. But there are huge consequences when we stop asking questions, and those consequences carry over into business. Somehow we have to reignite those fires that burned within us, reawaken our sleeping questioning child, and apply a questioning, open and honest attitude toward our lives, both personal and business. Questioning and honesty lead to growth and achievement. Silence just perpetuates the status quo.

2026 Commentary

This is probably more true now than ever. The only way the human race will advance is to keep asking questions, and to keep seeking better ways of life.

142. Why Upscale Hotels Charge for WiFi, and What We Can Learn from It

Originally published on JULY 10, 2012 in #LEADERSHIP, #MANAGEMENT, #PROJECTS, #STRATEGY

If you've traveled much, then you've probably noticed that the more expensive, "better" hotels charge for WiFi while the same service is often included in the room rate at less expensive hotels. This phenomenon probably seems odd to you—isn't this the opposite of what you should expect? In this article I'll provide an explanation, and I'll offer advice on how you can avoid the same problem in your own business.

Upscale Hotels were Early Adopters

The roots of the pricing problem go back to the early history of providing Internet access in hotels. The more upscale hotels were the first to offer Internet access, and because early Internet access required a wired connection, there was a considerable investment required to update a hotel's infrastructure. Ethernet outlets had to be provided in every room, cables had to run through the hotel walls to link the outlets, and routers and bridges had to be added throughout the building to link together all of this cable. This required an expertise that was beyond the capabilities of most hotels, and so independent companies jumped in to fill the gap.

The independent company would wire your hotel, service all of the equipment, provide an Internet link, supply a billing system for customers, and deliver ongoing troubleshooting and support. And based on the negotiated deal with each hotel chain, the independent company would keep a percentage of the revenues. For some chains, the hotel provided most of the investment and took most of the risk, but in many cases during the early years, the independent company would provide the equipment and labor in return for a long-term contract which split the revenue between the hotel and the independent company.

This outsourcing model for hotel services is fairly common in upscale hotels. The same financial approach is often used to outsource the management of many other services provided on a hotel property: snack machines, florist shops, shoeshine stands, cleaners, car rentals, and even in some cases gift shops, restaurants and bars.

Less Upscale Hotels and Motels were Later Adopters

Other hotels and motels added Internet access later. Some added wired Internet access, and some late adopters jumped over the wired Internet stage and went directly to WiFi, eliminating the need to put wired outlets in every room. By the time they adopted it, the less upscale hotels and motels viewed Internet access as a ticket-to-play item—not as a revenue generator. Not having Internet access was like not having cable TV; customers wouldn't stay at the hotel without basic services. And the less upscale hotels don't rely as much on the outsourcing model for hotel services, so even if they used an outside service to install the Internet equipment, they were less likely to negotiate a deal with their service provider to share revenue and cost.

Why Do Upscale Hotels Continue to Charge for Internet?

The practice continues because:

1. Contracts for outsourced Internet service are still in place in some hotels. In some cases, the hotel doesn't even own the equipment and cable—it's owned by the outsourcing company. So there is no financial incentive to switch to free Internet service.

2. Once the precedent was established, the practice of charging separately for Internet access was taken for granted by customers at the high-end hotels. I'm sure the customers didn't like it, but they still made the assumption that it would continue. And since the higher-end hotels rely more on business travelers and wealthy guests who are less cost sensitive, the practice has continued.
3. Keeping the Internet cost separate allows the hotel the appearance of a lower room rate so that the hotel can appear to be competitive on price while still getting additional revenue from Internet use.
4. Making Internet service free would eliminate a revenue stream that's already factored into the hotel projections.
5. The practice hasn't yet had a significant measurable negative impact on hotel business.

What Can We Learn from This Example?

There are a number of lessons here that can be applied to technology decisions in any business:

1. Financial decisions are often made with an unstated assumption that the technology underlying the decision will not change significantly over time. Sometimes it's easier to make the assumption of no change than it is to predict how the technology will change. But regardless of how difficult it might be to predict technology change, you have to anticipate that technology will change. Your best option is to make financial decisions based on the short-term: a payback in a year or two. The longer the payback of an investment, or the longer the commitment of a contract, the more you need to factor into the decision the possibility of technology change.

2. It's very difficult for a business to drop a revenue item, even in the face of customer dissatisfaction. And once a capital investment has been made, there's a tendency to stay with that capital investment even if there's now a better approach. Financial best practices call for depreciation of capital investment in order to apportion a one-time cost over the life of the investment. But that shouldn't mean that you're stuck with using that same asset over the anticipated asset life when better, more cost effective technologies become available. The best strategic decisions are not always the same as the best financial decisions, and good decisions should be made looking forward—not backward.
3. Outsourcing a part of your business usually requires a commitment to a certain way of operating your business for the duration of the outsourcing contract. The financial advantage of outsourcing depends on the guarantee that things will stay the same. Don't sign an outsourcing contract for a contract duration longer than you can vouch for such a guarantee. Outsourcing locks you in, and prevents change. Understand the consequences of that lock-in before you commit to a long-term contract.
4. The reasons something first happens are often different from the reasons something stays in place later. But habit and repetition keep things going the way they were until someone challenges a practice.

Conclusion

I believe that upscale hotels started to charge for Internet access because they locked themselves into the practice by accident. They continue to charge because they haven't yet seen an incentive to do otherwise. Where in your own business are you doing something because of a past decision that makes no sense in today's world? What are you going to do about it?

2026 Commentary

I wrote this fourteen years ago, when WiFi was just becoming a common technology in hotels. I don't know how many hotels continue the practice of charging separately for Internet access, but I suspect that the number has decreased.

I know that some hotels have adopted a tiered approach to pricing: providing limited-speed Internet access for "free" (included in the room rate), while providing higher-speed Internet access at extra cost.

Regardless of what hotels are doing with Internet access, however, the "lessons" in the article are still true. It's difficult to make high-dollar investments in technology that's rapidly changing unless there is a short-term payback. And these days, almost all technology is rapidly changing.

143. It's All About Trust

Originally published on JULY 24, 2012 in #ALIGNMENT, #LEADERSHIP, #MANAGEMENT, #PROJECTS, #STRATEGY

Trust is an important part of every relationship, but in some cases it's more important than others. It's easy to say you trust someone when you can watch their every move to see if they're doing what they said they would do. It's harder to trust someone when they're far away for extended periods of time and you can't see what they're doing.

Trust is a substitute for inspection. When you trust someone, it means that you have an inherent faith that they will do what they said they will do, and that they will abide by the principles to which you have both agreed. But even in a trusting relationship, it's good to do inspection every now and then, just to reassure both parties that the trusting relationship is continuing.

Trust is Between Individuals

Trust is between individuals, not organizations. You may say that you trust your IT organization, but what you really mean is that you trust the person in charge of your IT organization to make sure that everyone in their organization follows the agreements you have established. Or you may mean that you trust a specific individual in your IT organization—whether or not they're the manager—to follow up with other individuals in the organization who are doing work on your behalf. That trusted individual will either ensure that the work is being done properly or will report back to you immediately if there is a problem that needs to be solved.

In either case, if that individual leaves the organization and is replaced by someone else, then the trust is suspended and must be re-earned. You may give the replacement individual the benefit of the doubt and make it easier for the new person to earn your trust, but you will at the very least have a heightened sense of concern about the new relationship until the trust is proven.

Building and Losing Trust

When you start a relationship from a position of strong distrust, it may take a long time for that trust to build. If you have a history of being betrayed in your trust—trusting someone and then having the person violate the terms of your trust—then it will take even longer for you to trust someone new. The ease with which you build trust depends on the sum of the results of all of the trusting relationships you've ever had during your lifetime. For some individuals, the history of trust is so bad that they will never have a trusting relationship.

No matter how long it takes to build a trusting relationship, the trust can be lost in an instant. One betrayal can sabotage a trusting relationship and reset the trust level back to zero. Even a miscommunication that seems to temporarily imply betrayal can severely handicap a trusting relationship. In either situation the trust can usually be rebuilt, but it will take effort on the part of both parties in the relationship. Every betrayal of trust—no matter how small—makes the possible rebuilding of the relationship that much more difficult, and there is a point where the relationship can never be rebuilt.

Breaking Trust

Trusting relationships are between two human beings, and all human beings occasionally make mistakes. The difference between betrayal and a mistake is intent: if the violation of trust is accidental then it is more easily forgiven. But too many mistakes will destroy trust no matter what the intent. The perception of carelessness is in itself a betrayal.

Trusting relationships can be damaged due to causes outside the control of the two parties. If you trust your IT organization to keep your systems working but equipment has been purchased from an unreliable supplier, then repeated equipment failure can cause you to lose trust in the IT organization, despite the best intentions of the IT organization when choosing the supplier. In this situation a truly trustworthy IT organization would solve the problem, either by getting resolution through the supplier or by repairing or replacing the equipment. This situation falls into the category of a mistake, and the trust can be repaired. But repeated problems of this nature are an indication of incompetence, and the trust will be lost.

Trusting relationships can also be damaged due to organizational conflicts in the trusted organization. You may trust a particular individual in your IT organization to always help you with your system problems, and you may have built that trust with many years of a good working relationship. But if a new manager orders the trusted individual to stop "wasting time on support" and to spend time doing something else, then the trust will be lost because of this conflict in prioritization. This loss of trust is not only unfortunate, it's also demoralizing to the individual who had your trust. When there's a conflict of interest like this, the individual is being put in a tough position. Whose trust should they violate: their customer's or their boss's? There's not a correct answer, and the best course of action may be to bring the conflict of interest to the attention of the two trusting people, and let them work the conflict out between themselves.

The Necessity of Trust

Some relationships require more trust than others. When two individuals have the same level of knowledge and experience on a subject, then they can easily discuss different approaches to doing projects or solving problems in the subject area. When one of these individuals starts doing a project as a result of the discussion, it's easy for the other individual to assess progress and to inspect the result of the work while it's being done. The level of trust required in this relationship is small because both individuals have the same level of knowledge, and communication and prioritization is easy.

As an example, let's say you're hiring someone to mow your lawn. You've mowed your lawn by yourself many times, you know where the problem areas are, and you know the differences between a good job and a bad job. It's easy for you to talk about the job with someone new and to get an idea of their competence from the discussion. And it's easy for you to watch them as they work and see if they're doing a good job. You can stop them immediately if they deviate from your agreement, and so the level of trust required is small.

On the other hand, let's consider a situation where you have a medical problem which requires brain surgery. You talk with your surgeon about what needs to be done. The surgeon obviously has a lot more knowledge and experience in this area than you do. You can ask questions, and a good surgeon will do their best to provide answers and to make you feel comfortable that everything possible will be done to ensure a successful outcome.

In this situation, the level of trust required is large. Because you have no way of adequately assessing the surgeon's knowledge and competence, you judge the surgeon on aspects of behavior that probably have no relationship to competence: bedside manner, ability to explain things simply, how much the surgeon seems to care about your health, how distracted or focused the surgeon seems to be, and the surgeon's self-confidence. You look for indicators that help you build trust in the surgeon, because in this case trust is all you've got.

While I won't attempt to compare IT to brain surgery, there are similarities in the way that relationships develop between IT people and their customers. In most cases, the customers are dealing from a position of disadvantage: the customers don't have the same level of knowledge and experience that the IT people do. And because the customers don't have any real way to know whether IT is doing the best thing for them or making the right decisions, the customers have to rely on trust to a very great degree.

In some cases it's worse—the customers think they understand IT when in reality they only have a fundamental understanding of some of the technology, and they don't have any experience with the proper business use of that technology. In this situation the customer still has to rely on trust, but the relationship with IT starts out from a position of distrust. Business use of a technology will often differ from consumer use of the same technology due to the more stringent requirements of the business, and customers often see the additional safety measures as unnecessary. As a result, the IT people have to first educate the customers about the differences before trust can be established.

The Biggest Mistake

The biggest mistake that you can make in any relationship, whether personal or business, is to take trust for granted. Although trust between two parties is seldom a stated objective of any project, trust is essential to the success of every project. A project without trust is like a phone conversation with a bad connection. You can't quite understand what the other person is saying, and so you keep asking the other person to repeat things. You don't focus on what you're trying to communicate—you instead focus on being able to communicate at all. Eventually your level of focus shrinks to short intervals, and you reduce your expectations to just deal with the bare essentials.

It's the same way in a project without trust. You can't seem to get the other party to do what you expected and so you start repeating your expectations over and over. You shift your focus from the longer-term aspects of the project to instead deal with short-term project details, and so your conversations focus on tasks instead of strategies. This is the kiss of death for most projects, since it is bad planning and lack of attention to strategy that cause most projects to fail—not the execution of the details.

Conclusion

You can help build trust with your customers by doing things that will give them confidence in your abilities, and belief that you are committed to their satisfaction. You can build trust with your vendors by giving them a consistent set of requirements and by respecting their ability to deliver what you've asked for.

Trust is often an unspoken topic in most relationships, but it's the silent foundation upon which all relationships are built. If you don't believe that you have a mutually trusting relationship with your customers or with your vendors, then ask yourself why. What problems are being caused by lack of trust? What can you do to establish or reestablish trust? Your future success depends on the answers.

[For more on the topic of trust, and for specific recommendations on how to build trust in the IT/customer relationship, see my book, *Boiling the IT Frog*.]

2026 Commentary

It is interesting to me how little we talk about trust, and yet how truly important it is to any relationship—business or personal. We go through the trust-building process with every new vendor, every new friend, even with casual relationships. But we seldom think about it.

144. The 4 Things I Look for in a Prospective Employee

Originally published on AUGUST 14, 2012 in #CAREERS, #MANAGEMENT

When I interview prospective candidates, I look for four key attributes: enthusiasm, curiosity, insight, and perspective. Here's why:

Enthusiasm

Motivation is probably one of the most important attributes of a good employee, and the best kind of motivation comes from enthusiasm. Enthusiastic employees are eager to work. They volunteer for assignments. They work longer hours because they want to—not because they have to. They have an inner drive that keeps them going when other employees are running out of steam.

There are other types of motivation. An employee with a new baby at home may be motivated. An employee who is deeply in debt may be motivated. But enthusiasm is not the same as working for a good reason. Enthusiasm is a motivation that comes from inside rather than a feeling prompted by external factors. Most enthusiastic people are enthusiastic about everything. They live life just a little bit larger than everyone else.

Curiosity

A person who is naturally curious always wants to know more. Curious people don't take things for granted. If something doesn't work they want to know why. If something works, they want to know how, and then after they learn how, then they want to know how it could be made to work better. They dig deeper than other people to get at the underlying causes of things. If an organization isn't working well, then they'll figure out how to fix it. If a system or product is popular, then they'll learn what makes it popular so that they can make other systems or products popular too.

Curious people aren't satisfied with the status quo. To them, life is a series of questions to be answered and puzzles to be solved, and work is a process of continuous improvement.

Insight

Enthusiasm and curiosity aren't enough if a person learns things by rote, just memorizing the facts and not internalizing their meaning. Insight is required to actually do something with the information you've learned. Insightful people absorb facts and then arrange the facts in their heads to form mental models of reality. Insight allows a person to apply a physics model to a business problem, or to explain a computer design using a plumbing analogy.

Insightful people don't put rigid barriers around knowledge, seeing it only for its value within a narrow field. Instead, insightful people view the entire world as connected in a series of universal systems and principles. Learning from any area can be applied to all areas, and new learning is reduced to its fundamental elements and mentally saved for use in countless future situations.

Perspective

Insight without perspective is like textbook knowledge—technically correct but unusable because it lacks the practical elements that make it work in the real world. Yes, there might be a formula like A = B + C, but what about friction? What about the natural resistance of people to new ideas and ways of doing things? What about the difficulties of scaling a simple solution to a large-scale environment? What about the complications of dealing with environmental factors and unreliable equipment?

People with perspective have the practical experience required to know the difference between theory and reality, and to know how to adapt a theory to apply it in a real situation. Their experience isn't measured in years, but in the diversity of situations to which they've been exposed, in the difficulties that they've faced, and in the problems and challenges they've overcome. Perspective isn't the same thing as experience—perspective is what they've learned as a result of the experience. There are a lot of people with experience, but not so many with perspective.

How Do You Interview for These 4 Things?

You may say, "Well, that's an interesting list, but you can't determine these characteristics of a person in an interview."

I agree that you can't determine these characteristics in a *traditional* interview. The problem with traditional interviews is that they are based on asking questions and getting short answers, and they are usually focused on establishing facts rather than determining whether a candidate possesses some of these more complex personal attributes. And to make matters worse, candidates are usually reluctant to exhibit some of these characteristics in an interview, since they feel constrained to answer the questions without offering personal opinions.

The only way to get a candidate to reveal some of these characteristics in an interview is to turn the interview into a conversation. Get the candidate talking about something that they are enthusiastic about. Give the candidate a situation that provokes their curiosity. Ask the candidate to explain something abstract, and notice how insightful they are in using analogy or metaphor. Provide a real-world problem and see what kind of perspective they apply to a possible solution.

Uncover these four traits in an interview, and you get an additional benefit too. People who have these four traits love to use them. So doing an interview which allows the candidate to exercise these four characteristics is more likely to convince the candidate that your position is a desirable one. And that's important, because these people are scarce and in high demand. Hire them when you can.

2026 Commentary

I stand by this list, but I rarely see people interview this way. That's unfortunate.

145. 12 Problems in IT Caused by the E-Myth

Originally published on AUGUST 28, 2012 in #IT, #ALIGNMENT, #MANAGEMENT, #PROJECTS

In 1986 a book called *The E-Myth* attracted a lot of attention. The "E" in "E-Myth" refers to entrepreneur, and according to the myth in Michael Gerber's book, if you're good at a particular skill then you'll do well starting a business which requires that skill. So, for example, if you're a good cook then you'll do well starting a restaurant. Or if you take good photographs then you'll do well starting a photography business.

As you might expect from the use of the word "myth," the book presents a case that this myth is not true. The skills and inclinations of an artisan are not the same ones required of a businessperson, and most people don't realize that starting and running a business require only a relatively small percentage of time devoted to the skill that underlies the business. Very little of a restaurant owner's time is spent cooking; in fact, most restaurant owners hire someone else to cook. Very little of a professional photographer's time is spent taking pictures; more time is spent marketing and running the business. Even for the highly-paid professions like dentists, doctors or lawyers, you might spend a lot of time doing your craft if you work for someone else, but if you start your own business then you'll have to devote a lot of hours to basic business tasks that are unrelated to the profession itself.

The book goes on to say that when people who believe the E-Myth attempt to follow their dream and turn an amateur skill into an actual business, they either wake up to reality and scramble to learn the real skills required, or their business idea fails.

The E-Myth's Impact on Information Technology Organizations

In Information Technology, we often run into people who follow technology columns in the press and so they claim to understand IT. Or they take a programming course in college and so they claim to know how to develop software. Or they own an iPhone and use Siri, and so they claim to understand mobile technology and artificial intelligence.

These people have fallen victim to the same erroneous thinking as the E-Myth: they're equating an amateur level knowledge of a subject with the more extensive professional experience actually required to put that knowledge into real-life use. Fortunately, these amateur technologists don't attempt to start their own business based on their knowledge, but they still manage to cause a certain amount of business havoc when they become customers of an IT organization or—worse—when they're assigned to be the business project manager of an IT project.

Here are some of the problems these people cause:

1. They don't understand the level of effort required to create some of the products and services they're used to, and so they expect an IT organization to be able to create something equivalent to a high budget commercial product with an amateur budget and timescale.

2. They don't understand the way software is licensed. In particular, they don't understand the need for a licensing agreement that provides for growth and shrinkage of user seats as demand rises and falls.

3. They underestimate the amount of time, money and effort required to adapt off-the-shelf software to use existing company data, and to interface or integrate with existing company systems. One recent survey found that integration cost is often

seven times the cost of the off-the-shelf software being integrated.

4. They believe a vendor when the vendor says that their total cost is $x, when in reality, (a) the vendor is neglecting the internal company cost of installing, maintaining, adapting, and integrating the system, (b) the vendor is omitting the vendor consulting cost required to help with internal company effort, and (c) the vendor is low-balling the estimate by leaving out things like the license required for the test or backup system.

5. They overestimate the accuracy of company data. They propose systems which will automatically take action based on certain data values (e.g., automatically order parts when inventory is low) without having solid processes in place to ensure that the data is correct.

6. They don't realize that most programming effort is spent dealing with exceptions and that programming for the "normal" situation is the easy part. Their estimates for the programming time required are always too low.

7. They tend to leave testing and the subsequent corrective effort out of their estimates.

8. They tend to forget the need for training and for the learning curve associated with employees learning new systems and processes.

9. They don't understand the difference between a proven and reliable technology and an unproven state-of-the-art technology that has lower reliability. They tend to prefer the state-of-the-art technology, but aren't willing to budget for the extra testing, debugging and workarounds that are required to use it.

10. They don't understand the difference between a preliminary estimate and a guarantee, and they refuse to include contingency time in a project budget.

11. They don't understand the impact of changing the scope of a project, so they allow major scope changes without changing the budget or the schedule.

12. They don't understand the difference between a minimum infrastructure and a fault-tolerant, resilient infrastructure that's scalable to handle future volume increases, nor do they understand that software and hardware have to be designed and constructed to handle peak demand—not average demand. So they base the budget on the minimum infrastructure required to run a system during an average time of the day, then complain when hardware outages prevent use of a system, or when spikes in customer demand bring the system to its knees.

Conclusion

The E-Myth book is now 26 years old, but the myth persists, and it will continue to persist as long as human nature remains the same. The myth starts in childhood when we put a blanket over a card table and call it a house, or when we brandish a stick and call ourselves a warrior, or when we hold a toy bottle up to the mouth of a doll and call ourselves a parent. It's fun to make believe that the world is simple. It's nice to imagine that just because we know a little about a subject that the logic based on that little bit of knowledge will enable us to make big decisions. But it's not true. Nor is it true that having a knack for a particular skill like art or cooking or video games will enable us to successfully start or run a business that uses that skill.

But if we recognize the limitations imposed by the E-Myth, and if we get help from professional advisors who have the proper experience, then with enough motivation and resources we can accomplish anything we want. The secret is to take the myth into account and to include the extra resources in our time and budget estimates. Reality can be very rewarding, but unfortunately it's much more expensive and time-consuming than amateurs would expect.

2026 Commentary

As you might expect, technology has advanced since I wrote this article in 2012, but the problems persist. To use a more current example, AI is no different than programming projects in 2012. Just because you know how to word a few queries using ChatGPT doesn't make you an expert on how to implement AI for customer service in your business.

146. 2 Secrets of a Portable Expert

Originally published on SEPTEMBER 27, 2012 in #CAREERS, #MANAGEMENT, #PROJECTS, #STRATEGY

The traditional view of expertise is that you become an expert by spending many years working in a broad area. In gardening, for example, you gain expertise by working with different plants, experimenting with different nutrients and soils, and by making mistakes and then learning from your mistakes. Under this traditional view you are then labeled as an expert gardener, but not an expert on anything else. To become an expert on something else, you have to devote many more years. Thus you're unlikely to become an expert on more than one or two things during your lifetime.

But there's an alternate approach to expertise. Using the alternate approach, you can become a temporary expert on hundreds of different things during your lifetime. You can't remain an expert on many of these things at one time, but usually being an expert on one or two things at a time is enough.

Change has Accelerated

The pace of life has changed a great deal in the last fifty years. Up until the mid-1900s most people got educated in a certain field and then remained in that field their whole lives. Technology change was slow, and so the skills and expertise you developed were unlikely to become obsolete. It was fairly easy to keep up with technological developments in your field—just a small amount of reading and supplemental education would maintain your expertise.

But during the last fifty or sixty years, technology change has accelerated. There are very few fields in which you can maintain expertise without constant reading and reeducation. It's gotten to the point where entire concepts and approaches are replaced by new ones in the span of five years or so.

This new environment is awful for old-school experts—far more of your time is devoted to just maintaining your level of expertise. But the new world is wonderful for people who have a short-term need for expertise—people who want to become experts quickly in a narrow area, use the expertise for their own purposes, and then move on to become a different kind of expert to satisfy a different need. I call these people "portable experts."

An Analogy

Suppose I want to clear a path through a dense jungle to get to a certain location several miles away. The portable expert would use a map to figure out a path, then clear a path to the destination that's just wide enough for our needs. The more traditional expert would bulldoze all of the area between us and our destination, clearing far more area than is required.

In essence, the portable expert only learns what is needed to walk down the narrow path, while the more traditional expert attempts to learn everything there is to know about a broad subject area before embarking on the journey.

The Trade-off

The portable expert doesn't learn as much, but they learn exactly what is needed. The traditional expert learns a whole lot of irrelevant stuff along with the relevant things, on the assumption that all of that knowledge might be useful someday.

In the old days, the approach of the traditional expert would be superior. It's difficult to know exactly what knowledge is needed for a particular task, and knowing more about the "irrelevant stuff" might provide an advantage. But in today's world of rapidly changing technology, it's more difficult to keep up with a wide field of knowledge, so the traditional expert is less likely to be up-to-date on the latest advances in the particular subject area needed, simply because of the breadth of knowledge that the traditional expert is having to maintain. That's where the portable expert has the advantage: the portable expert can develop short-term expertise in the required narrow area of expertise, and then hold on to that expertise only as long as the expertise is needed for a specific project.

Two Secrets of Success

There are two secrets to making portable expertise work. The first secret is to be very sure of exactly what knowledge and wisdom is needed, and to make sure that the things you don't learn are truly irrelevant. That's the primary skill of a portable expert: They have gained the experience over the years to identify the learning that is required and the learning that isn't required, and know how to verify those learning decisions with instructors, mentors and other sources. This differentiation skill is difficult to master, and the skill is only acquired by becoming an expert in different specific areas over and over again.

The second secret is that the portable expert has to have enough of a basic foundation of knowledge in a field to be able to comprehend the subject area and understand its principles. The portable expert must be able to have significant conversations with leaders in the field, and must be able to learn from those conversations not just the core concepts but the subtle nuances that make all the difference in long-term success. In a sense the portable expert must be multilingual—they must develop the ability to carry on conversations in the technical language that goes along with each new area of expertise.

Samuel Johnson once said, "Knowledge is of two kinds. We know a subject ourselves, or we know where we can find information upon it." The portable expert is the ultimate implementer of this principle: someone who is able to supplement personal knowledge with rapidly acquired information from other sources.

To be continued

In my next article, I'll give you some examples of portable expertise from my own experience, I'll list four advantages of hiring a portable expert, and I'll give you some tips on how to create your own portable experts.

2026 Commentary

This “portable expert” idea is probably one of the more “outside the box” ideas that I wrote about. It worked for me, but it’s difficult to find portable experts, and even more difficult to be able to assess their trustworthiness.

It’s also worth contrasting this portable expert idea with the message from the prior article about the E-Myth. In a sense, the concepts are opposites. The E-Myth problem describes someone with narrow experience in an area who is attempting to generalize that narrow experience to a much larger area where the narrow experience doesn’t apply. The Portable Expert concept is for someone to quickly become a narrow expert just in a small subset of a topic area, and just for a short-term need. I think the examples in the next article will make this distinction clearer.

147. 4 Advantages of a Portable Expert

Originally published on OCTOBER 2, 2012 in #CAREERS, #ALIGNMENT, #LEADERSHIP, #MANAGEMENT, #PERSONAL, #PROJECTS

In my previous article I defined the term "Portable Expert" and I described the two secrets that make portable expertise possible. In this post I'll give you some examples of portable expertise from my own experience, I'll list four advantages of hiring a portable expert, and I'll give you some tips on how to create your own portable experts.

An Example of Portable Expertise

Back in the mid-1980s I led a group that was responsible for creating new systems for a tax-filing business. We were starting from scratch with no existing investment in hardware or software, so I first had to figure out what hardware and software to buy. I did enough research to determine the front-running candidates, and then did even more research to narrow down my choices. I decided to use an SQL database (somewhat leading edge in the 1980s) because I knew we needed flexibility in extracting information from huge amounts of data. I also went with a 4GL (fourth general language—a concept that was pretty advanced for the day) for construction of the user interface for our systems.

Next I had to train our programmers in the use of these techniques. I was very much a working manager—not an administrator—so I went through all of the training myself along with my team. I needed to understand the strengths and weaknesses of the software we were using, and I needed to know enough to match up team members with areas in which they would excel.

We needed someone to design the database, and I knew that extensive knowledge of the business was required, so I decided to do the database design myself. I went through advanced SQL training, getting an understanding of how the actual database engine worked in order to optimize the data structure to get maximum performance. I learned data normalization and did enough research to get a practical understanding of when normalization was good and when real world considerations suggested that extreme normalization should not be followed. I also spent many hours evaluating the existing data used in the tax filing business, getting an understanding of data relationships, and constructing a database design that gave us what we needed.

Once the database was designed, I worked with my team to construct a model for transactions. Our 4GL wasn't literally an object-oriented language, but in effect we designed objects to standardize interface design and to simplify screen construction for the programmers.

This sounds like it might have been a long drawn-out process, but in fact we went from purchase decision to a live system in about ten months. We started with basic transactions and then gradually worked up to the more complex ones, but the system was considered a success right from the start.

This was just one example. I later did similar rapid-development projects for a customer service system, a browser-based payroll system, and an intranet-based content management system. In each case I developed enough expertise to make purchase decisions, attained enough expertise in the software and tools to be able to direct designers and programmers, and then moved on to the next need.

The Background I Needed

Looking back at the technology-related decisions I had to make in these different examples, it's clear to me that someone without my base level of experience wouldn't have been successful. I had many years of software and hardware experience on multiple platforms as a knowledge foundation, and so it wasn't a huge leap for me to be able to become a temporary expert in the use of a SQL database or in a 4GL or in browser-based systems.

But I've also used my portable expertise in non-work-related fields, becoming a temporary expert in several different medical fields when expertise was needed to help make significant life decisions for loved ones. In these cases the most significant obstacle to temporary expertise was the language, but in each case there were probably no more than about fifty new words required to converse intelligently with medical experts in each field.

I found that my experience in applying portable expertise in IT was directly transferrable to the medical challenges I faced. The goal was to become a temporary expert in a narrow field, and to gain enough expertise to make the best-informed decisions. That goal was achieved, and my knowledge was respected by the doctors I worked with during my brief period of expertise.

Hiring a Portable Expert

It's common practice to hire an expert for certain projects which require experience in a specific area that is unknown to us. Most companies will hire someone who has done exactly what you want to do using the tools you've chosen to do it. The result will almost always be mediocre because you're hiring people who are in a rut. They have done the job so many times that they're on autopilot. They will do an adequate job but not an exceptional job, because they tend to force a standard solution on you instead of tailoring a solution for your unique needs.

I prefer a different approach. I look for people who have the right base level of knowledge together with portable expertise—people who are experienced in becoming experts in the area in which they need expertise at the moment. They don't just apply cookie-cutter knowledge, using a one-size-fits-all approach. Instead they look at the specific needs in each specific situation, do enough research to figure out what should be done, and then develop the expertise in that area to be able to optimally solve the problem.

When looking for a portable expert, I look for people who have skills and experience corresponding to the two secrets I mentioned in the prior post:

- People who have broad experience in the general area, using a number of different technologies in a number of different business situations
- People who have demonstrated portable expertise in the past, achieving temporary expertise in a technology and/or business area, putting that expertise to use, and then moving on to the next challenge

Advantages

I've found that I get better results hiring portable experts than I do hiring highly specialized experts. With a portable expert you have the advantage that:

- You won't try to fit a square peg into a round hole just because you have an expert in round holes
- You'll get a good solution even if requirements take you away from your original definition of need
- Your solution can evolve in a different direction rather than be constrained by the expertise limits of the original team
- You'll usually find that your result has a lower cost because you can exclude some components of the typical one-size-fits-all solution from your implementation and just do the things that are important

The overall key is flexibility: you're not locked into a specific approach just because you've hired a particular type of expert.

Build Your Own Portable Expert

Portable experts don't always come from outside your organization. You can build your own portable experts by implementing a few simple business philosophies:

- Rotate your people through different jobs in different areas. Give them a broad exposure to the business and to different technologies.
- Train your employees to recognize concepts in the work they do—not just to learn procedures by rote. Help them understand why things are done the way they're done, and what would happen if things were done differently.
- Teach people how to do technology transfer, and take learnings from other technologies or industries and adapt them for the use of your business. To understand technology transfer, they need to understand the concepts underlying the use of a technology. This means you need to hire thinkers—not just people who follow directions.

Portable expertise is like taking a single tool and replacing it with a Swiss army knife. It gives you a way to maximize your success without having to constantly switch from one hired consultant to another. It lets your business "travel light" and get the important things accomplished quickly. And using internal portable experts gives you continuity in your business processes, and lets you evolve your business in reasonable steps instead of jumping from one solution to the next, making drastic changes at every jump.

Conclusion

The next time you are about to bring in a specialized expert, consider whether you can use a portable expert instead—maybe even someone within your own organization. You might find that portable expertise gives you a more satisfactory solution at lower cost. And you'll certainly appreciate the flexibility that portable expertise gives your business.

2026 Commentary

As I said in my comment for the previous article, this "portable expert" idea worked for me, but it's difficult to find portable experts, and even more difficult to be able to assess their trustworthiness. It's critical to make sure that you and the portable expert agree on objectives.

148. Why Both Government and IT Need Transparency

Originally published on OCTOBER 15, 2012 in #IT, #ALIGNMENT, #MANAGEMENT, #STRATEGY

A lot of people in the United States are against what they call "big government." Yet most of those people are in favor of government services that they consider essential. In many cases, the concern about big government is not actually a concern about the size of government—it's more a concern about a government that has gotten out of control, a government that no longer seems to manage its expenses wisely, a government that seems to waste money on things that people consider foolish.

Much the same could be said about many IT organizations. When business people have a negative view of their IT organizations, it's not a statement that information technology shouldn't be used in their business—it's more of a concern that the company's investment in IT is not being used wisely.

In some cases, self-interest plays a role in criticism of government or IT spending. Some people consider an investment "wise" if it provides benefit to *them*, and unwise if it provides benefit to someone else instead.

But in most cases, a criticism of IT spending—just like a criticism of government spending—is based on a lack of understanding of exactly where the money is going. The popular term for this is a "lack of transparency."

What is Transparency?

Transparency can mean different things to different people, but the key elements of successful transparency are:

1. A simple, clear (thus the word "transparency") accounting of where the money is going, in high-level terms. For example, x% of the budget to infrastructure, y% of the budget to maintenance, z% of the budget to projects.
2. An explanation of the high-level terms, showing the next-level breakdown of why the money needs to be spent in those areas. I emphasize the word "why" because it's critically important for people to understand the business (or government) purpose of funding in an area. If people are going to justify the spending in their own minds, then it's key for them to understand what would happen if that funding didn't exist. What services would disappear? How would the world be different if that funding were discontinued?
3. For projects—which have a beginning, a middle and an end—a sense of where things are in each project. How long is the project supposed to take? What will it cost? How many people are working on the project? When will the project be completed? Will the project finish on schedule? Is the project over or under budget, and why? What has changed during the course of the project that is influencing its status? And most importantly, how will the business (or government) be different after the project is complete? What will be better?
4. An acknowledgement of current issues. Where are things not going as well as they could? Are certain areas underfunded and thus causing problems in the business (or government)? Are projects stalled due to lack of support or resources? How could additional investment help?
5. A recognition of opportunities. Are there areas where the need for funding is declining? Are there services which could be discontinued or replaced by something less expensive? Are there areas where outsourcing could help? Are there opportunities to

replace an old technology with a newer technology which would make things more effective and/or less expensive?

The Risks of Transparency

Many CIOs oppose transparency because they're afraid that providing transparency will result in loss of control. They think that if business people know exactly how the IT budget is being spent then the business people will try to micromanage the IT organization. In my personal experience, I have not found that fear to be justified, as long as budgets and project plans are presented in a logical manner that ties financial information to specific business benefits.

Financial transparency runs into trouble when budgets are presented based on expenditures rather than benefits. For example, technical training for IT employees is key to employee development and to successfully keeping up with technology advances. But you should never include technical training as a separate line item in a high-level presentation of IT costs, just as you never include employee retirement benefits as a separate line item. Instead, the technical training cost should be considered part of the cost of having an employee, and so the training cost should be included in the basic employee cost which is associated with each service provided by IT or each project done by IT.

If someone in the business wants to suggest specific changes to your IT budget, then let them make decisions on business benefit—not decisions on how efficient you are at delivering that benefit. So, for example, if they think an infrastructure expense for Internet servers is too high, then let them decide how important it is for the servers to be available all the time—not whether or not you could get by with fewer servers. And if they insist that your costs may be out of line with other providers, then let them suggest an IT project to do an analysis of whether there's a better way to provide the required capability. That analysis can look at the hardware and software being used and determine whether current technology offers a better, more cost-effective way of delivering that service.

The Key to Successful Transparency

The key to successful use of transparency is to connect every dollar being spent to a specific business benefit. Often that requires restating your financials in a way that doesn't correspond to your organizational structure. For example, you may have a help desk organization that provides support across a number of different application areas. The cost of that help desk organization should never be presented as a single line item because it will immediately become a target for cuts. Instead, the cost of the help desk should be apportioned out based on the areas of business benefit that it provides. For example, support for users of a Customer Relationship Management system should be rolled into the support cost for that system, and support for desktop applications should be rolled into the support costs for the specific business organizations being supported.

Just to be clear, I'm not talking about charge-backs here, where the cost of these services is cross-charged from IT to the other departments and the money is essentially taken out of the IT budget. I'm merely saying that the budget summary for the IT organization should account for the support costs under the line items for the individual systems or areas being supported. Don't lump support all together, but instead summarize the support costs under the categories where benefit is provided.

This approach may seem deceitful to some readers, but I'm not advocating this restatement of financials as an attempt to hide costs —the approach actually *clarifies* costs. If a help desk organization is considered a single expense and the budget is cut, then someone will have to make decisions about which user areas will get a lower level of service. And whichever areas get the lower levels of service, IT will get the blame. My suggested approach allocates the help desk costs to the business areas from the beginning, and lets the business areas themselves make the decisions about help desk cuts by tying those cuts to specific areas of support, thus avoiding IT blame and putting the responsibility where it belongs.

Conclusion

The secret to avoiding the stigma of big government or big IT or any out-of-control spending is to use transparency to make it absolutely clear where the money is going. While people tend to argue about generalities, they're seldom in doubt about specifics that affect them. By tying each dollar and resource of an IT organization to a specific business benefit, you can make it very clear to everyone that your organization is doing exactly what has been asked of it. If the business wants to cut spending, then the business should be prepared to live with fewer services. And if the business wants more out of IT, then it's clear that more investment is required.

Transparency gives you a better relationship with the business, and it lets you sleep better at night, knowing that your IT organization is doing exactly what you agreed to do.

2026 Commentary

I've got to wonder why this isn't done more often? Do people consider it too much trouble? For more insight, see the article, "124. Why IT Magic is Never Good."

149. Taking Shadow IT Out of the Shadows, Part 1

Originally published on OCTOBER 22, 2012 in #IT, #ALIGNMENT, #LEADERSHIP, #MANAGEMENT, #PROJECTS, #STRATEGY

Shadow IT is one of the names for the Information Technology work and expense that's done outside of the control of the formal IT organization and outside the formal IT budget. It's more prevalent in some companies than in others, and it often changes over time within a company. I've found that the amount of Shadow IT in a business usually varies in inverse proportion to the level of satisfaction the business feels for the formal IT organization. If business people feel that the formal IT organization is providing everything the business needs, then Shadow IT is virtually nonexistent. But if the formal IT organization is failing, then the expenditure on Shadow IT can exceed the formal IT budget.

To understand Shadow IT, imagine yourself as a business executive. You've got a job to do and business objectives to achieve. If you feel that the formal IT organization is standing in the way of achieving your goals, then you'll find another way to be successful, even if it means hiring consultants or utilizing your employees to create your own underground IT organization.

The Good News

Shadow IT has good points and bad points. One of the good points is that Shadow IT demonstrates a real business interest in information technology and in information technology projects. You know the interest is real when the business is willing to invest their own budget to get work done that they feel they can't get done through normal channels.

The Problem

But Shadow IT also has bad points. Business executives who make their own IT decisions optimize those decisions around their perception of their own short-term needs, and they underestimate the interdependency that their systems have with other corporate systems.

Let's break down that problem into its three components:

1. Perception of Needs

The focus of a business executive is on their own needs—not the needs of the business as a whole. For example, a sales executive will see the need for a sales tracking system or an order entry system or a customer relationship management system, but will not typically step back and look at the bigger picture. Yes, they realize that sales information is required by manufacturing or distribution or service providers downstream. But that won't be their focus, and so those other needs won't be a principal part of their system requirements.

For example, I worked with a company in the late 1980s where the sales organization made its own deal to buy notebook PCs for its sales force. The vendor was pushing WordPerfect word processing software, and so they bundled a copy of the WordPerfect software into the deal. As a result, the sales organization established WordPerfect as a standard for the organization, in spite of the fact that Microsoft Word was used everywhere else in the company. This worked fine for internal sales documents, but created havoc whenever anyone from outside the sales organization needed to review or share documents.

2. Short-term Emphasis

The business executive is usually focused on the current quarter or the current year. Goals are defined that way, and so the business executive is typically looking for quick results.

Rarely do business executives understand or consider all of the long-term implications of a systems purchasing decision or outsourcing decision. They don't see the problems of supporting a set of diverse software and hardware platforms. They don't consider the need for a resilient system that has to be available 24×7, and has to be able to handle peak user demand that may be a hundred times greater than the average demand. They don't see that the simple "off-the-shelf" software they're purchasing may not adapt to the evolving needs of the business.

The short-term emphasis also leads to quick decisions, and sometimes a desperate business executive will just buy the first solution that they see, without going through any sort of logical analysis of alternatives. Or in many cases the executive will just buy the same solution that they used at a previous employer, whether or not it's the right fit for the current business.

I've seen this happen several times. The most notable example was a department executive who was under budget at the end of the year, and ran into a pushy salesperson who offered a bottom-of-the-line enterprise software product at an end-of-quarter special price. Without looking at other alternatives, without doing a pilot of the software to assess its capabilities, and without even considering the need for hardware to run the software, the executive made a mid-six-figure dollar commitment to buy the software and use up his remaining budget for the year. The consequences of the decision would haunt the business for years to come.

3. Systems Interdependency

The business executive often makes simplistic assumptions about data. According to most executives, data is always accurate, consistent, timely, complete, and easily available. A manufacturing executive might buy a scheduling system and assume that existing order data will just flow right into the new system without any changes or validation. But the reality may be very different: data is only as good as the people who enter it, the system that validates it, the format it's stored in, and the management discipline of other organizations in the company. And feeding data from one system to another may require a tremendous amount of technical expertise and possibly an entirely new system just to do data conversions.

The Cost of Shadow IT

I had a neighbor once who, in preparation for selling her house, brought in a cheap painter to make the outside of the house look nicer. The painter skipped the normal steps of scraping, cleaning and priming the house, and just painted over whatever was there, including dirt and even leaves that adhered to the house. The neighbor was happy with the low price for painting the house, and the house looked good for a few months, but I'm sure the people who bought the house paid the price for my neighbor's shortcuts.

And so it is with Shadow IT. A professional IT organization has to look at all of the different aspects of information technology use: not just the initial cost of software and hardware, but all of the training, maintenance, customization, integration, infrastructure and support that continues for years to come. Shadow IT is often done following less than professional standards. While the result may be adequate in the short-term, the systems implemented using this approach often fail in the long-term, fall apart when critical people leave the business, or incur large costs later in their life cycle.

Often the Shadow IT projects end up being handed off to the formal IT organization. The formal IT organization then has to go back and fill in all of the gaps left by amateur IT people who took shortcuts, much like the shortcuts taken by my neighbor's house painter.

And since Shadow IT projects often use hardware and operating system platforms that don't follow corporate standards, the formal IT organization gets saddled with new platforms that require training and support resources that don't exist in their organizations.

Thus the short-term savings that may have accrued by using Shadow IT are more than offset by longer-term costs paid by the formal IT group. And ultimately those costs hurt the business bottom line.

Next

In Part 2 I'll describe the five most common approaches that formal IT organizations take in dealing with Shadow IT, I'll offer some comments and advice on each approach, and I'll give you my recommendations.

2026 Commentary

My background is in IT, and so Shadow IT is obviously an IT issue. But IT isn't the only business organization that can fall victim to "Shadow" work. A company can have shadow engineering that is performed locally instead of in a corporate engineering group. It can have shadow process engineering where employees in a single location alter a "standard" manufacturing process in an effort to improve it. It can even have shadow salespeople who bypass national account reps to try to get local business from subsidiaries of national companies. In most cases, the risks and rewards of shadow work are the same.

150. Taking Shadow IT Out of the Shadows, Part 2

Originally published on OCTOBER 29, 2012 in #IT, #ALIGNMENT, #LEADERSHIP, #MANAGEMENT, #PROJECTS, #STRATEGY

In my previous article I described Shadow IT and the problems it causes. In this post I'll describe some approaches that the formal IT organization can take to deal with Shadow IT, and I'll give you some recommendations.

5 Approaches to Dealing with Shadow IT

Most formal IT organizations take one of five different approaches in dealing with Shadow IT:

1. **Ignore it.** This can be dangerous for two reasons. First, since a business organization's use of Shadow IT is an indication that the business organization is not happy with what your IT organization is providing, ignoring the Shadow IT means that you're ignoring the fact that the customer organization is unhappy. And second, it's quite likely that some of the systems purchased or written by the Shadow IT group will end up being "thrown over the wall" into the formal IT organization, or at least require interfaces with formal IT systems. So if you ignore Shadow IT now, then you (or your replacement) will regret it later.

2. **Fight it.** This can work if the formal IT organization has enough support from the CEO. To fight Shadow IT successfully, you'll have to create explicit standards on what can or can't be purchased, built, or contracted for by the business organizations, and then you'll have to police those standards

rigorously. This works well if IT has a good relationship with the various parts of the business, so the approach's success is highly dependent on the individual people holding the various positions. But you can't let down your guard for a second—it's very easy for a business executive to sneak something in when you're not looking.

3. **Discourage it.** This is a slightly milder version of "Fight it," and it's often easier than the "Fight it" approach when IT doesn't have all the executive support you would like. You still need the explicit standards on what the business can or can't do, but you have to pick your battles on what you fight and what you ignore.

4. **Work with it.** This is a more proactive, positive version of alternatives #2 and #3. You define standards for what the business can or can't do, and then you define further standards for how the business can do the things that they're allowed to do. You can assign people to work with the Shadow IT groups to help them with technical decisions. And you can help them choose products and services which meet the business needs while conforming to the basic architectural standards you've established for the overall business. The approach is a tightrope walk, though: You're walking a fine line between control and lack of control, and it takes some serious relationship management to make this approach succeed.

5. **Integrate it.** This goes further than #4 in taking Shadow IT out of the shadows and making it an explicit part of your IT strategy. This is kind of like a decentralized approach to corporate IT where certain services and projects are done by the business IT organizations, and certain other services and projects are done by a corporate IT organization. Standards are decided jointly, and enforced by both the business and corporate IT organizations. This can be a very effective way

to share technology direction setting across a diverse group of business organizations. But if it's not done well, it can lead to more Shadow IT, or maybe we should now call it Shadow Shadow IT, since it's in the darker shadows beyond the shadows.

How an IT Organization can Derive Benefit from Shadow IT

Used properly, Shadow IT can be a supplement to your formal IT budget. Think of it like overflow seating at an event: all of the normal seating is taken, and the overflow seating provides some measure of benefit at a lower level of service. So if you've got rigid constraints on your formal IT budget, but you need to accomplish technology objectives that require more money than your budget, Shadow IT can provide a way to achieve some of those objectives using other people's money. This works best, of course, if you use alternative #4 or #5, and works least well if there is no communication whatsoever between the formal IT organization and the Shadow IT project teams.

It's easy to get overwhelmed by just managing your formal IT organization, and obviously the coordination and communication with Shadow IT projects will make your life more complex. But if you can't make a case for moving the budget to your own organization, or if your formal IT organization isn't trusted enough for the business organizations to give their Shadow IT budgets to you, then you'll have to do the best you can with the situation you're given.

Recommendations

No matter which of the five approaches you choose, consider these recommendations:

- Don't let the business disapprove a formal IT project and then let Shadow IT do it anyway. If there are business reasons why a project shouldn't be done then those business reasons should apply no matter what organization does the project.
- Provide IT advisors to Shadow IT organizations to establish realistic expectations on data availability, and to help steer Shadow IT projects toward appropriate standards for software and hardware. The key word here is "help." Be positive, not negative. You're an advisor in this case—not a manager. Pick your battles wisely.
- Provide a "quick response" capability to do short-term projects without all of the red tape of the bigger projects. Carve out a specific percentage of your overall formal IT budget to be used for these quick response projects. Many of the Shadow IT projects come from frustration with having to live with systems issues that could be fixed fairly quickly, or with systems limitations that could be eliminated with a small amount of resource.
- Alternatively, let the Shadow IT people be the quick response providers as long as they conform to the standards set by the formal IT organization. This can be win/win: they get satisfaction using their own people, and you can focus your formal IT resource on doing the bigger projects.

Conclusion

Shadow IT can be your enemy or your friend; it's all in how you deal with it. Start by getting it out of the shadows—if you don't know about it, then you'll never be able to cope with it, no matter which approach you take. Then choose your approach from the five approaches listed above, based on the trust level between the formal IT organization and the individual business organizations, and based on the support level that the formal IT organization is getting from the CEO. In many cases, you'll find that you need a separate approach for each business organization because your relationship with each business organization is different. But whichever approach you take, keep up the communication between the formal IT organization and the Shadow IT project resources. You may be reporting through different executives, but you're all trying to make things better for the same company.

2026 Commentary

In my commentary on the previous article (Part 1), I mentioned that shadow work isn't confined to the IT area. I believe that the alternatives and advice presented in this article (Part 2) apply equally well to shadow work in other organizations.

151. Is Your Strategy a Rifle or a Shotgun?

Originally published on NOVEMBER 12, 2012 in #LEADERSHIP, #STRATEGY

A strategy can be viewed as a way of achieving an objective. A coach's strategy for the football game might be to attack relentlessly on the ground. A general's strategy for the battle might be to feint an attack to the center while flanking from the right. A CEO's strategy for the business might be to become #1 or #2 in every market and to withdraw from markets where the company can't be #1 or #2. A corrupt politician's strategy to get elected might be to align with the "smaller government" movement to gain votes while secretly taking money under the table from businesses who need special favors.

The Common Thread

As diverse as these strategies are, they all share one thing: they deal with a limited amount of time and resources by focusing that time and those resources in one particular direction.

The football coach knows that a game only has 60 minutes of play, that their opposing team has the best pass defense in the league, and that the coach's own defensive team isn't as good as the opposing team's defense. So they play to their team's strengths, and focus as much playing time as possible to keep the ball on the ground and to chew up the clock by making short gains.

The general knows that their army is a pretty even match for the enemy forces, but rearranges the troops to move more of the soldiers to the right, hoping to keep the move secret until they can attack in surprise from that direction. By moving soldiers to the right side, the general is leaving the left side vulnerable. But the general is betting that the enemy won't realize the vulnerability until the right side troops overwhelm a significant number of the enemy.

The CEO has done enough research to show that, in these particular markets, only the #1 and #2 competitors get enough market share to be profitable. So by withdrawing from any market in which that #1 or #2 status can't be achieved, they can focus their investment on increasing the market share and profitability of the #1 and #2 products.

The corrupt politician wants fame and riches. By aligning with the "smaller government" camp, they get an instant bump in the polls without having to invest time and advertising money in building a name and position for themselves. By taking money under the table, they build a war chest to use for ads against their political opponents, and they lay the groundwork for a stream of income that will lead to the riches they're coveting.

The Road Not Taken

Every one of these strategies—even the strategy of the ethically challenged politician—deals with choices. And for every choice that's made, there's an alternative choice that wasn't made.

The football coach could have counted on surprise and tried mixing up their plays to catch the opposing team in a moment of weakness. They could have violated the spirit of the game and asked their players to play dirty and try to physically cripple key players on the opposing team.

The general could have counted on sheer will and patriotism to win in a direct confrontation. They could have ordered a full attack right down the middle of enemy forces.

The CEO could have reasoned that no one gets to be #1 or #2 without being #3 or #4 first. They could have invested heavily in advertising or new products for their lesser markets, hoping that they could break through into a #2 position in a few years.

The corrupt politician could have taken an honest and ethical route, counting on their own integrity to capture votes. They could have looked forward to income from the lecture circuit after they retired from a well-respected political career.

If Your "Strategy" Doesn't Reflect a Choice, It Isn't a Strategy

It's not a strategy to tell your team, "Do the best you can" because no trade-offs have been made. It's not a strategy to tell your army, "Keep calm and carry on" because what's the alternative, "Panic and run around in circles"? It's not a strategy to tell your employees, "Increase sales 20% this year" because even though you're being specific about a goal, you're not giving them any information about how to make choices. Which sales? How? By doing what?

And it's not a strategy for the politician to say, "Win the race, then get rich," because it doesn't give specifics on how they're going to win and how they're going to get rich.

If no choices are being made, then no strategy is involved.

Shotgun or Rifle?

Take a look at what you call the "strategies" for your business. Do they reflect a choice that has been made? Is there an alternative choice that could have been the desired strategy instead? Do the strategies focus resources on one area in preference to another, essentially shutting down one approach while investing resources in the chosen direction?

Committees tend to develop "shotgun strategies": a laundry list of items that spread effort in a number of different and diverging directions, without a clear preference for one objective over another. Strategies set by one person, or set by a focused group of people, tend to be "rifle strategies" instead. They maximize the time and resources of the business by focusing on a few key things.

Shotguns are used for small game that moves quickly. You cover a large area with the blast and hope that at least a few of the pellets hit a target. Sometimes you shoot at multiple birds in flight, and hope to bring down at least one.

Rifles are used for large game. It takes more power to bring down large game so you have to focus the entire bullet on one key spot. You have to pick a specific target, and concentrate on what it takes to deliver that one target.

If you know where you want to go, and you know how you want to get there, then use a rifle strategy. You can achieve big things for your organization.

If you don't know where you're going and you have no idea how to get there, then by all means use a shotgun strategy (although I hesitate in even calling it a strategy). Who knows? You might accidentally be successful. But since shotguns are used for small game, you won't achieve very much.

What strategy approach are you using? Do you know where you want to go? Have you made choices?

2026 Commentary

I’ve seen a lot more shotgun strategies than rifle strategies. And I’ve seen a lot more mediocre companies than great companies. There’s a connection.

152. Emergencies Aren't Strategy

Originally published on NOVEMBER 27, 2012 in #LEADERSHIP, #MANAGEMENT, #PROJECTS, #STRATEGY

In a previous article (#113), I talked about how there are two reasons for strategy: focus and communication. It seems like companies have a lot of trouble with focus. Particularly in large companies, there's a tendency to "focus" on everything at once: accomplishing all of your objectives, expanding in all of your markets, increasing revenues in all product lines at the same time that you're cutting costs in all departments. And of course this defeats the entire idea of focus, which is to concentrate your resources on the *few* things that are the most important. If everything is important, then nothing is *really* important.

Some companies and governments have attempted to get around their lack of focus by using emergencies. They declare a certain situation as an emergency, and then go into some sort of lock-down to concentrate all of their labor and money on solving the emergency. This can be successful but it's self-defeating. It leads to employee burnout, it's seldom cost-effective, it typically goes after symptoms instead of the underlying core problem, and it has a short-term focus with no support for long-term solutions.

Companies get into trouble when they replace their strategy process with a series of emergencies, jumping from emergency to emergency without ever putting things into perspective. This is everything that strategy shouldn't be: impromptu, short-lived, wasteful, pushy, and stressful.

A state of emergency should be reserved for those situations when an unexpected event catches you by surprise, or when things you anticipated as possible but improbable suddenly become certain. But probable things shouldn't be handled as emergencies—they should be handled as part of normal strategy.

If you can't use your strategy to focus your resources, then you don't have a strategy—you have a wish list. A wish list isn't a bad place to start when you put together your strategy, but it's not the end point. You still have to prioritize the objectives on your wish list, eliminate or defer all but the most important items, and then determine the specific ways you are going to accomplish the remaining important objectives. Then when you assign your resources to those remaining objectives, your strategy will be complete, at least for the moment. You still have to periodically review your strategy, of course. And you have to adjust it to take changing conditions into account. But it's still the way that you focus your resources at any point in time, and it's the way that you communicate what's important to the people with whom you work.

Don't use emergencies as a substitute for strategy. Do the strategy correctly from the start, and you'll have far fewer emergencies.

2026 Commentary

I've worked in a company where emergencies were a substitute for strategy. From personal experience, it's not an approach I would recommend. We would jump from one emergency to another, sometimes without any pauses in between. It's no wonder that the company wasn't ever very successful.

153. The "Emperor's New Clothes" Trick and How to Deal with It

Originally published on DECEMBER 11, 2012 in #IT, #ALIGNMENT, #LEADERSHIP, #MANAGEMENT, #PROJECTS

As a child, I read the story "The Emperor's New Clothes" by Hans Christian Andersen. The story is one of the earliest known accounts of a trick that technology salespeople use all of the time. Let me first recap the story, and then I'll explain the trick and how to deal with it.

The Plot of the Children's Story

According to the plot, a pretentious emperor always wants to be seen as superior to everyone else. Some con men convince him that they can make him a new set of clothes that are better than anything he has ever worn. The clothes are incredibly thin and luxurious, and so light that they feel like air. But the clothes have one interesting property: People who are unworthy can't see them at all.

So the con men get out their weaving equipment and go through the motions of making cloth. The emperor watches the early stages of manufacture, and notices that he can't see the cloth. But he keeps his mouth shut because speaking out would reveal the emperor as unworthy. The con men continue with their charade, pretending to sew the invisible cloth together until they have an invisible set of clothes.

The con men bring their fake clothes to the emperor in his bedroom. The emperor strips down, and the con men pantomime putting the clothes over the emperor's head and fastening them in place. "Are they not as light as air like we promised?" they ask. The emperor, too proud to admit that he can't see anything, agrees that they are definitely very light.

The emperor gets in his open carriage for a trip through the city. The story of the emperor's clothes has preceded the procession, so everyone watching the king figures they're just unworthy because they can't see anything. And no one watching wants to acknowledge this because then everyone would know that they're unworthy. But then a small child watching the procession cries out, "The emperor has no clothes!" Suddenly everyone, including the emperor, realizes that they've been conned.

The Modern Trick

The modern equivalent of this trick is to sell a technology that doesn't make sense by convincing you that the technology is so advanced that you can't understand it. The sales people start with a simple yet far-reaching premise, something like "we've invented a revolutionary new database technology that will completely eliminate fraud by providing full traceability of all transactions back to their original source." The target audience for the sales presentation is a group of senior executives in a company that has recently been victim to fraud or, even better, a group of senior executives who work for a competitor to the victim. The executives in the actual victim company will be more suspicious, but their competitors will be naive and yet eager to prevent a similar problem in their own company.

The salespeople schedule a meeting with the executives in the company. IT people are included, but they're outnumbered by operational executives with no technical background. The CIO will probably bring along one or two database experts, just to make sure there is technology expertise.

The presentation begins. The first part of the presentation attempts to scare the hell out of the audience, reeling off horror story after horror story about fraud and the cost to its victims. Statistics are cited to make it seem likely that there's no way to escape this problem—it will hit you sooner or later no matter what you do.

Next come some complicated diagrams and charts which "explain" the new innovation. To a non-technology person they look impressive. They're full of technical terms that are vaguely reminiscent of other technical words that have been thrown around a little bit by your IT department—things like relational database, multi-phase commit, fault-tolerance, multi-threading, and maybe even that most sacred of all current technology terms, "the cloud."

The appeal of the presentation is emotional—not logical. The salespeople want you to see their software as the only true solution to a problem that's infecting the entire world. Given the heavy executive weighting of the audience, it's unlikely that the company's technology experts will get much opportunity to ask questions. But if they do, the salespeople are well prepared with answers that sound good to the executives but don't actually address any of the real technical issues. The unspoken message to the technical people is, "this is technology so advanced that we don't expect a mere mortal like you to understand it. But believe me, it works."

By the end of the presentation, the executives are ready to sign up. The discussion is not about yes or no but instead about how fast the technology can be implemented. The CIO and the technical people know that there is too much momentum to stop this thing so they instead try to rein in the discussion, offering alternatives like a "proof of concept" or a "limited trial."

And this is where the most important issue comes to the forefront: *How much is the CIO trusted by the executives?* If the trust is there, then the momentum of the sales pitch will be broken, and a more rational approach will prevail. Maybe the CIO will assign someone to investigate this technology further, the IT organization will gather information from reputable technology assessment firms like Gartner, and a team will talk with current customers of the technology to get a feel for the real cost/benefit.

But if the trust between the business and IT *isn't* there, then there's a 90% chance that some form of contract will be signed before the end of the week. Some number of months later, the poor

CIO will be put in the unenviable position of having to explain to the business executives why this miracle technology couldn't deliver the promised results for their business. And there's a very good chance that the CIO won't survive in the job another two years. After all, if a miracle is promised and it's not delivered, then whose fault is it? It must be the CIO who didn't implement it correctly, since it obviously couldn't be the technology.

Lessons from the Parable

- If a concept can't be explained simply, don't buy it. Good products are sometimes complex, but there's no magic to them. Every good yet complex product has a simple explanation for its effectiveness that's understandable to any knowledgeable buyer. Put your vendors on the spot: Ask them for that simple explanation. And if they can't provide an acceptable answer, then throw the vendor out.
- There are a lot of reasons why it's important to have trust between your business executives and your IT executives. One of the most important reasons is so the non-technical executives have someone they can trust with technical decisions. If that trust isn't there, then you're going to have technical decisions made by non-technical people—not a good thing.
- Don't let the business buy a product whose advantage is high technology if your own technology people don't agree with the premise.
- If something sounds too good to be true, it probably is.

And although my example is an "advanced" database technology, I've seen this trick used equally effectively by people selling CRM systems, ERP systems, cloud solutions, and even accounting software. In each case the salespeople promise a magic solution but fail to disclose all of the behind-the-scenes hard work and changes in business processes that are required to successfully implement their product. Even the best products have no magic—the computer systems are just doing something that you could do manually, but the systems are doing it more reliably and millions of times faster. Ultimately it's not the product that produces the good result—it's the changes you make in your business to support the product. Don't let your business people forget that.

2026 Commentary

Fourteen years later, the practice continues. Most of the current sales pitches are using buzzwords like AI (Artificial Intelligence), LLM (Large Language Model), or the latest security buzzwords. But the same rules apply: see the list of lessons above.

154. How to Sell to a CIO

Originally published on JANUARY 7, 2013 in #IT, #ALIGNMENT, #PROJECTS, #STRATEGY

In 2006 I wrote an article explaining a recommended sales approach for selling to a CIO. The article has been available as a downloadable PDF file since that time, but I thought it might be a good idea to have the text of that article available directly on my blog as well. And after my last article on the Emperor's New Clothes, I figured this is the right time to share the original 2006 article. If you're in sales, then this is aimed directly at you. If you're a CIO (or someone else in IT) being bothered by salespeople, then have the sales reps read this article so they'll figure out how to meet your needs. Here's the article:

Chief Information Officers (CIOs) have responsibility for choosing the best information technology for use by their company, and then for implementing the technology and supporting it on an ongoing basis. Their job definition calls for them to pay attention to new technology appearing on the market so that they can determine whether use of the new technology will provide benefit to their company.

I've known a lot of CIOs over the years, and they all have one thing in common: they're very busy people. Most CIOs like to hear about new products and services as much as the next person, but because of their position they're bombarded by salespeople day and night. It gets to the point where they're afraid to go to industry meetings because they get attacked by salespeople; it's like movie stars and paparazzi. I know some CIOs who carefully register for industry meetings using a less prestigious job title just so their name tag doesn't identify them as a target.

The reason for this CIO behavior is obvious: there's a lot of money in technology sales, and CIOs are key decision-makers for most of those technology sales. If you can't sell the CIO, then you're not likely to sell a technology at all. And because there are so many more technology salespeople than CIOs, the CIOs are hopelessly outnumbered, and many of them become reclusive.

This article is written for technology salespeople to help them sell to CIOs. I'm writing the article because someone needs to define the rules of engagement for such a sale. I want to protect CIOs from being badgered, and I figure that if I give you a better way to sell to CIOs then maybe you'll be more likely to leave them alone when they're out in public. Most of the badgering comes from bad sales technique, so here's the right way to sell to CIOs.

The Psychology of a CIO

Let's start by getting an understanding of the type of person who becomes a CIO. There are two types of CIO: those who come from a technology background and those who don't. But whether the CIO comes from a technology background or not, the very fact that the person has achieved the office of CIO tells us that the person's primary motivator is the needs of the business. In very small companies someone may get the title of CIO as a perk when in fact the person just manages the technology infrastructure, but that's rare. Most CIOs are senior executives, and that means that their focus is on business needs: profit, revenue, productivity, business speed, and shareholder value.

I make this point because the most common mistake I see technology salespeople make is to try to interest a CIO in a new technology because it's "cool." Although you might get a brief smile of appreciation for the product's coolness (after all, CIOs were children once, so they appreciate toys), it's not going to be enough to make a sale unless there's an overwhelming business case to be made for the product. No CIO wants to bet their job on something just because it's cool.

So why would a CIO be interested in your product or service? Companies generally compete on one or more of three factors: product, price or relationship. Most technology companies compete on product: their product is what differentiates them from their competition. A few technology companies—mostly in services—compete on price: the product is pretty much a commodity and they compete by offering the commodity at a lower price than others. And once in a while I see a technology company that competes on relationship: they build a personal relationship with the client and then make sales based on that relationship.

Here's a warning for anyone trying to sell to a CIO: most CIOs don't make an initial purchase from a salesperson based on relationship; the initial contact has to be made based on product or price differentiation. That's partly because CIOs tend to be more tool-oriented than most senior executives (certainly the ones who come from a technology background will be that way) and partly because most CIOs are too busy to allow time for a relationship with a new salesperson to build. So right from the beginning you have to understand why your product or service is better than any other alternative or why it's comparable to other alternatives but more cost effective (Note that I said "cost effective"—not "cheaper." It could be more expensive than alternatives if its overall life-cycle cost is lower due to other factors, for example ease of training or low ongoing maintenance.).

Here's the bottom line on CIOs: They're way too busy to waste time on anything that won't provide benefit to their business. So you have to show them what the benefit of your product or service is, and you have to do it quickly, succinctly and convincingly.

Your Leverage: Fear

Experts in advertising often tell us that ads have to appeal to some personal emotion like guilt, flattery, greed, anger, sympathy or fear. From my own experience, fear is the most likely candidate for CIO advertising. CIOs live in constant fear that on the one hand they'll pin too much hope on an unproven technology or that on the other hand they'll be humiliated when a competitor gains an advantage by using a technology that the CIO hasn't approved. Although CIOs learn to live with their fears, they're constantly trying to strike a balance between reliability and forward thinking. Depending on the CIO and the business, the balance is struck in different places; some CIOs are early adopters, while others lag behind everyone else in implementing a new technology.

The most effective way to motivate a CIO to consider your product or service is to address a CIO's fear head-on. The message will depend on your product or service. Some products and services are specifically aimed at reducing risk, for example by increasing infrastructure up-time or by handling emergencies. Other products and services are useful in pursuing business objectives of increased performance, decreased time-to-market or lower cost. This second group attacks competitive fears rather than the fear of down time.

The biggest fear of CIOs is that they will run out of time; they won't be able to accomplish their goals before they are replaced. Thus showing a CIO how they can accomplish more in a fixed amount of time is a big attraction, but only if the implementation of the time-saver can be done quickly. It's virtually impossible to sell a CIO a product or service that takes multiple years to achieve benefit; the CIO will be long gone before the benefit is achieved, and any accolades will fall to the CIO's successor.

Find a Need and Fill It

This old business slogan is just as true for CIOs as it is for any other customer. But to sell to individual CIOs, you have a "chicken and egg" problem: you can't get any time in front of the CIO until you convince them that you can fill one of their needs, but you can't find out their needs without getting time in front of the CIO.

There are various approaches you can use to break through this obstacle. Here are a few:

1. Talk to lower-level people in the CIO's organization to find out what the organization needs. But be careful, since quite often the lower-level people—even if they report directly to the CIO—are more focused on *how* things are done than on *what* things should be done. You'll need to read between the lines, and you'll need to keep reminding yourself, "How does this need relate to an actual business benefit?" Lower-level people have their own agendas, but it's the CIO's agenda you should be interested in.

2. Talk to people who buy the products and services provided by the CIO's company. Find out how they compare the CIO's company with the competition, and see if you can discover a competitive disadvantage you can remedy or a competitive advantage you can strengthen.

3. Apply your knowledge of similar IT organizations in similar businesses. Do some research and find something you can offer a CIO that will bring their organization up to the level of excellence achieved by another IT organization.

4. Talk to vendors of products and services that are complementary to your own, looking for a vendor who already sells to the CIO's company. Gather information about the CIO's needs from these other vendors.

Once you've found a match between a specific CIO need and one of your products or services, write up a brief but compelling case for why the CIO should talk to you. Get specific about dollar benefits, and pay particular attention to how long it will take before payback is achieved. If you can go to a CIO and say something like, "My products can add $10 million to your company's bottom-line within one year," there is no way that the CIO will say no to a meeting.

Make the Approach

Ok, so you've identified a need, and you've developed a great pitch to explain to the CIO how your product or service will provide huge dollar benefit to the CIO's company within one year. Now how do you get time in front of the CIO to convey your message?

The answer depends on the CIO and the specific company. For smaller companies with only one business location, you can probably approach the CIO directly by sending a letter that outlines your case and that asks for a meeting to discuss your proposal further. A letter works better than an email or a phone call. Incoming calls to a CIO are usually screened by an administrative assistant, and you're not likely to get through without name recognition. And most people—CIO or not—get hundreds of spam emails a day, so an email out of the blue is not likely to be perceived as positive. Of course, letters are likely to be screened by an administrative assistant as well, but if it's a personal letter—not just a bulk mailing—then it will probably get through to the CIO.

After the letter is sent, wait a week or so and then call the CIO to set up a meeting. Refer to your letter in the call, and reference the bottom-line value that you can deliver.

For larger companies that have more than one business location, it's going to be more difficult to talk to the CIO. They'll be traveling a lot more, and they are likely to jealously guard their time at their home location. You can still try to contact the CIO directly using a personal letter; your degree of success will depend on how strongly you make your case in the letter. If you don't get a response, or if you don't think you'll have success with a letter, then consider the following alternative ways to get in touch with the CIO:

1. Find someone who knows the CIO, and ask to be introduced. You'll have a better chance of success if the person doing the introduction is a respected colleague or business associate. If

the person is a personal friend of the CIO outside of business, then the introduction will be more suspect, but it might still work out. Do not, however, go over the CIO's head and contact the CEO or President of the company, even if you know them. You'll get your CIO meeting, but the CIO will harbor a grudge against you, and you'll have to work even harder to make the sale.

2. Identify another CIO of equivalent rank (same size company) who knows your products and services and thinks highly of you. Have this other CIO make contact first in an equal-to-equal communication, and then have this other CIO introduce you. This works particularly well if your own company is about the same size (or larger) than the company to which you want to make the sale, and if you can have the CIO of your own company do the introduction. If that isn't appropriate, then see if you can persuade one of your other CIO customers to introduce you.

3. Sell up the ladder. Start lower down in the CIO's organization, and introduce yourself to the highest ranking IT person who will listen to you. Make the case for why they should consider doing business with your company, perhaps sell them some smaller product or service, and then persuade them (or their boss) to introduce you to the CIO.

4. Sell to the business users in the CIO's company. This is a tricky approach, since it can backfire on you just like going over the CIO's head. But if your company has a product or service which offers significant non-technical benefit, then it's perfectly valid to sell to the business users instead of the IT organization. However, make it clear from the beginning that you would like to involve the IT organization as well, and ask the business users to invite someone from IT to some of the meetings. Bend over backwards to show any IT attendees that you

have their best interests at heart. At an appropriate time, ask to be introduced to the higher management of the IT organization, and eventually the CIO.

Make Your Case

You've finally got that appointment with the CIO; now make the most of it. No matter how long the scheduled time is for the appointment, you've got no more than about fifteen minutes to convince the CIO that they should do business with you. This is your opportunity to convey the brief but compelling case that you've previously put together. Here are some tips:

1. As soon as introductions are done, start with the bottom line. If you can save the CIO's business $10 million, then say so to get their attention. Then gradually peel back successive layers of your proposal based on the CIO's interest and questions. Approaching the pitch this way guarantees that you'll get a strong message across no matter how little time you have. And don't be surprised if an interruption occurs and your time is cut short; that's to be expected in a meeting with a time-crunched CIO.
2. Don't start with background information on your company, except for maybe a sentence or two. The key message isn't a history of your company—it's the contribution that your products and services can make to the CIO's business. That's why you should start with the bottom line.
3. After presenting the bottom line, substantiate your claim. Provide the facts underlying your figures. Compare the CIO's situation to other customers or to other companies who have a similar challenge or problem. But be very careful not to reveal information about another customer that might be regarded as confidential by that customer; if the CIO sees you being cavalier with another customer's information, then they will expect

that you won't keep their own company's information confidential. That's a deal killer.

4. There are four messages that you should convey in your meeting:
   - My products and services can give your company bottom-line benefit.
   - You can receive the benefit quickly, ideally within the same fiscal year as the expense.
   - My company is trustworthy; we're solid, and you can count on us.
   - I will make your job easier; my products and services will solve more problems than they create.

If you can truthfully and convincingly convey these four messages, then you've laid a solid foundation for a sale.

Eliminate Risk

Ok, so now you've got the CIO's attention. You've conveyed the four messages, but the CIO will still have doubts; after all, it's their job that's on the line. Your next task at the first meeting, and at subsequent meetings, is to reduce or eliminate the risk for the CIO. There are four types of risk that you need to reduce or eliminate:

- Financial Risk for the CIO's company. There needs to be a win-win arrangement between your company and the CIO's company.
- Job Risk for the CIO. There can't be any way that this will affect the CIO's job other than positively.
- Technical Risk associated with your products and services. The CIO needs to be sure that the products and services will work as proposed.

- Future Risk associated with changes in the state of technology over time. The CIO needs to understand how future technology changes will impact the particular products and services that are being proposed.

There are a number of ways in which you can reduce or eliminate these four risks. The specifics depend on the product or services you offer, but here are a few ideas:

1. Offer a free trial of your product or service for a limited scope test (e.g., short-term software licenses for a small test organization, or a service provided to a small subset of the general company population). This reduces Technical Risk, since the CIO's company will see first hand how well your products and services work. It also reduces Job Risk for the CIO if the test can demonstrate the kinds of benefits that will come from using the product or service. And if the financial benefits can be demonstrated in the trial then it reduces Financial Risk as well.

2. Provide free training for the people involved in the test or trial. If you provide the product or service without the training then the test organization is not likely to use the product or service correctly. Furthermore, you'll find that once people are trained in the use of a good product or service, they'll want to continue using it after the trial. The test organization will actively lobby for the sale.

3. Provide reference information about CIOs in other companies who are successfully using your product or service. Contacting the other references will reassure the CIO and will reduce Job Risk and Technical Risk. And if these reference accounts can provide the CIO with specific financial data that show the benefit of the products and services to their companies, then it will reduce Financial Risk as well.

4. Provide white papers written by your company that show how your products and services fit into various IT architectures,

both current and future. Case studies with actual customer information (even if the customers are disguised to protect their confidentiality) will make more of an impact. This reduces both Technical Risk and Future Risk.

5. Have the CIO sign a nondisclosure agreement, and then share information with the CIO on future products and services from your company. If it makes sense, and if the size of the deal justifies it, then offer to have the CIO visit your development organization and/or talk to your own CIO, CTO or CEO. Sharing future direction will reduce Future Risk. Having the CIO sign a nondisclosure agreement will protect your company, and it has the added benefit that it makes the CIO feel like an insider; this makes it more likely that the CIO will buy.
6. Reduce Financial Risk by working out a financial proposal for your sale that provides a net positive impact in every fiscal budget year. Ideally the product or service will pay for itself in benefits in the first fiscal year. But if that isn't possible, then work out a way to finance the purchase (taking the financial accounting guidelines of the CIO's company into account) to generate positive cash flow for the CIO's company in every fiscal year. It's a whole lot easier for a CIO to sell a project when they can show that it impacts the bottom line positively in every upcoming year.
7. If you and your company are absolutely convinced that the bottom-line impact of your product or service will be positive, then contractually make that commitment. This eliminates the CIO's Financial Risk altogether and makes the purchase a no-brainer. But be careful to specify exactly how the bottom-line impact will be measured, and explicitly state any assumptions underlying the agreement. You need to protect your own company in case the CIO's company makes major changes in own-

ership, in commitment, or in underlying conditions that invalidate your guarantee.

Assuring Add-On Sales

Earlier in this article I said that CIOs don't make an initial purchase from a salesperson based on relationship. However, once the initial sale has been completed, and once the validity of your bottom-line promise has been established, then you're well on your way to establishing a trusting relationship with the CIO. Continuing to build this trusting relationship will enhance your ability to get add-on sales from the CIO's company.

But remember that the trusting relationship is very fragile, and it's based on your continued honesty and your continued ability to deliver bottom-line performance. You can't slack off just because you've got your foot in the door; you have to do your homework on every add-on sale just as you did for the initial sale. If you promise something and don't deliver it, then your trusting relationship will be over very quickly.

You also need to keep an eye on the politics in the CIO's company, arrange to meet additional executives, and build trusting relationships with them as well. Sooner or later, the CIO will leave, in spite of the support you've given them (or possibly because of it; the CIO may go on to a bigger company). You need to be prepared to start over in building a trusting relationship with the new CIO, aided by your history of success with the company.

Conclusion

The key to guaranteed CIO sales is to:

1. Determine the bottom-line value of your product or service for a CIO before you ever contact the CIO.

2. Get the CIO's attention by letting the CIO know how much monetary value you offer.
3. Take maximum advantage of the limited time you have with the CIO by focusing on the value, and on conveying that value to the CIO.
4. Reduce or eliminate all of the risks that will prevent the CIO from completing the sale.
5. Once you've gotten the sale, focus on building a trusting relationship with the CIO and with other company executives.

Always remember that you're not selling a product or service—you're selling the value of your product or service to the CIO's business. Don't ever confuse those two things; without business value your product or service is worthless.

2026 Commentary

I had positive feedback on this article from a number of technical salespeople. But if you read this and your reaction is that it's too much trouble, then you're obviously not selling the high-value technical products to which I refer.

155. Failed First-Time Managers: 2 Case Studies

Originally published on MARCH 19, 2013 in #CAREERS, #MANAGEMENT

I've promoted 20 or 30 people into their first manager jobs during my career. Two of those promotions were failures—they did so badly that I had to take them back out of the positions. Here's what I learned from that experience.

Background

In both cases there was ample evidence that the people were ready for a promotion. Both people had performed well as project leaders where they provided direction to others. Both were respected by their peers and customers, and had good communication skills. Both people had worked for me for a few years, so I had a pretty good idea of what their strengths and weaknesses were.

The First Failure

The first failed manager was promoted into a software development manager position. She had shown good ideas as a project leader, and I expected her to dig right in when she had her own people reporting to her. But I didn't recognize a problem she had with self-confidence and prioritization.

As a project leader she had to make decisions, and she seemed to cope well with the decision-making process. But apparently her confidence in her project leader decisions was bolstered by the fact that her manager had the final word on those decisions. When she took that manager role, she began to second guess herself. She couldn't adequately prioritize her time between the important decisions and the trivial ones, and she lost her self-confidence. She got to the point where she would decide something, tell her employees, then change her mind after talking to someone else. Many of those decisions were trivial enough that they didn't significantly affect project success, but it was unproductive to keep changing her mind, and it began to cause a lot of rework.

I talked to her numerous times about the problem, but she was unable to regain the focus she had as a project leader. Eventually her stress level got too high, and we jointly decided to take her out of the position.

The Second Failure

The second failed manager had been very successful leading infrastructure projects with well-defined objectives. We needed a services manager to lead an organization that was responsible for the overall infrastructure for a large department, and I promoted him into the position.

He had the technical background to understand the area he was responsible for, and he worked well with the employees who reported to him. He initially did well in the position, taking care of infrastructure needs, and prioritizing work based on the needs of the business.

Over a period of several months, however, he began to get overconfident. The customers for his services began to complain that he wouldn't listen to them, and when I discussed the complaints with him, he took the position that he understood their needs better than they did. I tried various approaches to improve his customer relationships. I tried working with him on defining customer expectations and on better communication. Ultimately, however, I couldn't get him to understand that he needed to redefine success to include customer perception. I removed him from the position, and transferred him to another area that was more project focused.

What I Learned:

1. Sometimes you can't tell in advance whether someone will do well in a management role (or any other role). And you can't wait until you're 100% positive, because no one is ever sure. There will be failures. You just need to stay on top of the situation and change things if necessary.

2. All first-time managers have to adjust to a new situation. All require support from their manager and some degree of hand-holding. It's important to meet regularly with a new manager to talk about any questions and difficulties. And it's important to get independent information from the new manager's customers, peers and subordinates to get a full picture of what's going on.

3. The biggest unknown about promoting someone into a first-time manager position is how it will affect the new manager's self-perception. Some feel inadequate and have trouble with stress, uncertainty and the reactions from their new subordinates. Some let the new implied status and power go to their heads, they become overconfident, and they don't pay enough attention to the views of their customers and employees. Both reactions can be temporary and can be corrected with time. But different situations demand different levels of tolerance. Sometimes the person has to be taken out of the position if it doesn't appear that the situation can be corrected in the time available.

4. Even if a first-time manager has to be removed from the management position, that doesn't mean that the person can't become a manager later. It's possible that the person will learn from their mistakes and do better in a few years.

5. Taking someone out of a manager position has to be done carefully and announced with tact. The action can be devastating to the ego of the person being removed from the position. If that person is going to continue in your organization then it's impor-

tant to explain the change in a way that minimizes the damage to the person's reputation.

2026 Commentary

Maybe the most important thing to understand about becoming a manager is that it's totally different from being an independent contributor or even a project leader. When you have the responsibility for other people and their day-to-day activities, it requires a different way of thinking about things. Some of the other articles in this book address those differences.

156. Secrets of the Hiring Process: How to Hire

Originally published on MAY 7, 2013 in #CAREERS, #MANAGEMENT

I've hired hundreds of people during my career, and my process has been pretty consistent for all of those hires:

1. Decide on the need for a person in a job.

Maybe it's a replacement for someone who left, or maybe it's a new position that I need for a project or for an expansion of a functional area.

2. Define the job.

I'm not a personal fan of job descriptions, but they're important for recruiting and for determining a fair level of pay. So I'll usually put one together, sometimes working with someone in Human Resources (HR). In many cases, the job is similar to another existing job, and so the job definition and salary range are based on that other job. Once in a while, I'll create a totally new position, and then it takes a bit more thought and effort to define the position and its salary level.

3. Get approval for the position if necessary (this depends on your level in the organization).

If the position is already in my budget then this is usually pretty easy. Otherwise the position approval will probably be part of a project proposal. If the project is approved, then the positions associated with the project are approved along with it.

4. Look internally.

When I'm creating a position or filling a vacancy, I usually have a pretty good idea whether or not there's someone already in the organization who can do the job—maybe as a lateral transfer or a promotion. I always look internally first to see if there's someone who would be good for the position. If not, then I'll look externally. If I'm not sure, then I'll look externally while keeping one or more internal candidates in mind.

5. Find candidates.

There are three primary sources for external candidates: agencies, direct responses from job postings, and referrals. Good agencies have some advantages: they pre-screen the candidates; if you do a lot of hiring then your agency rep already knows the kind of person you like to hire; and they're able to respond rapidly, sometimes with candidates who weren't even actively in the job market. Bad agencies are not worth dealing with—they throw resumes at you without regard to quality, and they're generally a waste of time. Worse, the bad agencies can try to sue you for fees if you hire someone for whom they sent a resume—even if you didn't hire because of that resume but got a different resume for the person through other channels (I've had it happen).

Direct responses come from job postings you put on your company web site or on public job web sites. Sometimes they're algorithmically filtered by job selection software, but in the years when I was hiring, the algorithm wasn't very good. Because the filters aren't as good as the human processes used by agencies, you'll get a large variation in the quality of the resumes submitted, and you'll have to spend more time going through the resumes yourself. You might save an agency fee (or pay a reduced fee to a job web site), but you'll spend more time on the resume search.

Referrals are a mixed bag. It's worth mentioning to your friends in other companies that you have a specific need, but in most cases your friends will hold on to their good people, and so the only referrals you'll get are friends of friends who are in a job search, or sometimes relatives of friends. I've had reasonable success getting referrals from my own employees, since they often know someone in another company who's looking. And it helps if your company pays a bonus for employee referrals who are hired; that cuts your agency fees and provides a good incentive to employees. (A Tip: Pay half the incentive when the person is hired, and pay the other half if the new person and the employee are still with the company a year later. That provides additional motivation for the employee to recommend someone who can do the job well, and it helps retain your employees too.)

6. Select viable candidates to interview.

Initial culling of external candidates is done by reviewing resumes. Some people look for a candidate who has done the exact same thing as the job for which you are hiring, but I don't favor that approach. I look for someone who shows the characteristics I want in an employee, and is positioned to do the job and be challenged by it (see other articles in this book on those subjects). Tom Peters calls this the "Hire for attitude, train for skills" approach. Being challenged by a job makes you want to do it well, and it makes you alert to new ways to do a job better. If you've done the same job over and over, then you're not likely to be very creative about doing it yet again.

7. Interview viable candidates.

Historically, I've found that I interview about 1 in 10 of the people for whom I get resumes if the resumes come from agencies, and I'll interview about 10 people for every hire. If the resumes haven't been pre-screened, then I might interview 1 in 20 or even 1 in 50. So I'll end up reviewing about 100–500 resumes for every hire. That may sound like a lot, but when you're looking at that many resumes, you get to be good at doing a quick first read to eliminate people who obviously don't fit. Then you go back and read the remaining resumes in more detail to select the best ones to interview.

The interview will usually be in person if the candidate is local, or by phone or Skype if the person is remote. I never hire without an in-person interview, but doing a first interview by phone or Skype saves travel expense and it avoids wasting the candidate's time on an unnecessary trip.

For me, the interview is the most important part of the hiring process. I eliminate a lot of candidates during the interview because of the attitude they show toward the job or the work. I want someone who can look forward to doing the job, not just do the work because they have to. I want someone who is eager to learn, no matter how advanced or high level the position is. I look for people who are open-minded and curious; responsible but willing to take risks when it's appropriate; focused on results, but understanding of the human issues that are so important for success. And for most jobs I only consider people who can communicate well.

Sometimes one interview is enough—I have had occasional situations where I made an offer right after the interview. But often I have a good feeling about a person for the job but want to see if any of the other candidates are even better. It's critical to schedule interviews for all of the candidates during a short period of time so that you can interview a candidate, interview them again if necessary in a subsequent visit, and then make an offer quickly. The really good people aren't on the market very long.

8. Have other people in the company interview each candidate.

Usually I'll arrange to have the candidate talk to other people right after I talk to the person. That way I can stop the process quickly if the candidate is an obvious non-fit, and save the time of the other interviewers.

Sometimes the interviews with other people are a way to get additional viewpoints on the person. Sometimes the other interviews are a way to expose the candidate to the business customer whom the candidate will support. Sometimes, depending on your level in the organization, it's necessary to have your boss interview the candidate. For "doer" jobs like programmers, I've often had the candidate spend some time with one of the other doers who is performing a very similar job. That has three advantages: it gives me some feedback on technical competency from someone who is doing the work day-to-day, it gives the candidate a view of how we actually work on a day-to-day basis, and it gives the doer a feeling of participating in the process.

When a candidate talks to multiple people in the organization in a single interview visit, I'll try to talk to the candidate again at the end of thc visit, with me being the last person to talk to the candidate before they leave for the day. That way I can get feedback from the candidate on how they view the position and the interview process. I can deal with any obstacles that might stand in the way of the candidate accepting the position. And it gives me one last opportunity to sell the position and the company to the candidate, even though I'm not sure at this point that I'll make an offer.

9. Discuss interview viewpoints with all of the people who interviewed the candidate.

I'll do this at least twice: once right after the interview, and again after we've had a chance to talk to all of the candidates. Even if the hiring decision is mine alone, I want to get input from other people, and I want them to feel supportive of the decision I'm making.

That's particularly true for business customers who will work with the candidate after the hire. The business customers need to agree with me that we're hiring the right person, so that the new hire will be supported by business customers when they start in the job.

10. Select a candidate and check references.

If the references hold up to scrutiny, then I'll move to the next step. If the references raise any questions, then I may need to contact the candidate to get answers or to clarify something I heard.

If I have any doubts about certain areas of the candidate's past or about their qualifications, I'll go deeper with references. Often the best approach is to ask the references for the names and contact information for other people who may be familiar with the candidate's work. Then I'll call these second-level references.

I do this because it is expected that the candidate will list references who will say nice things about the candidate. Otherwise why would this person be given as a reference? But by getting the reference to give me other contacts, I can get around this bias and get in touch with people who might be more likely to be honest and straightforward about the candidate. I've occasionally learned a few things this way that disqualified a candidate altogether.

11. Get together with HR and decide on a specific salary offer.

This is done carefully to ensure that we pay equitably across all employees—both new and long-term. It's ultimately my decision, but I appreciate the advice that HR can give me. And often HR has broader information on pay for this job level than I have from my own employees.

12. Make a job offer to the candidate.

Sometimes the offer is accepted as is. Sometimes the candidate attempts to negotiate on salary or on some other aspect of the job: start date, vacation time, signing bonus, etc. The policies on some of these things are often set company-wide and so I may or may not have the ability to modify the terms. And frankly, I've already put together an offer that I consider fair, and so I'm reluctant to negotiate. I'll try to be flexible on things like start date, but I rarely change anything other than salary, and I've only changed salary amount a few times. To me, the way that the candidate tries to negotiate can be a potential red flag for the way that the candidate will work after they are hired. I've withdrawn offers to a few people because they didn't seem to value the offer I made.

Exceptions to this Process

Sometimes the process varies. If a candidate is so far away that an interview trip will be very costly or time consuming, then I might check references after a phone/Skype interview before I schedule an actual physical interview. Or if the candidate comes to me as a highly-recommended referral from someone I trust, then I might schedule an interview before I see a resume.

I've also had a few situations where something in a resume was so different than the norm that I wanted to get an interview even though the resume itself was borderline in its statement of the candidate's credentials. I've interviewed and hired people who didn't graduate from college because of the amazing commitment they showed in other areas. And I've interviewed people with unique backgrounds or experiences (like a helicopter traffic reporter) just because I thought they might make an unusual contribution to the job.

Virtually all of my hiring has been for professional and management jobs. If you are hiring for a job that doesn't require good written communication (like a retail store clerk, a warehouse worker or a physical laborer), then you can't always use a resume to screen out bad candidates. In that case the candidate usually just fills out a job application with a basic job history, and then a preliminary interview is conducted to determine job suitability. Essentially the interview is used to capture the information that would normally be part of the resume for a professional person.

Conclusion

If you are a hiring manager, then I hope this description of my own process has provided you with some ideas or insight.

If you are someone looking for a job, then perhaps this gives you a better idea of what you need to do to be hired. In general, the resume gets you an interview, the interview gets you an offer, and the references just keep you honest. It's always nice to get a direct referral to a hiring manager from someone who knows the applicability of your skills for a position, but in my experience it's the exception—not the rule.

2026 Commentary

I've been away from corporate life for a long time now, so I don't know how much this process has changed. I've heard that AI screening of resumes is popular now, but I'm not sure I trust it. The problem with AI screening is that it only screens for specific work credentials, and it doesn't take into account the intangibles that I can get by reading between the lines in a resume. As you'll know by reading the other articles I've written about hiring and careers, I look for attributes that don't usually show up on resumes, and can't be recognized by AI. I'm afraid an AI screening would eliminate some of my best candidates.

The Future

Originally published on MARCH 13, 2024 #PERSONAL

I've been publishing this newsletter for twenty-one years, but I've decided to stop. I hope the newsletters have been helpful to you, and I plan to put most of the content into a book that will be published toward the end of 2024 or the beginning of 2025. It will be available for reference, and maybe for a few new readers who want to learn from some of my experiences.

Thanks to all the people who have commented or replied to my newsletters in the past, who have bought my book, *Boiling the IT Frog*, and who have attended my various speaking engagements over the years. I learned a lot from interacting with people on these topics, and I believe it was a valuable lesson for me to attempt to put my thoughts into words. You don't really understand a topic until you can explain it to other people.

2026 Commentary

You may have noticed two things. First, I say that I've been publishing the newsletter for twenty-one years, but the publication dates on the articles range from April 8, 2003 to May 7, 2013, which is only ten years. That's because I actually stopped publishing new material after the May 7, 2013 article, and the newsletters after that point were "reruns" of prior material. I had picked up a lot of new readers by that point, so the material was new to them, even though it wasn't freshly written. The newsletters continued like that for another eleven years until I decided to stop in 2024.

Second, you may have noticed that I've taken a couple of years longer than I had planned to put this into book form. My biggest challenge was how to present the material. I thought about breaking this into separate books, and I spent time trying to do that. I finally decided to just publish all the articles in a single book, but use hashtags to help people zero in on what they're really interested in. Regardless of how the material is presented, I feel good about saving these articles "for posterity." I hope the readers of this book will find some value in what I've written.

About The Author

I started working with computers during summers and holidays at the Center for Disease Control while I was still in high school, then I went to M.I.T. and majored in what is now called "Information Technology." I did a few part-time jobs while I was at M.I.T., including working at Project MAC using Multics (the predecessor to Unix) and doing some contract programming on an inventory management system.

After graduation, I went to work at Digital Equipment Corporation (DEC) as a programmer/analyst in the manufacturing area. Over the next few years I got promoted into project leader and then IT management positions.

After seven years at DEC, I went to work for The Service Bureau Company in Atlanta, which became part of Control Data Corporation, and then became Ceridian Corporation. I mostly worked in IT management positions in software development, and when I left in 2002 after twenty-two years, I was Vice-President of Technology Architecture and Strategy. In my last few years there, I spent most of my time doing due diligence on approximately 25 corporate acquisitions in the U.S., Canada, the U.K., and Switzerland.

After leaving Ceridian, I started my own consulting company, MakingITclear, Inc., specializing in helping companies improve the relationship between IT organizations and their business customers. I started out writing these articles in 2003 as part of a blog and newsletter that I created in an attempt to establish myself as a knowledgeable expert on IT strategy and the relationship between IT and the business. But as the years went by, I found that the articles that attracted the widest audience were those that talked more generally about being a manager and how to be a better one. In fact, my most read article, with more than a million views, was article 87: "How to Become a Manager—13 Skills You'll Need."

[illegible]

[illegible]

[illegible] apparently [illegible] being [illegible]

www.ingramcontent.com/pod-product-compliance
Lightning Source LLC
LaVergne TN
LVHW021738071126
851857LV00039B/536

* 9 7 9 8 9 8 6 1 0 0 9 2 0 *